P9-DUM-217

INTERNATIONAL

REVIEW OF CYTOLOGY

VOLUME 74

ADVISORY EDITORS

INTERNATIONAL

Review of Cytology

EDITED BY

G. H. BOURNE
St. George's University School of Medicine
St. George's, Grenada
West Indies

J. F. DANIELLI
Worcester Polytechnic Institute
Worcester, Massachusetts

ASSISTANT EDITOR
K. W. JEON
Department of Zoology
University of Tennessee
Knoxville, Tennessee

VOLUME 74

1982

ACADEMIC PRESS *A Subsidiary of Harcourt Brace Jovanovich, Publishers*
New York London
Paris San Diego San Francisco São Paulo Sydney Tokyo Toronto

ACADEMIC PRESS, INC.
111 Fifth Avenue, New York, New York 10003

United Kingdom Edition published by
ACADEMIC PRESS, INC. (LONDON) LTD.
24/28 Oval Road, London NW1 7DX

LIBRARY OF CONGRESS CATALOG CARD NUMBER: 52–5203

ISBN 0–12–364474–7

PRINTED IN THE UNITED STATES OF AMERICA

82 83 84 85 9 8 7 6 5 4 3 2 1

Contents

Morphological and Cytological Aspects of Algal Calcification

Michael A. Borowitzka

Naturally Occurring Neuron Death and Its Regulation by Developing Neural Pathways

Timothy J. Cunningham

The Brown Fat Cell

Jan Nedergaard and Olov Lindberg

List of Contributors

Numbers in parentheses indicate the pages on which the authors' contributions begin.

RICHARD D. BERLIN (55), *Department of Physiology, University of Connecticut Health Center, Farmington, Connecticut 06032*

MICHAEL A. BOROWITZKA (127), *Roche Research Institute of Marine Pharmacology, P. O. Box 255, Dee Why, N.S.W., Australia*

TIMOTHY J. CUNNINGHAM (163), *Department of Anatomy, The Medical College of Pennsylvania, Philadelphia, Pennsylvania 19129*

SIEGFRIED W. DE LAAT (1), *Hubrecht Laboratory, Embryological Institute, Utrecht, The Netherlands*

KARLEN G. GASARYAN (95), *Laboratory of Molecular Basis of Differentiation, Institute of Molecular Genetics, 123182, USSR Academy of Sciences, Moscow, USSR*

OLOV LINDBERG (187), *The Wenner-Gren Institute, University of Stockholm, S-113 45 Stockholm, Sweden*

JAN NEDERGAARD (187), *The Wenner-Gren Institute, University of Stockholm, S-113 45 Stockholm, Sweden*

JANET M. OLIVER (55), *Department of Physiology, University of Connecticut Health Center, Farmington, Connecticut 06032*

PAUL T. VAN DER SAAG (1), *Hubrecht Laboratory, International Embryological Institute, Utrecht, The Netherlands*

INTERNATIONAL REVIEW OF CYTOLOGY, VOL. 74

The Plasma Membrane as a Regulatory Site in Growth and Differentiation of Neuroblastoma Cells

SIEGFRIED W. DE LAAT AND PAUL T. VAN DER SAAG

Hubrecht Laboratory, International Embryological Institute, Utrecht, The Netherlands

I. General Introduction

Understanding the molecular regulation of cellular growth and differentiation is one of the fundamental problems of present developmental and cell biology. Knowledge of the underlying regulatory mechanisms is not only crucial for the developmental biologist studying the temporal and spatial controlled processes of cellular proliferation, diversification, and differentiation during embryonic development, but also for the tumor biologist, who tries to understand the origins of neoplastic transformation. It has become clear in recent years that in many respects the problems of developmental biology and tumor biology have a common basis, and the arguments hereto have been reviewed previously (Pierce, 1967; Pierce *et al.*, 1978). The use of intact embryos to study the molecular regulation of cellular growth and differentiation has, however, posed considerable problems because of the complexity and diversity of the processes taking place simultaneously

ISBN 0-12-364474-7

in the developing embryo and the difficulties in obtaining sufficient amounts of homogeneous material for certain biochemical assays and characterizations. The availability of a variety of cultured mammalian cell lines, mostly derived from tumors of embryonic origin, which have retained the capacity to undergo simultaneously and in practically unlimited populations of cells parts of a developmental program *in vitro* in an experimentally controllable way, with a strong resemblance to that occurring in certain phases of normal embryonic development, has opened new ways for the investigation of these processes in molecular detail. These features have attracted numerous investigators for exploiting such cell systems in the investigation of certain aspects of development at the molecular level, ranging from early development, as in embryocarcinoma cells (for a recent review, see Martin, 1980), to terminal differentiation as in neuroblastoma cells, the subject of the present article.

Neuroblastoma is one of the most common malignant neoplasms of childhood with roughly one in 10,000 live-born infants developing this tumor. Neuroblastomas may arise from any dividing neuroblasts, but most frequently they develop from sympathetic ganglia or the adrenal medulla. For this reason neuroblastomas are considered to be tumors originating from neural crest cells in which some unknown disturbance has induced unscheduled proliferation. Neuroblastoma is one of the few solid tumors which may occasionally regress spontaneously, probably by undergoing maturation (differentiation) to ganglioneuroma or ganglioneurofibroma. In some cases genetic factors operate in the pathogenesis of neuroblastoma and under these conditions it is inherited as an autosomal dominant trait with intermediate penetrance. It seems plausible that neuroblastomas result from an abnormal regulation of neural crest cell differentiation. Support for this view comes from the observation that many differentiated functions, characteristic for mature neurons, can be expressed in these tumor cells in a varying degree. Furthermore, fully differentiated neuroblastoma cells have lost their oncogenic capacity. The properties of the neuroblastoma system have previously been rewiewed extensively (Prasad, 1975; Pochedly, 1977; Schimke, 1980).

Since the establishment of murine neuroblastoma cell lines in culture (Klebe and Ruddle, 1969) a great number of cell lines derived from neuroblastoma of mouse, rat, and human origin have been described. They have gained widespread attention as model systems for studying neuronal differentiation *in vitro*. In these cell lines irreversible differentiation is usually induced by serum deprivation or intracellular cAMP-elevating agents. The cells then become arrested in the G_1 phase of the cell cycle, and express various differentiated functions of mature neurons: formation of neurites, development of electrical excitable membranes, and the induction of high activities of neuron-specific enzymes. The degree of expression of the various neuronal functions to a certain extent differs among the various cell lines. For an earlier review, see Prasad (1975).

The ability of appropriate external stimuli to cause a switch from a proliferative program to an irreversible differentiation program in the G_1 phase of the cell cycle implies to us that an important key to understanding the critical events involved in the initiation of cellular differentiation could be found through knowledge of the regulatory circuits that control the cell cycle, and in particular the G_1/S transition. For these reasons, a number of years ago we initiated a study of the regulation of the cell cycle and differentiation of neuroblastoma cells *in vitro*. In these studies emphasis is given to the possible role of the plasma membrane as a regulatory site. Such a role is evident since (1) the plasma membrane forms the barrier between the cell and its surrounding, thus every extracellular influence, e.g., hormones, growth, differentiation, and transformation factors, on intracellular processes is mediated and modified by the properties of the plasma membrane; (2) cell–cell interactions, modulating cellular growth, and differentiation are under direct control of the properties of the cell surface; (3) the plasma membrane is directly involved in the molecular interactions by which cellular growth and differentiation are regulated through its selective permeability to ions and nutrients, and through membrane-bound enzymes that control the intracellular levels of critical constituents. It is thus not surprising that considerable data are now available from the work of a great number of investigations on a wide range of cell types, which relate various membrane components and modifications with changes in growth and differentiation characteristics, e.g., expression and dynamic properties of membrane receptors, properties of membrane lipids, ion metabolism and electrical membrane properties, transport of nutrients, ultrastructure of the plasma membrane (for review, see, e.g., Nicolson, 1976; Bluemink and de Laat, 1977). Despite all of these studies, cause and consequence of the molecular events at the plasma membrane are as yet rather obscure. In our opinion, they can possibly be resolved by coordinated research in which the various biochemical, biophysical, and ultrastructural aspects are studied in parallel on suitable experimental systems.

In this article we will primarily summarize the studies, carried out along these lines in recent years in this laboratory, on the role of the plasma membrane in the regulation of growth and differentiation of murine neuroblastoma cells *in vitro*. They concern mainly the cell cycle and morphological differentiation (neurite extension) of Neuro-2A cells, and serum stimulation of growth-arrested N1E-115 cells. In addition we will review the literature concerning plasma membrane-mediated events involved in the initiation and expression of the differentiated phenotype of neuroblastoma cells, with emphasis on publications from 1975 onward. For a review of earlier work, see, e.g., Prasad (1975).

II. Cell Cycle of Neuro-2A Cells

A. Cell Culture and Cell Cycle Kinetics

In most of our cell cycle studies we have employed Neuro-2A cells, originally described by Klebe and Ruddle (1969), which were adapted to grow in Dulbecco's modified Eagle's medium without bicarbonate, buffered to pH 7.5 with 25 m*M* *N*-2-hydroxyethylpiperazine-*N*′-2-ethane sulfonic acid (HEPES). Optimized growth rates were obtained by addition to this medium of 0.4 m*M* of each of the amino acids alanine, asparagine, glutamic acid, proline, and cysteine, and supplementation with 10% fetal calf serum. Depending on the batch of serum and the type of culture substratum the generation time ranges from 7.5 to 11 hours under these conditions. They can be readily synchronized by selective detachment of mitotic cells, without the use of interfering drugs, whereby their relatively short generation time ensures a reasonable yield of mitotic cells (de Laat *et al.*, 1977). At a doubling time of about 10 hours the duration of M, G_1, S, and G_2 phases is about 0.5, 2.5, 5, and 2 hours, respectively.

The analysis of cell cycle related phenomena and of their regulatory relevance requires the use of quantitative descriptions of cell cycle kinetics. Traverse through the cell cycle of synchronized cell populations can be followed by determining cell number, mitotic index, and [^{3}H]thymidine incorporation (de Laat *et al.*, 1977), but descriptions of cell cycle kinetics in exponentially growing cultures can be obtained most readily by analyzing the distribution of intermitotic times (T_i) within a culture, as determined from time lapse films, according to appropriate mathematical models. All models proposed in recent years account for the general observation that genetically identical cells show great variability in T_i, predominantly by variability in their G_1 period. Together with the observation that this period is most susceptible to experimentally induced modifications in cycle times, this has led to the concept that cycle controlling events occur predominantly in G_1 (Pardee, 1974). The various kinetic models for the cell cycle have been based on two principally different views on the origin of variability in T_i. From the earlier deterministic perspective genetically identical individual cells show variability in the rates at which properties are acquired which levels determine cell cycle progression. The variability in T_i arises in this view from cumulation of the differences of an unknown number of rate constants (Pardee *et al.*, 1978, 1979; Rossow *et al.*, 1979; Castor, 1980; Yanishevsky and Stein, 1981).

Smith and Martin (1973) proposed an alternative basis for the description of cell cycle kinetics. In their view the cell cycle consists of a quasi-resting A state comprising part of G_1 in which cells can stay for a variable period of time, governed by a constant probability per unit of time (λ) of leaving the A state. On leaving the A state cells traverse a deterministic B phase of determinate duration (T_B) before their daughters reenter the A state. The average T_i and its sample variation are fully determined by the values of λ and T_B. In this type of analysis the data are usually presented as semilogarithmic plots of (1) the fraction of cells with T_i values larger than or equal to a certain time t, as a function of t (α curve), and (2) the fraction of siblings having an absolute difference in T_i (T_s) larger than or equal to an indicated time t, as a function of t (β curve). According to the original single random transition model α and β curves should yield linear and parallel lines, with slopes equal to $-\lambda$. The Smith and Martin model has stimulated many investigators to analyze in detail the growth kinetics of a variety of cell types and this has resulted in a number of proposed modifications to accommodate for experimentally observed discrepancies with the predictions of the model, in particular regarding the α curve (Svetina, 1977; Murphy *et al.*, 1978; Brooks *et al.*, 1980). The generally observed good linearity of β curves and, in particular, the equality between the mean value of $T_s(\bar{T}_s)$ and its standard deviation [SD(T_s)] as predicted by the occurrence of a random transition (Brooks *et al.*, 1980), still gives support for a description of cell cycle kinetics in which a major control over T_i and its variability is attributed to a random transition. It is beyond the scope of this article to evaluate in more detail the various contributions to this problem. Clearly, the continuing discussion will generate further experimentally based refinements of the various concepts and might be helpful in finding a physical basis for the postulated controling mechanisms. Until then the application of these models at least provides a mean for quantifying differences in cell cycle kinetic behavior.

Family tree analysis of the growth kinetics of exponentially growing Neuro-2A cells under the described culture conditions has revealed features which could not be explained by any of the earlier proposed models (van Zoelen *et al.*, 1981a). As observed in other cell types (van Wijk and van de Poll, 1979) the mean absolute difference in T_i between cell pairs was found to increase with descending family relationship between the cells, and consequently the slope of β curves determined for cell pairs was a function of their relationship, as can be seen from Table I. Taking also into account the positively skewed normal distribution of T_i, these observations could be understood within the concept of a single random transition model by additionally postulating that λ and T_B show a variability within the culture which increases with decreasing family relationship (van Zoelen *et al.*, 1981a).

TABLE I
FAMILY TREE ANALYSIS FOR NEURO-2A CELL PROLIFERATION[a]

Family relationship of cell pairs	Number of cell pairs	Slope of β curve (hr^{-1})	$\bar{T}_s$ (hours)	SD(T_s) (hours)
Siblings	56	2.03 ± 0.11	0.55	0.49
Cousins	112	1.84 ± 0.14	0.72	0.52
Second cousins	224	1.30 ± 0.06	0.95	0.61
Unrelated	56	1.29 ± 0.08	1.00	0.71

[a] For further explanation see text. Data taken from van Zoelen *et al.* (1981a).

Therefore, a mathematical description of the transition probability model was formulated which takes into account a normally distributed variability in λ and T_B, characterized by a mean value and a sample standard deviation, on the basis of which the shape of α and β curves, the nonexponential distribution of T_i and the behavior of the family tree could be explained (van Zoelen *et al.*, 1981a,b). This formulation also allows for the determination of the fraction of cells in the A state (ϕ_A), a useful parameter for comparing kinetic data of cultures under various experimental conditions. Clearly this model contains elements of both the probabilistic and deterministic view of the cell cycle.

An example of an analysis of the cell cycle kinetics of exponentially growing Neuro-2A cells under optimal culture conditions, based on this modified description, is given in Table II. These data show that (1) the β curve for

TABLE II
CELL CYCLE KINETICS OF NEURO-2A CELLS, ACCORDING TO THE MODIFIED DESCRIPTION OF THE TRANSITION PROBABILITY MODEL[a]

Parameter	$\bar{X}$	SD(X)	SEM($\bar{X}$)[b]
Intermitotic time, T_i (hours)	7.57	0.81	
Difference between T_i of sister cells, T_s (hours)	0.55	0.49	
Transition probability, λ (hr^{-1})	2.03		0.11[c]
Duration of A state, T_A (hours)	0.49	0.49	
Duration of B state, T_B (hours)	7.08	0.64	0.06
Fraction of cells in A state, ϕ_A	0.086		0.004

[a] Data were obtained from analysis of time lapse film of exponentially growing Neuro-2A cells, frame interval 0.2 hour, cell density at start of film about 10,000 cells/cm^2. In total, T_i of 56 pairs of siblings were determined. From van Zoelen *et al.* (1981a).

[b] Only for parameters obeying a normal distribution.

[c] Obtained from linear regression analysis of β curve.

siblings fulfills the requirements for a single random transition [$\bar{T}_s =$ SD(T_s)]; (2) $\bar{T}_i$ (7.57 hours) is very short when compared to other mammalian cells in culture; (3) λ (2.03 hr^{-1}) is very high compared to that of other cell types (0.3–0.9 hr^{-1}, see Brooks *et al.*, 1980). The distribution of T_i in these cultures resembles more a normal than an exponential one, whereas the distribution of T_s for siblings clearly obeys an exponential distribution. This is understood by realizing that at such a high value for λ the relatively small variation in T_i will come more from the variability in T_B than from the postulated random transition. However, as will be seen later, experimental manipulation of the growth characteristics can alter these conditions. In conclusion, the short generation time, the possibility of using a simple mitotic selection procedure for synchronization, and the availability of a satisfying kinetic description of their growth behavior make Neuro-2A cells a suitable system for further cell cycle studies.

B. Surface Morphology and Ultrastructure of the Plasma Membrane

The concept of the plasma membrane as a regulatory site for cell cycle traverse implies that modulations in critical membrane functions provide signals necessary for certain transitions in the cell cycle. Such functional modulations will be manifestations of structural and physicochemical alterations within the plasma membrane itself, or might result from cycle-dependent interactions between membrane components and cytoplasmic constituents. During its life history a replicating cell duplicates all its constituents, including its plasma membrane, to ensure phenotypic identity of its progeny. As being documented previously (Bluemink and de Laat, 1977), a possible molecular basis for the proposed functional membrane modulations could arise from asynchronous assembly of the various membrane components with the growth of the plasma membrane during the cell cycle. The data presented in this article provide strong support for this concept.

Time-lapse cinematography and scanning electron microscopy of synchronized Neuro-2A cells show that their shape and their surface architecture undergo dramatic alterations upon progression through the cell cycle, as illustrated in Fig. 1 (van Maurik *et al.*, unpublished observations). The spherical mitotic cells are covered by numerous blebs and microvilli (Fig. 1a). As cells enter G_1 they reattach to the substratum and flatten with a gradual disappearance of the surface irregularities (Fig. 1b). S phase cells become even more flattened and their surface very smooth (Fig. 1c). Premitotic rounding-off is initiated by a retraction of cellular extensions, and is associated with the reappearance of surface blebs and microvilli (Fig. 1d, e, and f). As found in other cells (Porter *et al.*, 1973), these morphological observations suggest that mitotic cells have a large supply of surface area, stored in blebs

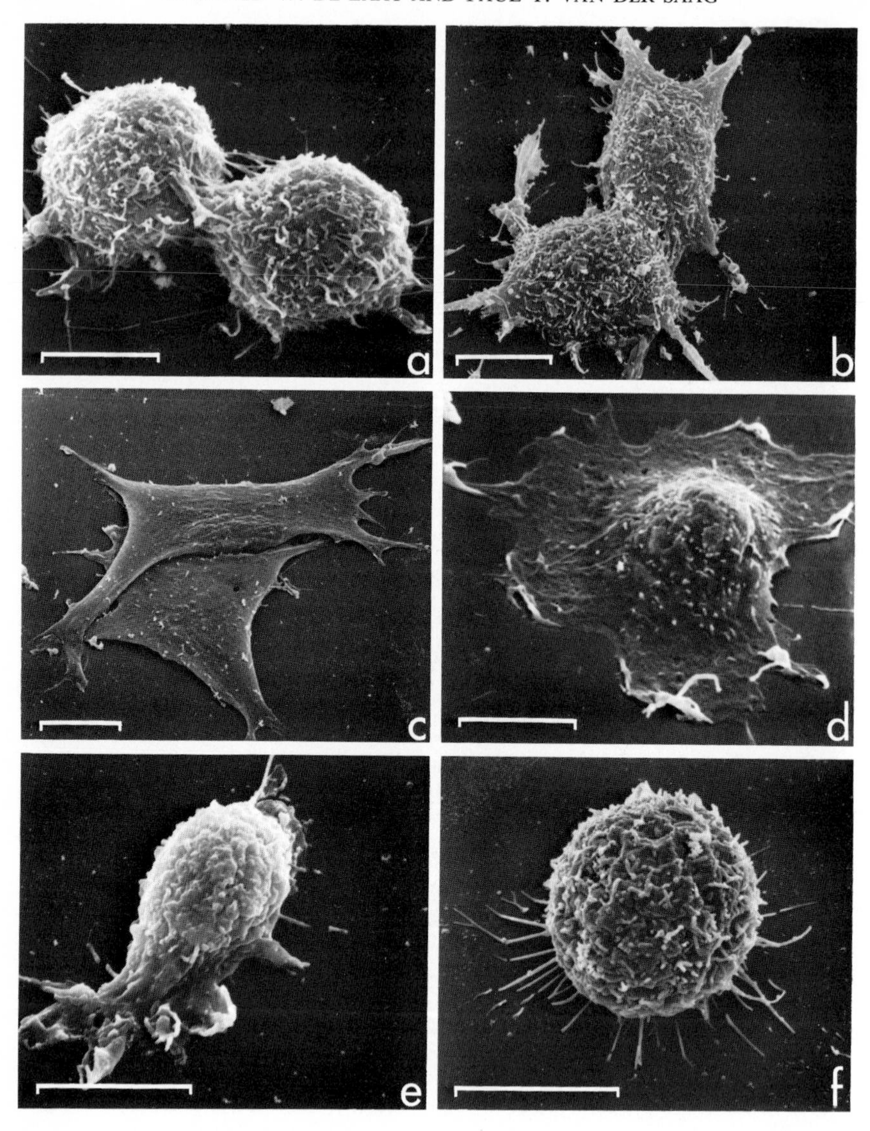

FIG. 1. Scanning electron micrographs of synchronized Neuro-2A cells at various time points after mitotic selection and replating in monolayers: (a) early G_1 (0.3 hour); (b) late G_1 (3.0 hours); (c) late S (7.0 hours); (d, e, f) premitotic rounding-off (9.5 hours). Bars represent 10 μm.

and microvilli, which is used during G_1 to accommodate the rapid increase in area/volume ratio as the cells flatten.

The alterations in surface morphology make it difficult to quantify the process of membrane growth from scanning electron micrographs. To obtain such data we determined the cell diameter during the cell cycle after maximal hypotonic swelling. These measurements showed that the increase in total cell surface area during the cell cycle is not linear (Boonstra *et al.*, 1981b), in contrast to the change in total cellular protein content (Mummery *et al.*, 1981a). The total surface area remains constant from mitosis until the G_1/S transition, then increases gradually until 30% greater in mid S phase, after which it remains constant till G_2. Prior to or during the next mitosis the cell surface area increases rapidly, roughly doubling relative to that in the previous G_1. As the cell surface area will be determined predominantly by the incorporation and turnover of plasma membrane phospholipids, these data indicate that net insertion of phospholipids is largely restricted to a short period around mitosis, and to a lesser extent to S phase. In addition, they provide direct evidence for a process of surface stretching during G_1.

A direct visualization of possible intrinsic modulations within the plasma membrane is given by an ultrastructural analysis using freeze-fracture electron microscopy (Branton, 1966). In the replica of the fracture plane through the membrane lipid bilayer the intramembrane particles (IMP) observed most probably represent integral membrane (glyco)proteins (Singer and Nicolson, 1972). Until now only a limited number of studies of the modifications in internal structure of the plasma membrane during the cell cycle of other cell types have been published. They were all restricted to an analysis of the total IMP density and gave contradictory results (Scott *et al.*, 1971; Torpier *et al.*, 1975; Knutton, 1976).

The development of an improved computer-assisted method to obtain a quantitative description of freeze-fracture images (de Laat *et al.*, 1981a) made it possible to carry out a detailed analysis of modulations in the internal structure of the plasma membrane of synchronized Neuro-2A cells (de Laat *et al.*, 1981b). The basic data for such a quantitative analysis are provided by the digitized relative coordinates (resolution, 10 Å) of the widest parts of the shadow of each of the IMP in selected electron micrographs (magnification, 252,000 ×). Simple counting procedures determine the numerical IMP-distribution in terms of (1) IMP density, as a function of the IMP diameter, and (2) IMP size distribution. These parameters provide information as to compositional changes of the plasma membrane. A differential density distribution analysis is applied to characterize the lateral IMP distribution, i.e., the spatial distribution of IMP within the plane of the membrane, of IMP populations selected on the basis of the IMP diameter. The important presumption of such an analysis is that the number of IMP (N)

per unit membrane area (U) will acquire a random (Poisson) distribution in the absence of significant directional constraints on the lateral mobility of the macromolecules represented by the IMP. A statistical measure for the deviation from a random distribution of the determined IMP density distribution is obtained by calculating the approximate normal deviate (Z) in a Poisson variance test. In other words, dynamic properties of membrane components, which result in significant directional constraints on their lateral mobility, can be detected by an analysis of their lateral distribution, whereby aggregative and dispersive forces can be distinguished. In addition, the nature of the lateral IMP distribution can be characterized at different levels of spatial organization by varying the size of the unit membrane area in the analysis. For full details, see de Laat *et al.* (1981a).

The ultrastructural features of the plasma membrane in Neuro-2A cells were analyzed by freeze-fracture electron microscopy of synchronized cells at 14 time points during the cell cycle. For each of these time points 10 replicas were selected, each representing 0.5 μm^2 of the P-face of the medium-exposed part of the plasma membrane, in which the coordinates of, in total,

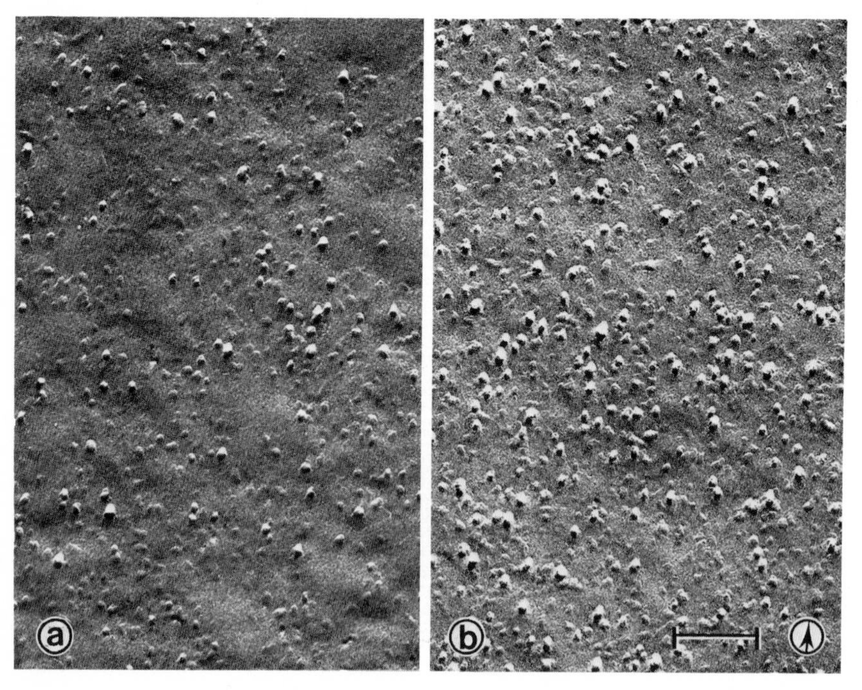

FIG. 2. Representative freeze-fracture electron micrographs of the P-face of the medium-exposed plasma membrane of Neuro-2A cells in late G_1 (a) and mid-S phase (b). Bar represents 0.1 μm.

72,462 IMP were digitized (de Laat *et al.*, 1981b). Visual inspection of the electron micrographs of replicas, representative of the various time points, clearly demonstrates a profound modulation of the total IMP density during the cell cycle, indicating the occurrence of cycle-dependent compositional changes in the plasma membrane (Fig. 2). This is further substantiated by quantitative determination of the IMP density, the IMP size distribution, and the lateral IMP distribution, of which the main features are presented in Figs. 3–5. Membrane areas void of surface protrusions were selected for the determinations. This introduces in particular a gross overestimate of the measured IMP density in mitotic cells (Fig. 3), since the surface blebs and the tips of microvilli described earlier appear as IMP-free (lipid) domains (not shown). This in itself suggests that the surplus surface area formed in this period is merely of lipid nature. Furthermore, it is evident that in mitotic cells also outside these protruding areas large local variations in IMP density occur which express themselves in a significant IMP aggregation, independent of the size of the unit membrane area selected (Fig. 5), indicating the

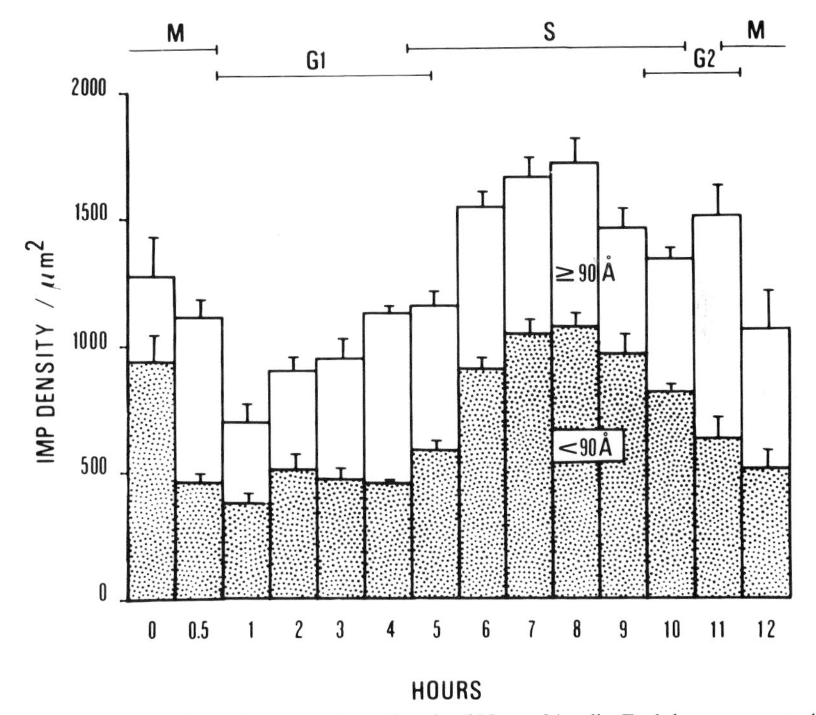

FIG. 3. Total IMP density during the cell cycle of Neuro-2A cells. Each bar represents the mean ± SEM of the number of IMP per μm^2 at a given time point, subdivided over IMP with diameters <90 Å and ≥90 Å, respectively.

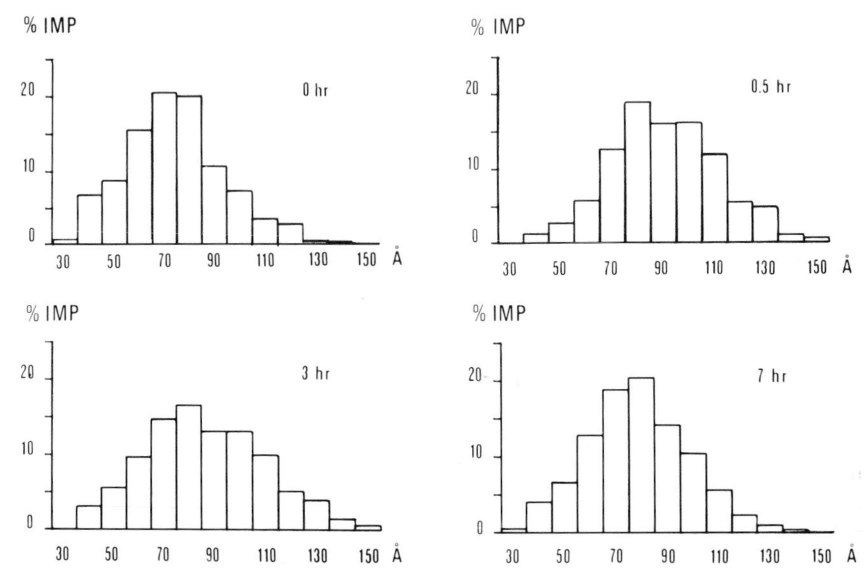

FIG. 4. Relative IMP distribution per diameter-class for mitotic, early G_1, late G_1, and S phase Neuro-2A cells.

action of significant directional constraints on the lateral mobility of the molecules represented. It is tempting to speculate that these constraints are related to or caused by the introduction of the lipid domains. The plasma membrane of mitotic cells is characterized not only by its aggregative IMP distribution, but also by the IMP size distribution which shows a number of unique features. In particular the IMP diameter in Neuro-2A cells ranges from 30 to 170 Å, whereas the smallest mean diameter (73 Å) is found in mitotic cells (Fig. 4).

As cells enter G_1 a radical reorganization within the plasma membrane is observed, which is associated with, or directly related to, the rapid changes in cell shape and surface morphology. In the membrane areas analyzed the density of small IMP (diameter <90 Å) rapidly decreases whereas the number of large IMP (diameter ≥ 90 Å) per unit membrane area increases, leaving the total IMP density nearly constant (Fig. 3). The IMP size distribution shifts accordingly toward larger diameters, so that the mean IMP diameter rapidly increases to 91 Å (Fig. 4). At the same time the IMP lose their aggregated distribution and take random positions within the plasma membrane (Fig. 5). All these data indicate that in this period the directional constraints on the mobility of the observed membrane molecules are released, so that the heterogeneities within the plasma membrane disappear. At the same time new macromolecular complexes are formed from smaller subunits,

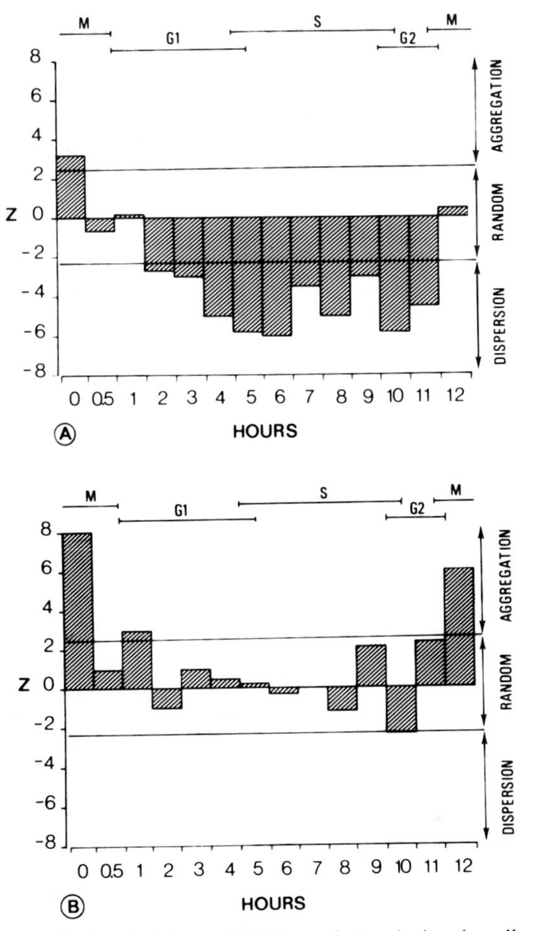

FIG. 5. Lateral distributional of the total IMP population during the cell cycle of Neuro-2A cells as determined by differential density distribution analysis, with unit membrane area U equal to 510 × 510 Å (A) and 1020 × 1020 Å (B), respectively. The horizontal lines indicate the 1% significance levels for the approximate normal deviate Z, as given by the Poisson variance test.

or a displacement of the IMP with respect to the fracture plane occurs (Borochov and Shinitzky, 1976), which is reflected in the rapid change in IMP size distribution.

Upon further progression through G_1 the density of small IMP remains constant but new large IMP are preferentially formed (Figs. 3 and 4). Macroscopic heterogeneities in the lateral IMP distribution remain negligible as seen by the random distribution at large value of U in the differential density distribution analysis (Fig. 5B). However, at a small value of U new molecular interactions become apparent, manifested by a significant dispersed IMP distribution (Fig. 5A), which suggest the formation of some sort of anchorage of membrane proteins.

The structural changes described so far have occurred at a constant total surface area reflecting structural modifications associated with the stretching of the cell surface and the disappearance of IMP-free areas. From the G_1/S transition onward the cell surface area increases. The gradual net insertion of phospholipids is accompanied by a preferential formation of small IMP (Fig. 3). Since the density of large IMP remains constant the IMP size distribution gradually shifts during S phase toward smaller diameters (Fig. 4), so that by the end of S the mean IMP diameter is 77 Å. The lateral IMP distribution remains similar to that of late G_1 cells.

A second phase of radical reorganization within the plasma membrane starts just prior to the next mitosis although the partial desynchronization of the synchronized cell populations at the end of the cell cycle restricts the possible time resolution of these events. The rapid increase in total cell surface area in this period is associated with the formation of lipid domains located in surface protrusions and with a corresponding decrease in the density of small IMP. The large IMP, however, show a transient increase in density (Fig. 3). Concomitantly, the dispersive constraints on the lateral mobility of the IMP are released and overruled by aggregative forces, in particular on a macroscopic scale (Fig. 5), probably resulting from the introduction of new lipid domains.

Table III summarizes the main features of the alterations in surface morphology and plasma membrane structure during the cell cycle of Neuro-2A cells. These data strongly support the concept of asynchronous assembly of various membrane components during cell cycle related membrane growth (see also Bluemink and de Laat, 1977). Characteristic alterations in the

TABLE III
MAIN FEATURES OF THE ALTERATIONS IN SURFACE MORPHOLOGY AND PLASMA MEMBRANE STRUCTURE DURING THE CELL CYCLE OF NEURO-2A CELLS

Phase of cell cycle	M	G_1	S	G_2/M
Surface morphology	Blebs, microvilli	Stretching	Smooth	Blebs, microvilli
Net lipid insertion	High	No	Moderate	High
Predominant size of formed IMP	Small	Large	Small	Large
Lateral IMP distribution at large scale	Aggregated	Aggregated/random	Random	Aggregated
Lateral IMP distribution at small scale	Aggregated	Random/dispersed	Dispersed	Random/aggregated

membrane properties investigated coincide with the main transitions in the cell cycle, so that, based upon this structural analysis, distinct static and dynamic properties can be attributed to various phases in the life history of the cell. It seems very likely that these structural alterations are reflected in modulations in the physicochemical and functional properties of the plasma membrane, which could provide essential signals for the cell cycle progression.

C. Dynamic Properties of Plasma Membrane Components

The ultrastructural analysis presented above (de Laat *et al.*, 1981b), and in particular the outcome of the analysis of the lateral IMP distribution, provides evidence for cell cycle-dependent modulations in the dynamic properties of membrane components, originating from compositional changes within the plasma membrane, and possibly also from stage-specific interactions between membrane components and cytoskeletal components. It might be expected that changes in the constraints on the dynamic properties of membrane components will influence, e.g., the conformation and functioning of membrane-bound enzymes and transport systems, and the expression of membrane receptors for hormones and growth factors. Through such mechanisms the cell surface might exert its proposed key role in growth control.

Two independent fluorescence methods have been applied to Neuro-2A cells to measure possible cell cycle-dependent modulations in the dynamic behavior of plasma membrane components directly. The rotational mobility of the fluorescent hydrocarbon 1,6-diphenyl-1,3,5-hexatriene (DPH) was determined by steady-state fluorescence polarization measurements (Shinitzky and Barenholz, 1978) on synchronized populations of intact cells, as a measure for the microviscosity ($\bar{\eta}$) of the hydrophobic region of the plasma membrane lipid matrix (de Laat *et al.*, 1977). These measurements have demonstrated that substantial changes in membrane lipid fluidity occur during the cell cycle, in particular around mitosis. During this period $\bar{\eta}$ rises transiently to a maximum of 3.5 poise. As cells enter G_1, $\bar{\eta}$ decreases rapidly about twofold to 1.9 poise, a value which is maintained during the remaining part of interphase.

The lateral mobility of fluorescent plasma membrane lipid probes and fluorescently labeled membrane proteins was determined during the cell cycle of Neuro-2A cells by the fluorescence photobleaching recovery (FPR) method (de Laat *et al.*, 1980). This method allows for the quantitative determination of the lateral motion of fluorescently labeled membrane components over macroscopic distances (a few micrometers) in a localized membrane area of an intact cell (Koppel *et al.*, 1976; Axelrod *et al.*, 1976). Usually, this method is based on single photon counting of the fluorescence intensity of a defined membrane area (radius 1 μm), using a modified epiluminescence fluorescence microscope with a focused attenuated laser beam as the excitation source. The mobility properties of the fluorophores are detected by partial photobleaching of this area by a short pulse of intense focused laser light and subsequent monitoring of the motion of unbleached fluorophores from the surrounding plasma membrane into the bleached region. Such motion leads to a recovery of the fluorescence intensity in the illuminated area. The kinetics and degree of this recovery are used to determine the nature and the rate of motion, as well as the fraction of mobile fluorophores. In Neuro-2A cells we have employed the lipid analog 3,3′-dioctadecylindocarbocyanine iodide (diI) and a fluorescein-labeled ganglioside (F-GM1) as probes for the lateral mobility of membrane lipids. Membrane proteins were labeled with rhodamine-conjugated rabbit antibodies (Fab′ fragments) against mouse El_4 cells ($RaEl_4$), which showed reactivity with an unknown cross-section of the surface antigens of Neuro-2A cells.

In correspondence with the earlier data on the modulation of membrane microviscosity, the lateral mobility of both lipid probes showed a two- to threefold change in diffusion coefficient (D) during the cell cycle, being minimal in mitosis at 1.8×10^{-9} and 3.1×10^{-9} cm^2/second for diI and F-GM1 (Fig. 6), respectively. After an initial increase during early G_1 phase

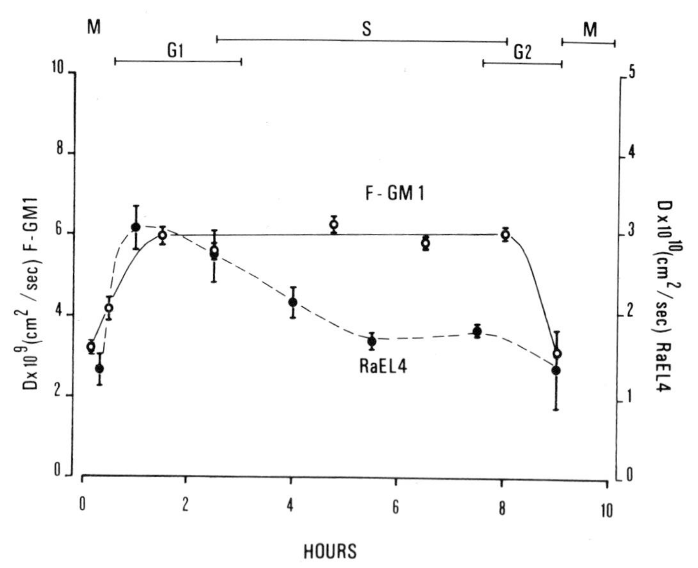

FIG. 6. Lateral mobility of membrane lipids (F-GM_1) and membrane proteins ($RaEl_4$) during the cell cycle of Neuro-2A cells. Diffusion coefficients (D) are given as mean ± SEM. From de Laat *et al.* (1980).

D remains constant for both probes at about 6×10^{-9} cm^2/second during further cycle progression, and decreases again just prior to the next mitosis (de Laat *et al.*, 1980). These results provide further evidence for the occurrence of substantial changes in the dynamic properties of plasma membrane lipids, associated particularly with the process of cytokinesis, and suggest that the lateral mobility of membrane lipids is primarily controlled by the fluidity of the membrane lipid matrix.

The antigenic sites recognized by $RaEl_4$ show, in part, a different behavior (Fig. 6). The lateral mobility of these membrane proteins is also minimal in mitosis ($D = 1.3 \times 10^{-10}$ cm^2/second) and rises rapidly in early G_1 to a maximum value of $D = 3.1 \times 10^{-10}$ cm^2/second. However, in contrast to the constant mobility properties of the membrane lipids during the remainder of interphase the diffusion coefficient of $RaEl_4$-binding sites shows a gradual decrease during S phase reaching a minimum again at the next mitosis. Although these experiments cannot exclude that qualitatively different sites are detected by $RaEl_4$ at different stages of the cell cycle, they suggest that during M and G_1 protein mobility is primarily controlled by membrane lipid fluidity, whereas other constraints, e.g., interactions with cytoskeleton elements, become effective in S phase.

It is interesting to consider these data in view of the changes in membrane structure described previously (de Laat *et al.*, 1981b, see Section II,B). Evidently, the plasma membrane of mitotic and early G_1 cells not only exhibits unique structural features, but also shows unique dynamic properties. In this period rapid lipid insertion, detectable by the formation of IMP-free domains (blebs) and by the sudden increase in total cell surface area, is associated with a change in the IMP size distribution, an aggregated lateral IMP distribution, and relatively low lipid and protein mobility. The transient formation of different membrane domains, as seen in freeze-fractured membranes of mitotic cells, poses, however, a problem in interpreting the spectroscopic data on the rotational and lateral mobility of membrane lipids and protein, since it is as yet unknown whether the probe molecules have preference for certain domains. As all probes show a similar behavior it seems nevertheless plausible that the rapid assembly of (specific?) lipids is the cause of the observed alterations in mobility properties, and the source of aggregative constraints on the IMP distribution. In addition, it could well be that the sudden change in the IMP size distribution during mitosis is partly, or fully, due to vertical displacement or conformational changes of membrane proteins induced by altered membrane lipid fluidity (Borochov and Shinitzky, 1976). During the other cell cycle phases a striking correlation also exists between the dynamic aspects of the ultrastructural analysis and membrane protein mobility. The lateral IMP distribution is only random under all conditions during a certain period of the G_1 phase.

As this indicates the absence of significant directional constraints on the represented macromolecules, this corresponds well with the observed transient maximum in protein mobility. Conversely, dispersive constraints become apparent in S phase, as protein mobility gradually decreases while lipid mobility is not affected. Anchorage of membrane proteins to cytoskeletal elements (Edelman, 1976) could explain this non-lipid-mediated restriction of protein motion.

Are these modulations in the mobility properties of membrane components general to cycling cells? This important question is as yet difficult to answer. There is general agreement from a number of studies on various cell types that stimulation of quiescent cells leads to an increase in membrane lipid fluidity (Inbar and Shinitzky, 1975; Collard *et al.*, 1977; Chen and Levy, 1979), which could be correlated in normal regenerating and malignant hepatocytes to a decreased cholesterol/phospholipid ratio in isolated plasma membranes (Chen and Levy, 1979). This suggests that the $G_1(G_0)/S$ transition is associated with an increased membrane fluidity, in accordance with our findings on synchronized Neuro-2A cells (de Laat *et al.*, 1977, 1980). On the other hand, fluorescence polarization measurements on synchronized L1210 mouse leukemic cells did not reveal significant membrane fluidity changes in relation to the cell cycle (Obrénovitch *et al.*, 1978), but it should be noted that the degree of cell synchronization was relatively poor. More recently Lai *et al.* (1980) reported an electron spin resonance study on the changes in membrane fluidity during the cell cycle of CHO cells. As for Neuro-2A cells, these cells were synchronized by mitotic selection, but the synchronized populations were cultured in suspension. A decreased fluidity was observed in G_1 but, in contrast to Neuro-2A cells, an increase occurred in M as well as in S phase. Whether technical differences, cell type specificity, or substrate attachment versus suspension conditions are the origin of the discrepancy with respect to the properties of mitotic cells is unclear, and further studies will be required to answer this question.

D. Cation Transport

So far we have presented experimental evidence for cell cycle-related ultrastructural and physicochemical modulations within the plasma membrane of Neuro-2A cells. In this section attention is given to functional membrane changes during the cell cycle, in particular related to the transport of cations across the plasma membrane.

All cells have developed mechanisms for maintaining ionic gradients across their plasma membranes. In eukaryotic cells the plasma membrane bound Na^+,K^+-ATPase is primarily responsible for setting intracellular Na^+ at a low level and K^+ at a high level with respect to the extracellular ionic conditions. The enzyme activity depends on substrate conditions like K^+ at the extracellular and Na^+, Mg^{2+}, inorganic phosphate and ATP at the intracellular side of the membrane (Robinson and Flashner, 1979), and on the physicochemical properties of the lipidic environment of the enzyme (Kimelberg, 1976; Mandersloot *et al.*, 1978; Sandermann, 1978). The enzyme can be inhibited specifically by the cardiac glycoside ouabain.

The resulting ionic gradients and electrical gradients, caused by the charge separation involved, are used in transmembrane signaling and in transporting other ions and solutes across the plasma membrane. Evidence has been presented that basic cellular processes, like protein synthesis, are dependent on the intracellular Na^+ and K^+ concentrations (Lubin, 1967; Christman, 1973). In addition, the involvement of the cation transport system and related electrical potentials in the regulation of cell proliferation has been demonstrated (Jung and Rothstein, 1967; Cone, 1969, 1971; Sachs *et al.*, 1974), but the mode of action in this context is far from understood.

In a recent series of studies from our laboratory two strategies were adopted to gain more insight in the role of the cation transport systems in growth control. First, the passive and active transport systems for K^+ and Na^+ have been characterized in synchronized Neuro-2A cells (Boonstra *et al.*, 1981a,b; Mummery *et al.*, 1981a,b; van Zoelen *et al.*, 1981b), and second, the early ionic events upon serum stimulation of serum-deprived, quiescent N1E-115 cells have been investigated (Moolenaar *et al.*, 1979, 1981b) (see Section III). In both cases, the relevance of the observed alterations is determined by studying the effects of selective manipulation of the various transport systems.

Combining electrophysiological, tracer flux, and biochemical methods a relatively comprehensive picture of the cell cycle related modulations in K^+ and Na^+ transport was obtained. A first indication as to such modulations was found by measuring the membrane potential (E_m) at various times in the cell cycle by conventional microelectrode techniques (Boonstra *et al.*, 1981b). E_m is maximally hyperpolarized in mitotic cells (-45 mV), depolarizes rapidly in G_1 to -23 mV, and rises gradually from the onset of S to reach again a value of -45 mV in mid S, after which it remains constant (Fig. 7A). The membrane potential is determined predominantly by the transmembrane K^+ gradient, maintained by the Na^+,K^+-ATPase, and the relative permeabilities for K^+ and Na^+ (Boonstra *et al.*, 1981a). Direct intracellular measurements of the K^+ activity (a_K^i) by K^+-selective, liquid ion-exchange type, microelectrodes demonstrated that a_K^i also shows a substantial modulation during the cell cycle (Fig. 7B), largely in parallel with E_m, with a maximum of 127 m*M* in mitotic cells and a minimum of 80 m*M* in G_1. As the changes in the transmembrane K^+ gradient contribute only partially to the magnitude of observed changes in E_m, also permeability

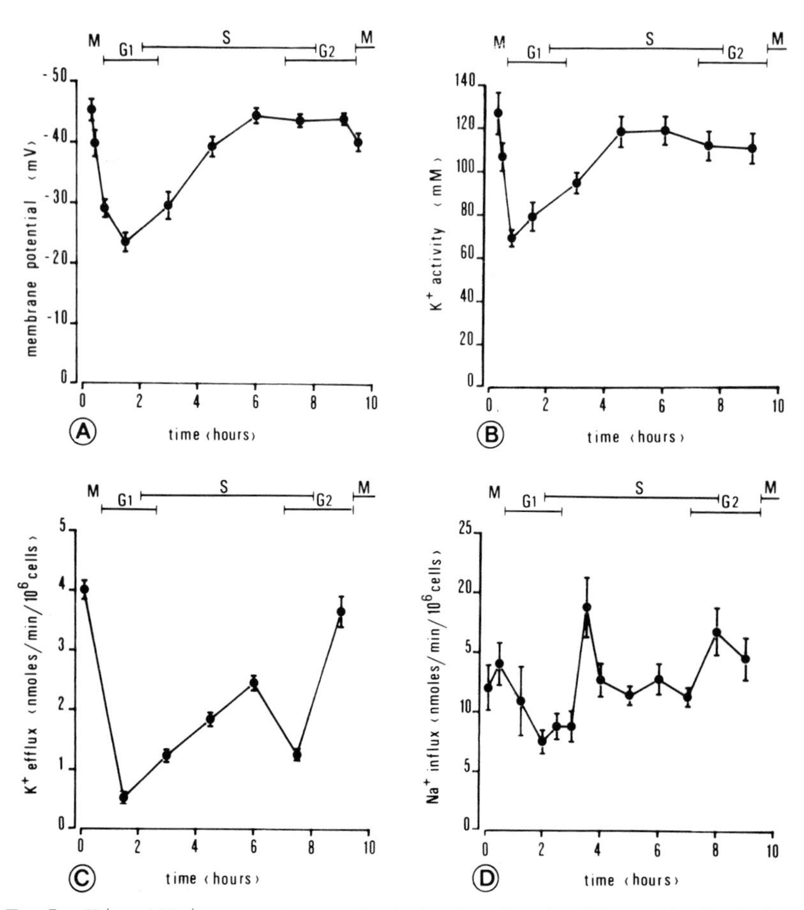

FIG. 7. K^+ and Na^+ transport properties during the cell cycle of Neuro-2A cells. (A) Membrane potential, as determined by intracellular microelectrode measurements. (B) Intracellular K^+ activity, as determined by intracellular measurements with K^+-selective microelectrodes. (C) K^+ efflux rate, as determined by $^{42}K^+$ efflux measurements. (D) Na^+ influx rate, as determined from the initial uptake of ^{24}Na during 3 minutes, in the presence of 5 mM ouabain. (E) Na^+,K^+-ATPase-mediated K^+ influx, as determined by measurements of the ouabain-sensitive ^{42}K influx. (F) Na^+,K^+-ATPase activity, as determined by the ATP hydrolysis under optimal substrate conditions in cell homogenates. In all cases the data are presented as mean ± SEM. From Boonstra *et al.* (1981b) (A, B, C) and Mummery *et al.* (1981a) (E, F), with permission of Alan R. Liss, Inc., New York.

changes must be involved. From the electrophysiological data the relative permeability for electrodiffusional transport of Na^+ and K^+ can be inferred. This permeability ratio changes during the cell cycle in association with the modulation in E_m, with a minimum of 0.16 in mitosis and a maximum of 0.26 in G_1 (Boonstra *et al.*, 1981b).

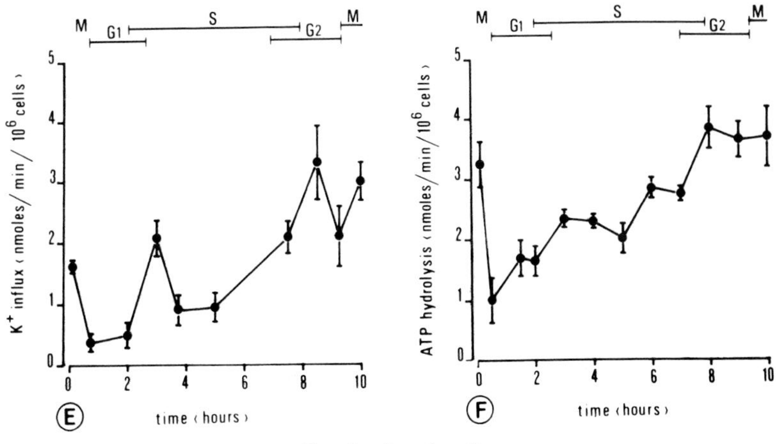

FIG. 7. (*continued*)

Further details were obtained by radioactive tracer flux analysis, using $^{42}K^+$ and $^{24}Na^+$. The unidirectional K^+ efflux rate, shown in Fig. 7C, is again at maximum during mitosis (4 nmoles K^+/minute/10^6 cells), decreases eightfold upon entering in G_1 to 0.5 nmole K^+/minute/10^6 cells, followed by a gradual five fold increase during S to 2.5 nmoles K^+/minute/10^6 cells. A large increase is seen before or during the next mitosis. The K^+ permeability (P_K), depending on flux rate and E_m, follows qualitatively a similar pattern during the cycle (Boonstra *et al.*, 1981b). Subsequently, the permeability for electrodiffusional transport of Na^+ (P_{Na}) can be determined from the values of P_K and P_{Na}/P_K. P_{Na} changes qualitatively according to a pattern similar to P_K but the relative change upon entry in G_1 is smaller than for P_K, leading to the increase in P_{Na}/P_K mentioned earlier (Boonstra *et al.*, 1981b).

To maintain its relatively high intracellular K^+ level the cell has to take up K^+ from the medium against its concentration gradient. This K^+ influx consists of at least two components: an ouabain-sensitive part, mediated by the membrane-bound Na^+,K^+-ATPase, and an ouabain-insensitive part, through, e.g., exchange mechanisms. Whereas the latter component rises roughly in a linear way during the cell cycle, a characteristic modulation of the Na^+,K^+-ATPase mediated K^+ influx is observed, as presented in Fig. 7E (Mummery *et al.*, 1981a). This component decreases more than fourfold as the cells enter into G_1, and, most characteristically, shows a transient sixfold increase on entry into S. Subsequently, a gradual increase is seen until a small but significant decrease becomes apparent just prior to the next mitosis.

As expected already from the changes in a_K^i, K^+ transport is not in a steady state during the cell cycle. Taking the difference between the total K^+

influx and efflux shows that certain periods in the cell cycle are characterized by a net efflux of K^+ (early G_1) and other phases by a net K^+ influx (late G_1 to early S; late S). These net fluxes are in the right order of magnitude to account for the observed changes in the intracellular K^+ content (Mummery *et al.*, 1981a).

The major alterations in cation permeabilities occur around mitosis, and involve intrinsic changes in plasma membrane properties. They coincide with radical changes in the ultrastructural and dynamic properties of the plasma membrane (see Sections II,B,C), suggesting a common molecular basis. Support for this hypothesis was obtained recently (Bierman *et al.*, unpublished observations) by studying the effects of the incorporation of fatty acids of varying degree of saturation on membrane lipid fluidity, membrane potential, and cation permeability of exponentially growing Neuro-2A cells, cultured in a serum-free chemically defined medium, which sustains continuous growth (Bottenstein and Sato, 1979). Supplementation with unsaturated fatty acids (oleic acid, linoleic acid) specifically induces a parallel increase in the lateral mobility of membrane lipids, a depolarization of E_m, and an increase in P_{Na}/P_K, thereby mimicking part of the events normally occurring upon entry into G_1.

The observed modulations in Na^+,K^+-ATPase mediated K^+ influx, or functional Na^+,K^+-ATPase activity, can be influenced, however, by membrane-bound and extrinsic mechanisms: (1) the number of active pump sites per cell, (2) the intrinsic properties or conformation of the active pump sites (e.g., substrate affinities), and (3) the availabilities of intracellular substrates, Na^+ being the most important one. In contrast to substrate availability the other mechanisms are related to intrinsic properties, controlled by assembly and turnover of the enzyme and interactions between the enzyme and its immediate environment.

A first distinction between these extrinsic and intrinsic modes of regulation is obtained by comparing the activity of the enzyme in cell homogenates under optimal substrate conditions (Fig. 7F) with its functional pump activity *in situ* (Fig. 7E). Obviously, the common characteristics are most probably due to intrinsic membrane changes. Such a comparison leads to a number of conclusions (Mummery *et al.*, 1981a). First, based on the assumption that two K^+ ions are being translocated for each hydrolyzed ATP molecule, the cell utilizes only part of the maximum pumping activity in all phases of the cell cycle. van Zoelen *et al.* (1981b) showed recently that addition of 3.5 m*M* ATP to intact cells induces a more active conformation of the Na^+,K^+-ATPase, as measured by the ouabain-sensitive K^+ influx. This ATP effect certainly will contribute to this discrepancy, since similar levels of ATP are present on both sides of the plasma membrane in the ATP hydrolysis assay. Second, a drastic decrease in optimal as well as in functional activity is observed as the cells enter into G_1. Preliminary results of measurements of the affinity for [^{3}H]ouabain indicate that a conformational change of the enzyme is associated with the entry into G_1 (van Zoelen *et al.*, unpublished data). Measurements of the number of [^{3}H]ouabain binding sites are in progress to see whether changes in the number of active pump sites are also involved. Third, optimal and functional activity do not change in concert as cells proceed from G_1 through the cell cycle, the most prominent difference being the substantial, transient increase in ouabain-sensitive K^+ influx at the G_1/S transition which is not reflected in the enzyme activity under optimal substrate conditions. Subsequent measurements of ^{24}Na influx (Fig. 7D) demonstrated that this transient increase in Na^+,K^+ pump activity is associated with a transient twofold increase in Na^+ influx rate (Mummery *et al.*, 1981b). It should be noted that the total Na^+ influx results from the sum of a number of different components, e.g., electrodiffusional transport, Na^+–Na^+ exchange, and solute-linked transport (Mills and Tupper, 1975).

Further characterization of the nature of the transient Na^+ influx at the G_1/S transition is currently in progress. So far, the following properties can be ascribed to it: (1) it is an electroneutral flux, i.e., it is not accompanied by changes in E_m, suggesting an ion-exchange process, (2) it is completely and specifically inhibitable by the diuretic amiloride (0.3 m*M*), a known inhibitor of Na^+,H^+ exchange in epithelial cells and, as shown recently by Moolenaar *et al.* (1981a,b), also in N1E-115 neuroblastoma cells (see Section III), (3) Na^+ influx in early G_1 cells is not amiloride sensitive, (4) in normal medium without serum the transient increase in amiloride-sensitive Na^+ influx is absent, and (5) inhibition of this Na^+ influx by amiloride prevents the transient stimulation of the Na^+,K^+-ATPase mediated K^+ influx, but the Na^+,K^+-ATPase activity as such is not amiloride-sensitive.

All together these properties suggest that we are dealing with a serum-induced activation of an amiloride-sensitive Na^+,H^+ exchange mechanism, which is silent in early G_1 cells. As a result of this activation, Na^+ enters the cells and is extruded again as it enhances the Na^+,K^+ pump activity, thereby inducing an increased K^+ influx. Furthermore, it seems likely that as a result of this sequence of ionic events intracellular pH will become more alkaline at the G_1/S transition, a possibility which is presently being studied.

Are these modulations in ionic transport properties relevant for cell cycle progression? The analysis of the effects of sublimiting concentrations of ouabain and amiloride on cell cycle kinetics of nonsychronized cultures demonstrates that in particular the ionic conditions set by the described transport changes at the G_1/S transition are required for normal progression through the cell cycle (Table IV). Partial inhibition of either the amiloride-sensitive Na^+ influx or the Na^+,K^+-ATPase activity results in a more

TABLE IV

EFFECTS OF OUABAIN AND AMILORIDE ON CELL CYCLE KINETICS OF NEURO-2A CELLS[a]

		Ouabain (mM)		Amiloride (mM)	
	Control	0.2	0.5	0.1	0.2
Generation time, T_g (hours)	10.3 ± 0.4	15.2 ± 0.8	20.3 ± 0.9	12.7 ± 0.3	26.0 ± 0.03
Transition probability, λ (hr^{-1})	1.39 ± 0.06	0.66 ± 0.06	0.35 ± 0.01	0.66 ± 0.13	0.11 ± 0.01
Duration of B state, T_B (hours)	9.6 ± 0.3	13.7 ± 0.7	17.5 ± 0.7	11.2 ± 0.3	17.8 ± 0.3
Fraction of cells in A state, ϕ_A	0.09 ± 0.04	0.13 ± 0.01	0.18 ± 0.01	0.15 ± 0.02	0.39 ± 0.03
Number of cells	20	34	32	52	18

[a] Data obtained from an analysis of time lapse films, according to van Zoelen *et al.* (1981a), of exponentially growing cultures in the presence or absence of sublimiting concentrations of ouabain and amiloride. From Mummery *et al.* (1981a,b).

pronounced effect on λ than on T_B, and consequently the cells accumulate in the A state, i.e, before entry into S.

In conclusion, profound changes in the properties of the passive and active transport systems for K^+ and Na^+ occur during the cell cycle of Neuro-2A cells, the major changes being associated with mitosis and the G_1/S transition. The significance of the mitosis-related alterations has yet to be established, but the sequence of ionic events in late G_1 creates conditions which are a prerequisite for further progression through the cell cycle. The large changes in K^+ and Na^+ permeabilities and Na^+,K^+-ATPase activity observed around mitosis reflect intrinsic membrane changes, which coincide with radical changes in the ultrastructural features and the dynamic behavior of the plasma membrane (Section II,B,C). It will be a challenge to investigate the possible common molecular basis and biological significance of these events. Great similarity is seen between the ionic changes in late G_1 in synchronized Neuro-2A cells, cultured in the continuous presence of serum, and those occurring upon serum addition to serum-deprived, quiescent N1E-115 cells (Moolenaar *et al.*, 1981b; see Section III). The present data favor the hypothesis that cells are not committed to respond properly to external growth factors in early G_1, but acquire this capacity in late G_1. Whether this is due to modulations in the number or expression of membrane receptor sites for growth factors, or to the formation of a functional coupling between the ligand–receptor complex and the described transport systems is yet unclear. However, the described unmasking of nerve growth factor (NGF) membrane-specific binding sites in neuroblastoma cells during G_1

(Revoltella *et al.*, 1974) is suggestive for the first mechanism. Finally, in view of the proposed role of cytoplasmic alkalinization in switching on DNA synthesis in sea urchin eggs (Johnson *et* al., 1976) and N1E-115 cells (Moolenaar *et al.*, 1981b), the effects of extracellular acidification on cell proliferation and DNA synthesis in Neuro-2A cells (van der Saag *et al.*, 1981), and the evidence summarized here for the turning-on of a Na^+,H^+ exchange system at the G_1/S transition, it seems likely that intracellular pH could play a general key role in growth regulation.

III. Growth Stimulation and Cation Transport in N1E-115 Cells

In the previous section we have described a number of cell cycle-dependent modulations in structural, dynamic, and functional plasma membrane properties, using synchronized Neuro-2A cell cultures, and we have analyzed to some extent their significance for cell cycle transition. Cause and effect of growth-related phenomena have been studied successfully in a variety of cells by monitoring the sequence of physiological and biochemical events leading to the initiation of DNA synthesis and cell division upon growth factor or serum addition to quiescent (G_0) cells. It is generally accepted that growth factors initiate their action by binding to specific plasma membrane receptor sites. Among the first detectable events following serum stimulation of quiescent animal cells are marked changes in cation transport properties, in particular a stimulation of the Na^+,K^+-ATPase pump activity (Rozengurt and Heppel, 1975; Smith, 1977; Tupper *et al.*, 1977; Kaplan, 1978). Evidence has been presented that the enhanced pump activity results from a rapid increase in Na^+ influx rate (Smith and Rozengurt, 1978; Mendoza *et al.*, 1980). Thus, great similarity seems to exist between the events evoked by growth factor–receptor interaction in quiescent G_0 cells and those occurring in late G_1 in synchronized Neuro-2A cells, cultured in the continuous presence of serum (Section II,D).

The presence of serum in the growth medium seems to suppress the expression of the differentiated phenotype in a number of neuroblastoma cell lines, e.g., N1E-115 cells, since serum withdrawal is often a sufficient trigger for growth arrest and initiation of differentiation (see Section IV). Knowledge of the serum-induced molecular processes is therefore pertinent to the understanding of the initiation of differentiation.

Mouse N1E-115 neuroblastoma cells have a generation time of about 24 hours under normal culture conditions. Serum withdrawal leads to an accumulation of the cells in G_1 and almost complete cessation of DNA synthesis within 24 to 48 hours. After this period the first signs of morphological differentiation are apparent, but the differentiation process is still reversible. Readdition of 10% fetal calf serum (FCS) or dialyzed FCS, lacking the low-molecular-weight components (MW < 1200), results in a synchronous reentry into S after an 8 to 10 hour lag period, and cell division 10 hours later. The Na^+ influx inhibitor amiloride completely suppresses this serum-induced DNA synthesis and cell multiplication (Moolenaar *et al.*, 1981b), as it does in other cultured cells (Koch and Leffert, 1979; Rozengurt and Mendoza, 1980) and fertilized sea urchin eggs (Johnson *et al.*, 1976).

The early ionic effects upon growth factor membrane interaction in N1E-115 cells have been studied in detail by electrophysiological and tracer flux methods (Moolenaar *et al.*, 1979, 1981b). Adding dialyzed FCS to 24 hour serum-deprived cells evokes within seconds a characteristic response in membrane potential (E_m) and resistance. Before stimulation E_m has a mean value of -38 mV. Serum addition results in a transient rapid hyperpolarization (H phase), mainly due to an increase in K^+ permeability, followed by a gradual depolarization (D phase) of 5 to 8 minutes duration, reflecting an increase in the permeability for several ionic species, including Na^+ and K^+ and possibly also Ca^{2+}. Subsequently, E_m reaches a new stable level of approximately -30 mV. A striking decrease in membrane resistance accompanies this electrical response. Growth factor-depleted FCS neither induces DNA synthesis nor induces comparable electrical effects. In addition, ouabain or amiloride has no effect on this electrical response, indicating that (1) the response is not mediated directly by the Na^+,K^+-ATPase, and (2) amiloride acts on electroneutral ion fluxes (Moolenaar *et al.*, 1979, 1981a,b).

Further details were obtained from measurements of the unidirectional Na^+ influx and the Na^+,K^+-ATPase-mediated K^+ influx (Fig. 8). Addition of dialyzed FCS to N1E-115 cells after 24 hours of serum starvation results in a twofold stimulation of the Na^+ influx within minutes, and a slightly delayed stimulation of similar order of the Na^+,K^+-ATPase pump activity. Growth factor-depleted FCS has no such effects, and amiloride blocks these events at concentrations that inhibit DNA synthesis (0.4 m*M*) without affecting the basal rates of Na^+ influx and ouabain-sensitive K^+ uptake. Taking also into account that (1) the activation of these transport changes is dependent on extracellular Na^+, and (2) various Na^+ translocators stimulate pump activity in the absence of FCS, we conclude that serum growth factors activate an amiloride-sensitive Na^+ transport system, which is not or hardly functional in quiescent cells, and that the entry of Na^+ into the cell in turn enhances the Na^+,K^+ pump activity (Moolenaar *et al.*, 1981b). Whether and how these ion fluxes are coupled to the changes in electrical membrane properties remain to be established.

What is the nature of this amiloride-sensitive Na^+ transport mechanism? As mentioned above, the activated Na^+ entry is electroneutral, suggesting a direct coupling of Na^+ influx to anion influx or cation efflux. In a subsequent

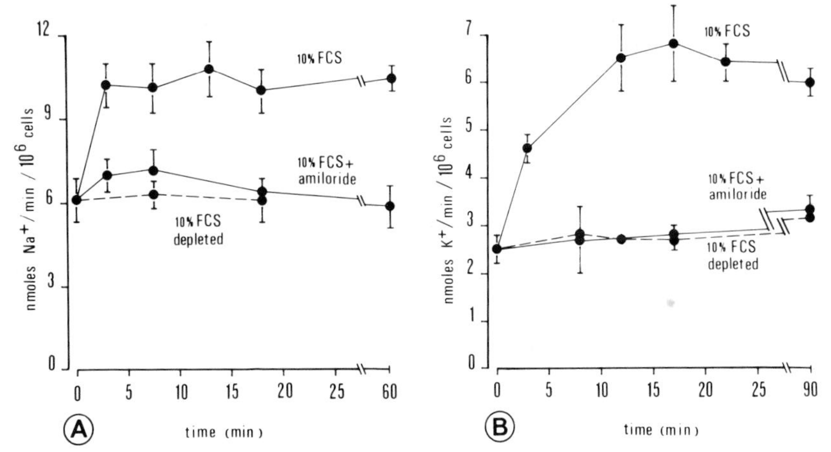

FIG. 8. Serum stimulation of N1E-115 cells after 24 hours of serum deprivation. Dialyzed fetal calf serum (FCS), growth factor-depleted serum (FCS depleted), and amiloride (0.3–0.4 m*M*) were added at time zero, and were present continuously. (A) Kinetics of stimulation of Na^+ influx, as determined by the initial rate of ^{24}Na uptake during 5-minute intervals in the presence of 5m*M* ouabain. (B) Kinetics of stimulation of the Na^+,K^+-ATPase-mediated K^+ influx, as determined by the ouabain-sensitive ^{86}Rb influx. The data are presented as mean ± SEM. From Moolenaar *et al.* (1981b) with permission of M.I.T. Press, Cambridge, Massachusetts.

series of experiments (Moolenaar *et al.*, 1981a,b) it was shown that the amiloride-sensitive Na^+ influx component is not of simple electrodiffusional origin. Furthermore, it was demonstrated that the entry of Na^+ is associated with a release of H^+, distinct from metabolic acid production, and a rise in intracellular pH. Conversely, Na^+ uptake can be stimulated by lowering the intracellular pH with externally applied acetate. Amiloride inhibits all these processes (Moolenaar *et al.*, 1981a).

It seems justified to conclude that serum growth factors activate a Na^+,H^+ exchange process in these cells. The enhanced Na^+ influx then results in the observed stimulation of the Na^+,K^+-ATPase activity while the simultaneous H^+ efflux may raise intracellular pH. The resulting intracellular ionic conditions are a prerequisite for the initiation of DNA synthesis and cell proliferation. These results are consistent with the idea that intracellular pH fulfills an important second messenger function in growth regulation.

Finally, we would like to emphasize the striking similarity between the early ionic events upon serum stimulation of quiescent N1E-115 cells and those occurring at the G_1/S transition in synchronized Neuro-2A cells (see Section II,D). After 24 hours of serum deprivation the plasma membrane of N1E-115 cells is apparently committed to respond to growth factor binding, but in the absence of such stimuli and thus of the resulting ionic conditions, the cells switch their fate from a proliferative program to a differentiation pathway. In the early phases of the cell cycle of Neuro-2A cells, cultured in the continuous presence of serum, these cells are apparently not able to bind serum growth factors or to respond properly to such binding. They acquire this capacity only in late G_1. Together these results indicate that neuroblastoma cells modulate their plasma membrane properties such that the absence or presence of appropriate external stimuli in certain stages of their life cycle can determine their fate: continued proliferation or the initiation of differentiation.

IV. Initiation of Differentiation

A A. Introduction

In this section we will focus on the role of the plasma membrane in the sequence of events initiating the transition of neuroblastoma cells from the proliferatory to the differentiating state. Differentiation of neuroblastoma cells is ultimately characterized by the irreversible acquisition of functional neuronal properties, such as an excitable membrane and the expression of high levels of specific neuronal enzymes. Morphological differentiation, i.e., neurite outgrowth, has often been taken as the main criterion for differentiation because it is most readily observable. Although a period of neurite outgrowth is indispensable for and highly characteristic of a maturing functional neuron, taking it as the only criterion for differentiation can lead to unjustified conclusions.

A number of reviews have been published on neuroblastoma cells (Prasad, 1975; Prasad and Sinha, 1978) and on neuroblastoma–glioma cell hybrids (Hamprecht, 1977), but only the review by Prasad (1975) dealt specifically with differentiation in any detail. Since then, considerably more information has become available, much of which could be relevant to the role of the plasma membrane in the differentiation process. In particular, the recent application of new techniques and technology (see also Section II,C) has made significant contributions, a number of which are covered by the present review.

B. Triggers for Differentiation

A strikingly wide diversity of differentiation-inducing procedures has been described for neuroblastoma cells. Based on their primary site of action two main classes can be distinguished among the inducing triggers: (1) those acting primarily on the plasma membrane, or on plasma membrane controlled cytoplasmic signaling pathways while bypassing the plasma membrane (Table VA), and (2) those acting primarily at the level of the genome (Table VC). The primary site of action of a number of other differentiation-inducing procedures is, however, less clear (Table VB), although some of them, nonpolar compounds (DMSO, HMBA) (Tsiftsoglov *et al.*, 1981) and low pH (van der Saag *et al.*, 1981), might also act through changes at the plasma membrane level. We will restrict ourselves here to a discussion of possible membrane-mediated triggers.

Transmembrane signaling usually involves (specific) membrane receptors and the participation of second messenger molecules. As can be seen from Table V only a limited number of studies have provided direct evidence for their involvement in initiating differentiation of neuroblastoma cells. Membrane-mediated initiation of differentiation could in theory result from three types of interaction with the appropriate signaling mechanism: (1) direct action at the receptor level by application of effective ligands (e.g., NGF) to committed cells, (2) direct regulation of second messenger production, while bypassing the ligand–receptor interaction (e.g., cAMP addition), and (3) indirect control of receptor expression or second messenger production through modified molecular interactions within the plasma membrane (e.g., liposomes). We will attempt to distinguish through what mechanisms the various membrane triggers act. Some speculation cannot be avoided due to insufficient experimental knowledge, but this may stimulate new approaches to gain more insight in this field.

If we consider neuroblastoma cells as an expression of neoplastic transformation of sympathetic neuroblasts, then nerve growth factor (NGF) certainly would be the more physiological trigger for differentiation. NGF receptors have been identified in mouse neuroblastoma cells (Revoltella *et al.*, 1974), and in particular a number of human neuroblastoma cell types respond to NGF addition by morphological and functional differentiation (Waris *et al.*, 1973; Kolber *et al.*, 1974; Perez-Polo *et al.*, 1979). In contrast to rat pheochromocytoma cells (PC12), the action of NGF has not been studied in any detail in neuroblastoma cells. Where appropriate we will therefore include in the following some studies on PC12 cells to substantiate certain lines of evidence.

Among the possible second messengers cAMP has been studied most extensively, and some of its effects have been traced down to the nuclear level (Prasad and Sinha, 1978). The response to NGF could be mediated by cAMP but studies on PC12 cells are not unequivocal at this point (for reviews see Greene and Shooter, 1980; Vinores and Guroff, 1980). In only one case a reduction in the level of intracellular cAMP was found upon differentiation of neuroblastoma cells (Kimhi *et al.*, 1976). These authors

TABLE V

INITIATION OF DIFFERENTIATION OF NEUROBLASTOMA CELLS[a]

Trigger	Second messenger	Growth inhibition	Morphological differentiation	Functional differentiation	Clone	Reference
A. Plasma membrane level						
1. Serum deprivation	cAMP↑	+	+	n.d.	N18	Seeds *et al.* (1970)
	cAMP↑	+	+	+	C1300	Schubert *et al.* (1971)
	cAMP↑	+	+	+	42B	Kates *et al.* (1971)
	cAMP↑	+	+	n.d.	B103, 104 (rat)	Schubert (1974)
2. Dibutyryl-cAMP	cAMP↑	+	+	+	NB60	Furmanski *et al.* (1971)
	cAMP↑	+	+	+	C1300	Prasad and Hsie (1971)
	cAMP↑	+	+	+	C1300	Prasad and Vernadakis (1972)
	cAMP↑	+	+	n.d.	B103, 104 (rat)	Schubert (1974)
3. PDE inhibitors	cAMP↑	+	+	+	C1300	Prasad and Sheppard (1972)
4. Prostaglandins						
PGE_1/PGE_2	cAMP↑	+	+	+	N4TG1, S20, N10, N18	Gilman and Nirenberg (1971)
					NBA2, NBA5	Prasad (1972)
PGA_2	n.d.	+	+	n.d.	N2A	Adolphe *et al.* (1974)
5. Butyrate	cAMP↑	+	−	+	C1300	Prasad and Sinha (1976)
6. Nerve growth factor	n.d.	n.d.	+	n.d.	Human	Waris *et al.* (1973)
	n.d.	−	+	n.d.	Human	Kolber *et al.* (1974)
				+	Human	Perez-Polo *et al.* (1979)
	n.d.	+	+	n.d.	NB6R	Revoltella and Butler (1980)
7. Dexamethasone	n.d.	+	+	+	NBP2	Sandquist *et al.* (1978, 1979)
8. Substrate/cell interaction						
Collagen	n.d.	+	+	n.d.	N2A, N4, N18	Miller and Levine (1972)
Silicon monoxide	n.d.	+	+	n.d.	NB41A	Cooper *et al.* (1976)
Cell variants	n.d.	−	+	n.d.	B104 (rat)	Culp (1980)

9. Gangliosides	n.d.	+	+	n.d.	B104 (rat)	Morgan and Seifert (1979)
10. Liposomes	n.d.	+	+	n.d.	41A3	Chen *et al.* (1976)
					N3P-2	Sandra *et al.* (1981)
11. Delipidated serum	n.d.	–	+	n.d.	NB2A	Monard *et al.* (1977)
12. Urea herbicides	n.d.	n.d.	+	+	41A3	Erkell and Walum (1979)
13. Interferon	n.d.	+	+	n.d.	41A3	BaldeKier Jaffe *et al.* (1979)
14. Hypertonic medium	n.d.	+	+	n.d.	N2A	Ross *et al.* (1975)
15. Carboxymethyl cellulose	n.d.	+	+	n.d.	N18	Koike and Pfeiffer (1979)
16. Valinomycin	n.d.	+	+	+	N1E115	Spector *et al.* (1975)
					N18	Koike (1978)
					NB2A	Hinnen and Monard (1980)
17. Ca ionophore	n.d.	n.d.	+	n.d.	NB2A	Hinnen and Monard (1980)
B. Level of action to be determined						
18. Dimethyl sulfoxide	cAMP↓	+	+	+	N1E115, NS20, N18	Kimhi *et al.* (1976)
19. HMBA	n.d.	+	+	+	N1E115	Palfrey *et al.* (1977)
20. Low pH	n.d.	+	+	+	NBA2	Bear and Schneider (1977)
21. Glial factor	cAMP: no change	–	+	n.d.	NB2A	Monard *et al.* (1973)
						Monard *et al.* (1975)
						Schuerch-Rathgeb and Monard (1978)
22. Quinidine, dinitrophenol, dicoumarol	n.d.	+	+	n.d.	N1E115	Egilsson (1977)
23. Hemin	n.d.	n.d.	+	n.d.	N2A	Ishii and Maniatis (1978)
24. CCA	n.d.	+	+	n.d.	N1E115, N1A103	Croizat *et al.* (1979); Portier *et al.* (1980)
25. Vitamin E	cAMP: no change	+	+	n.d.	NBP2	Prasad *et al.* (1979)
26. Oxygen	n.d.	+	+	+	41A3, N18, N1E115	Erkell (1980)

(continued)

TABLE V (*continued*)

Trigger	Second messenger	Growth inhibition	Morphological differentiation	Functional differentiation	Clone	Reference
C. Genome level	cAMP: no change	+	+	+	C1300	Prasad (1971); Prasad and Vernadakis (1972)
27. X-Rays						
28. 5-Bromo-2′-deoxyuridine	cAMP↑	+	+	+	NBA2	Schubert and Jacob (1970); Prasad *et al.* (1973a)
29. Fluorodeoxyuridine	n.d.	+	+	+	42B	Klebe and Ruddle (1969); Kates *et al.* (1971)
30. Thioguanine	cAMP: no change	+	+	+	NBA2, NBE	Prasad (1973); Prasad *et al.* (1973a)
31. Cytosine arabinoside	n.d.	+	+	+	42B	Kates *et al.* (1971); Byfield and Karlsson (1973)
32. DTIC	cAMP: no change	+	+	+	NBP2	Culver *et al.* (1977)
33. Methotrexate	n.d.	+	+	n.d.	C1300	Byfield and Karlsson (1973)
34. MNNG	n.d.	+	+	+	N18	Yoda and Fujimura (1979)
35. Sulfur mustards	n.d.	+	+	+	N18	Lanks *et al.* (1975)

[a] cAMP, Adenosine 3′:5′-cyclic monophosphate; CCA, 1-methyl-cyclohexane carboxylic acid; DTIC, 5-(3,3-dimithyl-1-triazeno)imidazole-4 carboxamide; HMBA, hexamethylene bisacetamide; MNNG, *N*-methyl-*N*′-nitro-*N*-nitroso guanidine; n.d., not determined; cAMP↑, cAMP level increase; cAMP↓, cAMP level decrease.

reported a small decrease (<10%) in intracellular cAMP after 2 days of treatment with 2% DMSO, and a more substantial decrease (20%) after 5 days of treatment. It is not likely that these changes are involved in the initiation of differentiation by DMSO, since no changes in cAMP content of N1E-115 cells were found during the first hours of exposure to DMSO (A. Feyen, unpublished data).

Another class of possible second messengers is formed by changes in the cytoplasmic ionic conditions, controlled mainly by the plasma membrane-bound transport systems. The Ca^{2+} ionophore A23187 can induce differentiation (Hinnen and Monard, 1980), and neurite extension induced by glia-conditioned medium is dependent on external Ca^{2+} (Hinnen and Monard, 1980). In addition, agents like quinidine, dinitrophenol, and dicoumarol could exert their effect through the release of Ca^{2+} from mitochondria (Egilsson, 1977). At present it is not yet clear whether Ca^{2+} ions can act as a primary trigger for differentiation or interact with cAMP metabolism.

In view of the studies reported in the previous section it is of particular interest to consider the possible role of monovalent cations in the initiation of differentiation. Such a role is indicated by the effects of the K^+ ionophore valinomycin on several clones (Spector *et al.*, 1975; Koike, 1978; Hinnen and Monard, 1980) and of acidification of the growth medium (Bear and Schneider, 1977). These studies did not include the determination of secondary effects on cyclic nucleotide metabolism, but there are indications that cAMP could be involved also in these cases. Valinomycin most likely causes a change in membrane potential, which in its turn could induce a shift in intracellular cAMP content (Pall, 1977). Similarly, we found recently in Neuro-2A cells that acidification of the medium leads within a few hours to an increase in membrane fluidity and cAMP production by adenylate cyclase, upon which growth is inhibited. However, in this cell line no differentiation is detectable under these conditions (van der Saag *et al.*, 1981). Moreover, it could be speculated that serum deprivation induces differentiation through similar mechanisms. Serum growth factors activate upon interaction with their membrane receptors a series of ion transport systems of which one ultimate effect probably is an alkalinization of the cytoplasm (Moolenaar *et al.*, 1981) (see Section II,E). If serum withdrawal would have the reverse effect, which still has to be demonstrated, then it could be speculated that the enhanced cAMP production and the initiation of differentiation are also being mediated here by a shift in intracellular pH. Although no such data have been published yet for neuroblastoma cells, studies on NGF-induced differentiation of primary sympathetic ganglia (Skaper and Varon, 1980, 1981) and PC12 cells (Boonstra *et al.*, 1981c) also suggest that changes in monovalent cation transport properties are part of the NGF-triggering pathway.

Some evidence exists that the initiation of differentiation is dependent on the properties of the plasma membrane lipid matrix. Treatment with liposomes composed of egg lecithin and phosphatidylserine (6.6:1, w:w) induced morphological differentiation in 41A3 cells (Chen *et al.*, 1976). Differentiation of Neuro-2A cells by cAMP-elevating treatment could be inhibited reversibly by incubating the cells with phospholipid vesicles composed of (saturated) dipalmitoyl phosphatidylcholine, while vesicles composed of (unsaturated) dioleoyl phosphatidylcholine had no such effect (de Laat *et al.*, 1978). Both studies indicate that a more fluid lipid matrix potentiates, or is required for the expression of morphological differentiation (see Section IV,D). In contrast, Monard *et al.* (1977) have reported the inhibitory effect of oleic acid on morphological differentiation following the addition of delipidated serum. It is not clear whether these treatments interact indirectly with second messenger production, or can be a sufficient trigger by itself. Furthermore, it should be considered that these lipidic agents might interfere with new membrane production, as required for neurite outgrowth.

A rather similar situation exists with respect to the possible active role of cell surface-associated enzyme activities or structural proteins. In a human neuroblastoma cell line (SK-N-SH) fibrinolytic activity resulting from the liberation of a plasminogen activator was demonstrated (Wachsman and Biedler, 1974). Growth of the cells in a plasminogen-deficient medium (under conditions of reduced general proteolytic activity) increases the number of morphologically differentiated cells considerably (40%) although the total number of differentiated cells reached a maximum value of only 27%. Under these conditions the amount of extracellular plasminogen activator per cell is increased fivefold (Becherer and Wachsman, 1980). Similarly uncloned mouse C1300 neuroblastoma cells markedly increased plasminogen activator upon treatment with differentiation promoting prostaglandins and cAMP (Lang *et al.*, 1976), again suggesting a positive correlation between the production of this specific protease and differentiation of neuroblastoma cells. More studies will be needed to characterize further the possible significance of these findings. In particular possible target molecule require more systemic identification at the level of the plasma membrane or the cell surface. It can be speculated that the generally accepted concept of increased cell–substratum interaction being a crucial condition for differentiation of neuronal cells (Schubert *et al.*, 1971) under *in vitro* conditions is partially achieved by proteolytic modification of the cell surface. The first example in which cell surface-associated proteins have been identified was given recently: laminin and type IV procollagen, both characteristic of basement membranes, were demonstrated in clone NB41A3 (Alitalo *et al.*, 1980).

An intriguing alternative regulatory mechanism is the autoregulation of differentiation through the production of differentiation-inducing (co)factors by the cells themselves. Nerve growth factor, for example, is produced by a variety of cells under tissue culture conditions (Greene and Shooter, 1980). Murphy *et al.* (1975) have discovered that neuroblastoma cells (N2A) also produce and secrete NGF in culture. They found, however, that this clone is not biologically responsive to NGF. Subsequently, it was shown that a cAMP-elevating differentiation trigger such as prostaglandin E_1 (PGE_1) will increase the NGF content of the cells and of the medium twofold (Schwartz and Costa, 1978). Apparently, the cAMP content can regulate NGF production of these cells.

A similar situation with respect to the production of prostaglandins was first described by Hamprecht *et al.* (1973) in neuroblastoma cells (clone N4TG3). In the presence of dibutyryl cAMP the production of PGE_1 was considerably enhanced, suggesting a positive regulatory function for cAMP in the synthesis of prostaglandins. More recently, prostaglandin D synthetase activity was demonstrated in cell homogenates of NS20 and N1E115, while the product PGD_2 is a potent activator of adenylcyclase in these cells (Shimizu *et al.*, 1979). Finally, the synthesis of predominantly $PGF_{2\alpha}$ was demonstrated in cell homogenates of four different mouse clones and one human clone of neuroblastoma cells (Tansik and White, 1979). These examples clearly illustrate the possibility of a positive feedback between different differentiation trigger molecules or systems. Moreover, it seems more than coincidental that this phenomenon has been described only for physiological plasma membrane active molecules or NGF and prostaglandins.

In conclusion, the number of differentiation-inducing agents seems at first sight even to exceed the number of different neuroblastoma cell lines described so far. However, certain common characteristics among various agents can be distinguished. In view of the resemblance between neuroblastoma cells and undifferentiated neuroblasts, the action of hormone addition (NGF) or withdrawal of serum is probably the more physiological mode of initiation, and certainly plasma membrane mediated. Differentiation can also be initiated by direct experimental interaction with the membrane-transducing pathways or second messenger production (cAMP, ions). It is possible that the applied membrane lipid modifications fall into the same category. As one would expect changes in gene expression upon differentiation, it is not surprising that direct interaction with the process of transcription can induce a differentiated phenotype. More detailed analysis at the molecular level of the effects of other agents, the actions of which are so far not well understood, will certainly help to unravel cause and effect in the initiation of differentiation. Finally, it should be realized that there exists a large variation in the response to various agents among the established neuroblastoma cell lines. This implies that care must be taken in generalizing the results obtained from a particular cell line. Furthermore, this could indicate that various neuroblastoma clones are inhibited in their normal differentiation pathways at different points as has been proposed for myeloid leukemic cell differentiation (Sachs, 1980).

C. Growth versus Differentiation

As shown in Table V, growth inhibition has been observed simultaneously with, or soon after the onset of morphological differentiation under most of of the differentiation-inducing conditions, but growth inhibition per se is not a sufficient trigger. In this respect neuroblastoma cells and also pheochromocytoma cells behave similarly to nontransformed neuroblasts. Some reports, however, have indicated no such mutual exclusion of growth and differentiation in neuroblastoma cells (Monard *et al.*, 1973, 1977; Waymire *et al.*, 1978; Culp, 1980). It would be interesting to see whether under these conditions only neurite-like cellular processes are being found, or also functional differentiation markers become detectable.

Information is limited regarding the phase of the cell cycle in which differentiated cells are arrested. Prasad *et al.* (1973b) reported that NBE^- cells, differentiated by dibutyryl-cAMP, PGE_1, or Ro20-1724 (a phosphodiesterase inhibitor) accumulate in G_1 on the basis of their reduced DNA content. The same clone showed, however, no change in its DNA content after serum withdrawal (Prasad *et al.*, 1973a), while another clone (N1A) did accumulate in the G_1 phase as measured by flow microfluorometric analysis (Baker, 1976). In addition, Moolenaar *et al.* (1981b) have shown that serum-deprived, early differentiating N1E-115 cells respond to the readdition of fetal calf serum by a synchronous reentry into S phase after 8 hours and cell division 10 hours later, indicating that the cells were arrested in G_1. Although more studies are definitely needed the frequently aneuploid character of neuroblastoma poses a problem. The human cell line SK-N-SH and its subclone SY-5Y is an interesting exception, since it has a nearly diploid karyotype (Perez-Polo *et al.*, 1979). The possibility that the degree of aneuploidy is negatively correlated with the capacity to react to physiological trigger molecules in a specific part of the cell cycle (G_1) should be considered.

In rat pheochromocytoma cells (PC12) plasma membrane receptors for NGF (Levi *et al.*, 1980) as well as functional receptors for the mitogen epidermal growth factor (EGF) (Huff *et al.*, 1981) have been demonstrated. Upon differentiation of these cells with NGF the binding capacity for EGF declines and is virtually abolished after 3 days indicating the disappearance or masking of EGF receptors in the plasma membrane as a result of differentiation. This important finding clearly illustrates the regulatory role of the plasma membrane under conditions in which cells are susceptible to two types of opposing signals. If we assume that cells under the action of NGF accumulate in the G_1 phase of the cell cycle then this specific growth arrest could lead to a subsequent decrease in the expression of growth factor receptor sites. Similar mechanisms could be operational in neuroblastoma cells since it has recently been found that these cells also carry EGF receptors (Bhargava *et al.*, 1980), and are biologically responsive to EGF under certain conditions (Mummery and Moolenaar, unpublished observations).

D. Dynamic Properties of Plasma Membrane Components

As described in Section II,C, pronounced changes in the dynamic behavior of the plasma membrane lipids and proteins have been demonstrated during the cell cycle of Neuro-2A cells, in particular at times of new plasma membrane formation (de Laat *et al.*, 1977, 1980). In these studies two independent methods were applied which gave essentially similar results: (1) determination of the rotational mobility of the lipid probe DPH by steady state fluorescence polarization measurements, and (2) determination of the lateral mobility of the lipid probes diI and F-GM1 and of a protein probe $RaEl_4$, a fluorescent Fab′ fragment against mouse El_4 cells.

Obviously, neurite extension will also involve the production of new membrane. To establish whether the plasma membrane changes its physicochemical properties upon differentiation, as compared to the proliferative state of the cells, we monitored the dynamic properties of membrane lipids and properties during early differentiation of Neuro-2A cells employing the same methods as before. Morphological differentiation was induced by dibutyryl cAMP and 3-isobutyl-1-methylxanthine (IMBX: a phosphodiesterase inhibitor), or by prostaglandin E_1 in the presence of R020-1724, another phosphodiesterase inhibitor. The fluorescence polarization measurements of the microviscosity ($\bar{\eta}$) of the membrane lipid matrix of intact cells on a glass substratum demonstrated a substantial decrease within 1 hour after the addition of the differentiation-inducing agents. A further progressive decrease in $\bar{\eta}$ was observed during the next 5 hours to a level never observed during the cell cycle ($\bar{\eta} = 1.4$ poise at 37°C). It was concluded that (1) the plasma membrane of differentiating cells has a more fluid character than that of proliferating Neuro-2A cells, and (2) the membrane fluidity changes are not simply the result of neurite extension but are associated with the initiation of morphological differentiation. As mentioned before (Section IV,C), we have shown in the same study that incubating the cells with phospholipid vesicles composed of saturated phospholipids (dipalmitoyl phosphatidylcholine) prevented the increase in membrane fluidity as well as neurite formation completely and reversibly, whereas (unsaturated) dioleoyl phosphatidylcholine vesicles had no effect. These results strongly suggest that an increased membrane fluidity is a prerequisite for neurite extension (de Laat *et al.*, 1978).

Subsequently, we analyzed in more detail the dynamic properties of the plasma membrane during neurite extension in this cell type by the fluorescence photobleaching recovery (FPR) method, using a fluorescein-labeled analog of ganglioside GM1 (F-GM1) and rhodamine-conjugated Fab′ fragments against mouse El_4 cells ($RaEl_4$) for the determination of the lateral mobilities of membrane lipids and proteins, respectively (de Laat *et al.*, 1979). These probe molecules were also applied in the FPR study on the cell cycle, described earlier (Section II,C). The great advantage of this method is its capacity to detect local differences in the mobility properties of membrane components within the plasma membrane of a single cell, and we exploited this capacity to detect possible differences between the plasma membrane of the cell body or perikaryon region on the one hand and that of the extending neurite on the other. As shown in Fig. 9 the lateral diffusion coefficients of both F-GM1 and $RaEl_4$ remain unchanged upon differentiation in the perkaryon region, but show a significant increase specifically in the neurite membrane. This demonstrates the appearance of a topographical heterogeneity within the plasma membrane of a differentiating Neuro-2A cell, with more fluid membrane domains being located in the extending neurites. It is very possible that these local differences in membrane properties are related to the process of membrane growth, since Pfenninger (1979) demonstrated by freeze-fracture electron microscopy of developing neurites of the fetal rat spinal cord that membrane growth occurs through a predominant insertion of membrane lipid in specialized areas at the growth cone.

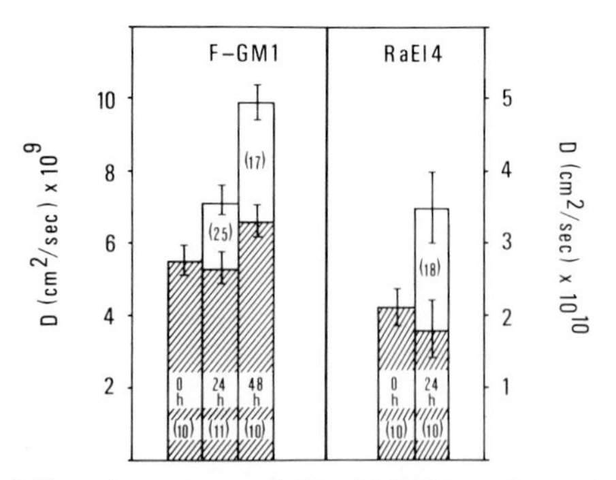

FIG. 9. Lateral mobility of membrane lipids ($F\text{-}GM_1$) and membrane proteins ($RaEl_4$) upon morphological differentiation of Neuro-2A cells. Diffusion coefficients are given as mean ± SEM (number of measurements). Where appropriate, distinction is made between perikaryon region (shaded area) and neurite (total length of bar).

A few other studies concerning the dynamic behavior of the plasma membrane upon differentiation of a number of neuroblastoma cells have been published. Struve *et al.* (1977) found in N18 cells by electron paramagnetic resonance measurements a slightly enhanced membrane fluidity upon differentiation induced by serum withdrawal. Correspondingly, an increased lateral mobility of Con A receptors was observed in morphologically differentiated S20 cells, in comparison to round cells without neurites (Zagyanski *et al.*, 1977). In contrast Koike (1978) could not detect any significant differences in N18 cells between control cells and cell differentiating upon low serum + valinomycin treatment, using 1-anilino-8-naphthalene sulfonic acid (ANS) fluorescence polarization measurements. Several authors have presented evidence as to changes in the dynamic organization of the plasma membrane upon attachment of cells to a substratum (Erkell, 1977; de Laat *et al.*, 1977; Kawasaki *et al.*, 1978). The latter authors unfortunately have identified the attachment of N18 cells in the presence of 10% fetal calf serum as the initiation of morphological differentiation, although no difference in cell proliferation was observed between suspension and attached cells. Furthermore, fluorescence polarization measurements, using a number of different probe molecules, were carried out on sonicated cells after 13–15 hours of labeling, thus reflecting most likely properties of the total cellular membrane pool and not of the plasma membrane.

An interesting observation was recently published by Fishman *et al.* (1981). DMSO-induced differentiation of N1E-115 cells is associated with a substantial increase in the number of immobile receptor sites for Con A but not for WGA. Serum withdrawal which also induces neurite formation, had no such effect. Furthermore, this receptor-specific immobilization by DMSO could be reversed by treatment with the microfilament-disrupting agent cytochalasin B, suggesting that specific microfilament–membrane interactions are involved in this process. FPR measurements did not reveal any significant changes in the lateral mobility of membrane lipids upon DMSO-induced differentiation. This study clearly shows that various triggers for the initiation of differentiation can have different effects at the level of the plasma membrane even in one particular cell line.

In conclusion, several independent techniques have demonstrated characteristic changes in the dynamic properties of plasma membrane components upon morphological differentiation, and at least in one case (de Laat *et al.*, 1979) these changes were found to be neurite-specific. It is yet unclear whether the induced changes merely provide necessary conditions for the morphological changes as such, or also are part of the differentiating-initiating regulatory mechanism. Finally, it will be clear from this discussion that care must be taken in generalizing the results obtained from one particular cell type or one particular differentiation-inducing treatment.

E. Plasma Membrane Composition

In this section we will discuss the evidence available regarding differentiation-dependent alterations in the composition of the plasma membrane and cell surface-associated molecular complexes, with some emphasis on the possible correlations with the dynamic membrane changes described in the previous section. Methodologically such studies can be distinguished into two classes: (1) the characterization of plasma membrane composition is being made without the isolation and purification of plasma membranes, which most readily can be done for (glyco)proteins, and (2) the characterization requires plasma membrane purification, e.g., for the determination of changes in lipid composition.

Among the first type of studies, Brown (1971) and Glick *et al.* (1973) have reported already quantitative and qualitative differences, respectively, in the patterns of tryptic cell surface-derived glycopeptides upon differentiation. Using antibodies against intact cells a group of neural-specific antigens was found to be associated with the cell surface of differentiated cells only (Akeson and Herschman, 1974a,b). In the near future monoclonal antibody techniques will probably open the way for more detailed analysis of the antigenic differences and will permit a more detailed temporal and spatial resolution of the proteins involved. The isolation of individual membrane proteins will be possible by monoclonal affinity chromatography and has already been applied with success for the isolation of a relatively small glycoprotein (MW 20,000) present in small amounts at the cell surface of growing IMR 5 human neuroblastoma cells (Momoi *et al.*, 1980). In general not much is known about individual membrane proteins of neuroblastoma cells. An interesting observation was recently made in relation to the Thy-1 cell surface antigen in neuroblastoma cells (Morris *et al.*, 1980). This extensively characterized antigen occurs in many cell types, but in the central nervous system only on neurons. Upon addition of dibutyryl-cAMP to the clones N18 and N1E115 the number of detectable antigenic sites increased from less than 1.2×10^3 per cell (undetectable) to $5–8 \times 10^5$, which is quite a remarkable increase. Although a number of rat and mouse cloned nerve cell lines also respond with an increase in Thy-1 sites, the response of the neuroblastoma cells was by far the most impressive.

The phenomenon of plasma membrane shedding, particularly by tumor cells (for a review see Black, 1980), can also be used to detect differences in plasma membrane composition. It is clear that this method also has inherent disadvantages, since the mechanism of shedding is not understood at all. Growing neuroblastoma cells (clone N2aE), prelabeled with glucosamine or fucose, released glycoproteins with a heterogeneous molecular weight distribution, while cells differentiating in the presence of dibutyryl-cAMP released preferentially a glycoprotein of about 66,000 daltons (Truding *et al.*, 1975). In a different approach it was observed that growing iodinated cells release glycoproteins centering in a peak of 87,000 molecular weight. This peak disappeared within 2 hours upon exposure of the cells to dbcAMP or bromodeoxyuridine, both differentiating agents in this clone (Truding and Morell, 1977). It is evident that these changes could be explained by changes in cell surface-associated or released proteolytic enzyme activity upon the initiation of differentiation.

Fucose-containing glycoproteins were also found in the culture medium of growing N1E115 and N1A103 cells (Littauer *et al.*, 1980). In this case a particular glycoprotein (200,000 dalton) possibly associated with the voltage-dependent sodium channel, disappeared from the medium upon differentiation by DMSO to increase its presence in the plasma membrane. A neuronal-specific glycoprotein of a similar molecular weight was reported at the surface and in the medium of growing N18 cells by immunological techniques (Akeson and Hsu, 1978).

Despite the foregoing it is necessary for a better understanding of structure–function relationship to obtain isolated plasma membranes from neuroblastoma cells. A compilation of studies in which plasma membranes have been isolated is presented in Table VI. The high indidence of the method first described by Warren and Glick (1969) in which Zn^{2+} is used to stabilize plasma membrane is striking. The number of instances in which the isolated membranes have been characterized to some extent, however, is disappointingly low. Only in three cases (Garvican and Brown, 1977; Charalampous, 1977, 1979) have nonplasma membrane localized enzymes (negative markers from mitochondria, microsomes, and lysosomes) been determined in the preparations, while only Charalampous (1979) has included phospholipids in his characterization. The same author is also unique in the fact that he employs two alternative isolation methods. The latex beads method (Charalampous *et al.*, 1973) is in particular an unorthodox approach but gives homogeneous membranes by interiorization of the latex particles covered with an inside-out membrane, which seems to reflect the enzymatic composition of the primary plasma membrane (Charalampous *et al.*, 1973). Plasma membranes isolated with the $ZnCl_2$ method were purified about 10 times on the basis of Na^+,K^+-ATPase activity, both from cells in suspension and under monolayer culture conditions. From the data obtained by Charalampous (1977) with the latex beads method one must conclude, however, that the same enzyme is excluded from the interiorized membrane of beads taken up by monolayer cells, while this is not the case with suspension cells. For this reason the results obtained with this method from monolayer cells must be interpreted with caution. A complicating factor in the comparison between suspension and monolayer cells is, moreover, that

TABLE VI

PLASMA MEMBRANE ISOLATION AND CHARACTERIZATION FROM GROWING AND DIFFERENTIATING NEUROBLASTOMA CELLS[a]

Clone	Growing/differentiating	Method	Characterization			Reference
			Enzymes	Purification factor	Phospholipids	
N2aE	+/+	$ZnCl_2$[b]	AchE	4–5×	n.d.	Truding *et al.* (1974)
N2A	+/+	$ZnCl_2$[b]	n.d.	—	n.d.	Matthews *et al.* (1976)
NB41A	+/+	Hypotonic lysis	5′-Nucleotidase	15.3×	n.d.	Garvican and Brown (1977)
N2A	+/+	$ZnCl_2$[b] + TPPS[c]	Na^+,K^+-ATPase	10×	+	Charalampous (1977, 1979)
N2A	+/+	Latex beads[d]	Na^+,K^+-ATPase	5×	+	Charalampous (1977, 1979)
N18	+/+	TPPS[c]	n.d.	—	n.d.	Kawasaki *et al.* (1978)
N1E115	+/−	$ZnCl_2$[b]	n.d.	—	n.d.	Strange *et al.* (1978)
N2A	+/+	$ZnCl_2$[b]	n.d.	—	n.d.	Milenkovic *et al.* (1978)
N18	+/−	$ZnCl_2$[b]	n.d.	—	n.d.	Milenkovic and Johnson (1980)
N18, N1E115, N1A103	+/+	$ZnCl_2$[b]	n.d.	—	n.d.	Littauer *et al.* (1980)

[a] AchE, Acetylcholinesterase; n.d., not determined; TPPS, two phase polymer system.

[b] Warren and Glick (1969).

[c] Brunette and Till (1971).

[d] Charalampous *et al.* (1973).

suspension cells have a 40% shorter generation time than monolayer cells (Charalampous, 1977).

It is clear from Table VII, where a survey is presented of studies concerning differences between the plasma membranes of growing and differentiating cells, that quantitative changes in protein composition have been described predominantly. In the two studies in which phospholipids and cholesterol have been analyzed (Kawasaki *et al.*, 1978; Charalampous, 1979) membranes of suspension and attached cells were compared. This type of comparison is, however, not an adequate equivalent for initiation of differentiation since differences in growth kinetics might also contribute in this case.

As has been shown, the regulatory role of phospholipids as an important constituent of the plasma membrane in growth and differentiation certainly deserves a more detailed study than has been undertaken so far. For this purpose isolation of plasma membranes is a prerequisite. In our laboratory plasma membranes were isolated from growing and differentiating Neuro-2A cells by hypotonic lysis in 1 m*M* sodium bicarbonate without divalent cations. Subsequently, pellets obtained from the lysate at 27,000 g_{max} were analyzed on continuous sucrose gradients. In these studies Na^+,K^+-ATPase was taken as the major marker enzyme for the plasma membrane. Moreover, marker enzymes for endoplasmic reticulum and mitochondria were included in the analysis. Plasma membranes of both growing and differentiating cells as measured by Na^+,K^+-ATPase activity were found at a buoyant density of 1.112 gm/cm^3. In these fractions the cholesterol/phospholipid ratio of differentiating cells is considerably lower (0.69) than in homologous fractions of growing cells (0.98) indicating a lower cholesterol content of plasma membranes of differentiating cells. These findings are only a part of an extensive study of phospholipid and fatty acid composition of isolated plasma membranes and will be published elsewhere (Nelemans *et al.*, in preparation). The observed difference in cholesterol content is a sufficient basis to explain the differences in microviscosity and lateral mobility of lipids and proteins previously described. So far, no other studies have reported results supporting this finding using comparable methods (Table VII).

As stated before, it is important to extend similar studies as described to different neuroblastoma clones and to different modes of triggering differentiation in the same clone to obtain a better idea of the molecular changes involved at the level of plasma membrane. Such studies are in progress in our laboratory.

To substantiate further these findings on isolated plasma membranes experiments have recently been carried out to measure lipid synthesis in growing and differentiating cells. Under conventional tissue culture conditions (with serum) cells obtain their necessary supply of cholesterol via a

TABLE VII

COMPOSITIONAL DIFFERENCES BETWEEN PLASMA MEMBRANES OF GROWING AND DIFFERENTIATING NEUROBLASTOMA CELLS[a]

Mode of differentiation	Isolated membranes	Type of membrane molecules described	Marker	Technique	Quantitative changes	Remarks	Clone	Reference
1. Suspension/monolayer ± dbcAMP	+	Proteins	Leucine	PAGE	–		N2aE	Truding *et al.* (1974)
	+	Glyco-proteins	Glucosamine	PAGE	+	MW 105K↑		
	+/–	Proteins	Iodination	PAGE	+	MW 78K↑		
	–	Tryptic peptides	Iodination	PAGE	+			
2. Suspension/monolayer ± dbcAMP (BUdR)	+	Proteins	Amino acids	PAGE	+		NB41A	Garvican and Brown (1977)
	+	Glyco-proteins	Fucose	PAGE	+	MW 70K↑		
3. Suspension/monolayer	+	Enzymes	—	—	+		N2A	Charalampous (1977)
	+	Proteins	—	PAGE	–(+[b])			
Monolayer ± dbcAMP	+	Enzymes	—	—	+			Charalampous (1977)
	+	Proteins	—	PAGE	+			

4. Suspension/monolayer	+	Phospholipids/cholesterol	—	—	+	Ratio↓	N18	Kawasaki *et al.* (1978)
5. Monolayer ± dbcAMP	+	glyco proteins	Fucose	PAGE	–	turnover↓	N2A	Milenkovic *et al.* (1978)
6. Suspension/monolayer	+[c]	Phospholipids/cholesterol	—	TLC	+	Ratio↑	N2A	Charalampous (1979)
Suspension/monolayer	+[b]	Phospholipids/cholesterol	—	TLC	+	Ratio: no change	N2A	Charalampous (1979)
7. Monolayer ± DMSO	–	Protein	Iodination[d]	PAGE	+	MW 95K↑	N1E115, N18	Zisapfel and Littauer (1979)
8. Monolayer ± DMSO (HMBA)	+ +	Glycoproteins	Fucose	PAGE	+	MW 200K↑	N1E115, N18, CHP134 (human)	Littauer *et al.* (1980)

[a] BUdR, 5-Bromo-2′-deoxyuridine; dbcAMP, dibutyryladenosine 3′:5′-cyclic monophosphate; DMSO, dimethyl sulfoxide; HMBA, hexamethylene bisacetamide; MW, molecular weight; PAGE, polyacrylamide gel electrophoresis; TLC, thin layer chromatography.

[b] Plasma membranes isolated with $ZnCl_2$ and two phase polymer system.

[c] Plasma membranes isolated with latex beads method.

[d] Also cell surface labeling with *N*,*N*,*N*-[^{3}H]trimethylamino-*β*-alenyl-*N*-hydroxysuccinimide ester.

specific receptor-mediated uptake of high-density lipoprotein from the medium. Therefore, it is clear that for the study of cholesterol and phospholipid synthesis culture in serum-free defined media is a prerequisite. Recently, such media have been described for a number of cell culture systems (Barnes and Sato, 1980a,b). In fact, the first instance in which such a medium was formulated was for rat neuroblastoma cells (clone B104) (Bottenstein and Sato, 1979). This medium was subsequently supplemented with compounds known to increase the interaction between cell and substratum, such as fibronectin and polylysine (Bottenstein and Sato, 1980) and under these conditions it could support continuous growth, including subculturing of B104 cells by trypsin. The medium does not include any lipid components but contains hormones (insulin and progesterone), the iron-carrying protein transferrin, putrescine, and selenite. Similar results have been described for the human neuroblastoma cell line LA-N-1 (Bottenstein, 1980). On the other hand, such possibilities have not yet been reported for the most commonly used C1300-derived mouse cell lines (Bottenstein *et al.*, 1979). We have been able, however, to sustain the continuous growth of Neuro-2A cells in a medium of comparable composition, consisting of Dulbecco's modified Eagle medium and Ham F-12 medium, supplemented only with transferrin and selenite (van der Saag and de Laat, unpublished results). Under these conditions cells have generation times comparable to those in serum-containing media, and no changes in cell morphology were observed. In other cases, however, more significant morphological changes resembling morphological differentiation have been described of growing cells in serum-free defined medium (Bottenstein, 1980; Bottenstein and Sato, 1980).

Morphological differentiation can be induced in Neuro-2A cells in serum-free defined medium as described above in the presence of dibutyryl-cAMP and 3-isobutyl-1-methylxanthine. These conditions were used to study the synthesis of phospholipid and cholesterol in differentiating Neuro-2A cells (Nelemans, unpublished results). In Fig. 10 a time course is given of the synthesis of cholesterol/phospholipid determined from the incorporation of [^{14}C]acetate pulses, as measured in whole cell extracts. After an initial increase in the relative incorporation of cholesterol/phospholipid a steady decrease occurs to reach values of around 50% as compared to the values of growing cells. In combination with total phospholipid and free cholesterol determinations under identical conditions (not shown) these results suggest a preferential inhibition of cholesterol biosynthesis in differentiating Neuro-2A cells. Measurements of the rate-limiting enzyme in the biosynthesis of cholesterol, 3-hydroxy-3-methylglutaryl coenzyme A reductase, must provide further support for this suggestion, but the results obtained so far give independent support to the compositional data of isolated membranes, in which a lower cholesterol/phospholipid ratio was measured in differentiated cells. The regulatory role of cAMP in cholesterol biosynthesis has been

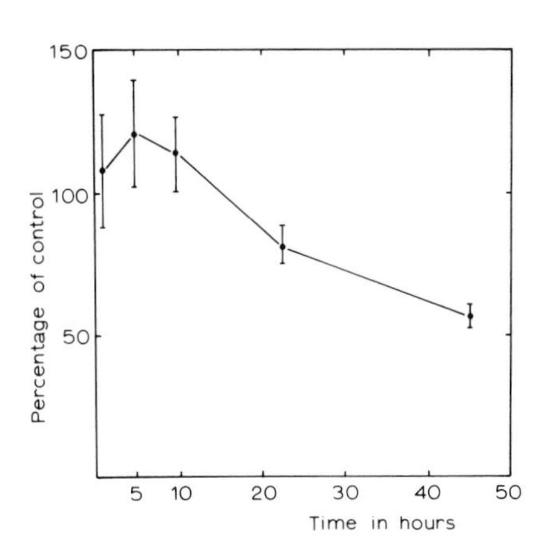

FIG. 10. [^{14}C]Acetate incorporation into free cholesterol and phospholipids of growing and differentiating Neuro-2A cells. Cells were plated in medium containing 10% (v/v) fetal calf serum, which was replaced after 5 hours by defined medium (without serum) as described in Section IVE. Differentiation was induced 24 hours after plating by the addition of dibutyryl-cAMP (1 m*M*) and 3-isobutyl-1-methylxanthine (0.2 mg/ml). Pulses of 60 minutes were given at the indicated times with [1-^{14}C]acetate (0.001 mCi/ml). Extracted lipids were analyzed by thin layer chromatography (solvents: heptane/ethyl ether/acetic acid, 90/15/1, v/v) and analyzed for radioactivity of collecting the appropriate spots. Results are expressed as the percentage of the control value (mean ± SEM) of the free cholesterol/phospholipid ratio of differentiating cells.

documented in other cell types before (Volpe and Marasa, 1976; Goodwin and Margolis, 1978; Respess *et al.*, 1978; Edwards *et al.*, 1979). Moreover, Maltese and Volpe (1980) investigated the role of cholesterol synthesis during neurite outgrowth and did not observe any effect of inhibition of this synthetic pathway by cholesterol analogs under serum-free culture conditions. This suggests that cellular cholesterol levels can be diminished considerably without interfering with morphological differentiation.

In conclusion, sufficient evidence is available demonstrating that compositional changes of the plasma membrane are associated with the early differentiation of neuroblastoma cells. Further studies are required to determine these changes in more detail, to separate cause and effect, and, in particular, to establish whether they are involved in the physiological triggering of neuronal differentiation.

V. Concluding Remarks

In this article a survey is given of the experimental evidence regarding the possible role of the plasma membrane as a regulatory site for the control of growth and initiation of differentiation of neuroblastoma cells *in vitro*, with emphasis on recent studies carried out in our laboratory. Ultrastructural, physicochemical, and physiological features of the plasma membrane change in association with cell cycle progression, growth stimulation, and the initiation of differentiation. Experimental manipulation of a still limited number of these membrane properties demonstrates their regulatory function, and sheds light on the molecular interactions involved. From these data a picture begins to emerge in which the cell continuously modulates the composition and properties of its plasma membrane during its life history. Such modulations originate most likely from the biosynthesis and assembly of the various plasma membrane constituents which do not occur in concert during the cell cycle. As an important result of this process the competence of the plasma membrane for external signal reception and transduction could be restricted to certain periods in the cell cycle during which the presence of appropriate ligands can determine the ultimate fate of the cell, i.e., cell multiplication, growth arrest, or differentiation. Obviously, at the molecular level a cell has a choice out of a variety of mechanisms to regulate the reception and transduction of external signals as well as the production of intracellular messengers by the plasma membrane, as number of receptor sites or enzyme copies, affinities for ligands and substrates, substrate levels, interactions between receptor or enzyme molecules with their lipidic environment, or with cytoskeletal elements, etc. So far, information as to the regulatory modes employed by the cell is still very limited, not the least due to the limited experimental approaches applied in studies related to this problem. In our opinion further coordinated research of the various possible regulatory pathways in a suitable experimental system is the proper way to clarify how the plasma membrane exerts its role in the regulation of growth and differentiation.

In this review we have restricted ourselves to one particular cell system, neuroblastoma cells *in vitro*, with emphasis on a limited number of cell lines. As emphasized before, it should be realized that it is yet unclear to what extent generalizations can be made to other neuroblastoma cell lines or even to embryonic neuroblasts or other committed, but not fully differentiated, cells. The section on triggers for differentiation (Section IV,B) illustrates not only that such generalizations should be made with great caution, but also that the number of possible experimental manipulations leading to the differentiated state even outnumbers the cell types studied. In part this can be understood in view of the complex and diverse ways by which the plasma membrane controls cell functioning, but cell type-dependent variations as to the mechanisms employed could as well be involved. We hope that the given review of data and the expressed views on the problems discussed will stimulate future research in this area, and will contribute to the development of new experimental strategies.

ACKNOWLEDGMENTS

We are grateful to our colleagues at the Hubrecht Laboratory: Drs. J. G. Bluemink, J. Boonstra, W. H. Moolenaar, C. L. Mummery, S. A. Nelemans, and E. J. J. van Zoelen for the contributions reported in this article, and to Drs. M. Shinitzky and J. Schlessinger (The

Weizmann Institute of Science, Rehovot, Israel) and E. L. Elson (Washington University, St. Louis, Missouri) for their cooperation in carrying out parts of this research. We thank A. Feyen, W. Miltenburg-Vonk, W. A. M. van Maurik, P. Meyer, and L. G. J. Tertoolen for for their excellent assistance and E. C. Ekelaar for preparing the manuscript. This work was supported in part by the Koningin Wilhelmina Fonds (Netherlands Cancer Foundation), Shell International Research Corporation, and the Foundation for Fundamental Biological Research (BION), which is subsidized by the Netherlands Organization for the Advancement of Pure Research (ZWO).

References

Adolphe, M., Giroud, J. P., Timsit, J., Fontagne, J., and Lechat, P. (1974). *C. R. Soc. Biol.* **168,** 694–698.

Akeson, R., and Herschman, H. R. (1974a). *Proc. Natl. Acad. Sci. U.S.A.* **71,** 187–191.

Akeson, R., and Herschman, H. R. (1974b). *Nature (London)* **249,** 620–623.

Akeson, R., and Hsu, W.-C. (1978). *Exp. Cell Res.* **115,** 367–377

Alitalo, K., Kurkinen, M., Vaheri, A., Virtanen, I., Rohde, H., and Timpl, R. (1980). *Nature (London)* **287,** 465–466.

Axelrod, D., Koppel, D. E., Schlessinger, J., Elson, E. L., and Webb, W. W. (1976). *Biophys. J.* **16,** 1055–1069.

Baker, M. E. (1976). *Biochem. Biophys. Res. Commun.* **68,** 1059–1065.

Bal de Kier Jaffe, E., Puricelli, L., and de Lustig, E. S. (1979). *Cell. Mol. Biol.* **24,** 257–264.

Barnes, D., and Sato, G. (1980a). *Anal. Biochem.* **102,** 255–270.

Barnes, D., and Sato, G. (1980b). *Cell* **22,** 649–655.

Bear, M. P., and Schneider, F. H. (1977). *J. Cell. Physiol.* **91,** 63–68.

Becherer, P. R., and Wachsman, J. T. (1980). *J. Cell. Physiol.* **104,** 47–52.

Bhargava, G., Bak, M., Levy, E., and Makman, M. H. (1980). *Fed. Proc. Fed. Am. Soc. Exp. Biol.* **39,** 593.

Black, P. H. (1980). *Adv. Cancer Res.* **32,** 75–199.

Bluemink, J. G., and de Laat, S. W. (1977). *In* "Cell Surface Reviews" (G. Poste and G. L. Nicolson, eds.), Vol. 4, pp. 403–461. North-Holland Publ., Amsterdam.

Boonstra, J., Mummery, C. L., Tertoolen, L. G. J., van der Saag, P. T., and de Laat, S. W. (1981a). *Biochim. Biophys. Acta* **643,** 89–100.

Boonstra, J., Mummery, C. L., Tertoolen, L. G. J., van der Saag, P. T., and de Laat, S. W. (1981b). *J. Cell. Physiol.* **107,** 75–83.

Boonstra, J., van der Saag, P. T., Moolenaar, W. H., and de Laat, S. W. (1981c). *Exp. Cell Res.* **131,** 452–455.

Borochov, H., and Shinitzky, M. (1976). *Proc. Natl. Acad. Sci. U.S.A.* **73,** 4526–4530.

Bottenstein, J. E. (1980). *In* "Advances in Neuroblastoma Research" (A. E. Evans, ed.), pp. 161–169. Raven, New York.

Bottenstein, J. E., and Sato, G. H. (1979). *Proc. Natl. Acad. Sci. U.S.A.* **76,** 514–517.

Bottenstein, J. E., and Sato, G. H. (1980). *Exp. Cell Res.* **129,** 361–366.

Bottenstein, J. E., Sato, G. H., and Mather, J. P. (1979). *In* "Hormones and Cell Culture" (G. H. Sato and R. Ross, eds.), pp. 531–544. Cold Spring Harbor Lab., Cold Spring Harbor, New York.

Branton, D. (1966). *Proc. Natl. Acad. Sci. U.S.A.* **55,** 1048–1056.

Brooks, R. F., Bennett, D. C., and Smith, J. A. (1980). *Cell* **19,** 493–504.

Brown, J. C. (1971). *Exp. Cell Res.* **69,** 440–442.

Brunette, D. M., and Till, J. E. (1971). *J. Membr. Biol.* **5,** 215–224.

Byfield, J. E., and Karlsson, U. (1973). *Cell Diff.* **2,** 55–64.
Castor, L. N. (1980). *Nature (London)* **287,** 857–859.
Charalampous, F. C. (1977). *Arch. Biochem. Biophys.* **181,** 103–116.
Charalampous, F. C. (1979). *Biochim. Biophys. Acta* **556,** 38–51.
Charalampous, F. C., Gonatas, N. K., and Melbourne, A. O. (1973). *J. Cell Biol.* **59,** 421–435.
Chen, J. S., Del Fa, A., Di Luzio, A., and Calissano, P. (1976). *Nature (London)* **263,** 604–606.
Chen, S., and Levy, D. (1979). *Arch. Biochem. Biophys.* **196,** 424–429.
Christman, J. K. (1973). *Biochim. Biophys. Acta* **294,** 138–152.
Collard, J. G., de Wildt, A., Oomen-Meulemans, E. P. M., Smeekens, J., and Emmelot, P. (1977). *FEBS Lett.* **77,** 173–178.
Cone, C. D. (1969). *Trans. N.Y. Acad. Sci. Ser. 2* **31,** 404–427.
Cone, C. D. (1971). *J. Theor. Biol.* **30,** 151–181.
Cooper, A., Munden, H. R., and Brown, G. L. (1976). *Exp. Cell Res.* **103,** 435–439.
Croizat, B., Berthelot, F., Ferrandes, B., Eymard, P., and C. Sahuquillo (1979). *C. R. Hebd. Seances Acad. Sci. (Paris) Ser. D* **289**, 1283–1286.
Culp, L. A. (1980). *Nature (London)* **286,** 77–79.
Culver, B., Sahu, S. K., Vernadakis, A., and Prasad, K. N. (1977). *Biochem. Biophys. Res. Commun.* **76,** 778–783.
de Laat, S. W., van der Saag, P. T., and Shinitzky, M. (1977). *Proc. Natl. Acad. Sci. U.S.A.* **74,** 4458–4461.
de Laat, S. W., van der Saag, P. T., Nelemans, S. A., and Shinitzky, M. (1978). *Biochim. Biophys. Acta* **509,** 188–193.
de Laat, S. W., van der Saag, P. T., Elson, E. L., and Schlessinger, J. (1979). *Biochim. Biophys. Acta* **558,** 247–250.
de Laat, S. W., van der Saag, P. T., Elson, E. L., and Schlessinger, J. (1980). *Proc. Natl. Acad. Sci. U.S.A.* **77,** 1526–1528.
de Laat, S. W., Tertoolen, L. G. J., and Bluemink, J. G. (1981a). *Eur. J. Cell Biol.* **23,** 273–279.
de Laat, S. W., Tertoolen, L. G. J., van der Saag, P. T., and Bluemink, J. G. (1981b). *J. Cell Biol.* (submitted).
Edelman, G. M. (1976). *Science* **192,** 218–226.
Edwards, P. A., Lemongello, D., and Fogelman, A. M. (1979). *J. Lipid Res.* **20,** 2–7.
Egilsson, V. (1977). *Cell Biol. Int. Rep.* **1,** 435–438.
Erkell, L. J. (1977). *FEBS Lett.* **77,** 187–190.
Erkell, L. J. (1980). *Exp. Cell Biol.* **48,** 374–380.
Erkell, L. J., and Walum, E. (1979). *FEBS Lett.* **104,** 401–404.
Fishman, M. C., Dragsten, P. R., and Spector, I. (1981). *Nature (London)* **290,** 781–783.
Furmanski, P., Silverman, D. J., and Lubin, M. (1971). *Nature (London)* **233,** 413–415.
Garvican, J. H., and Brown, G. L. (1977). *Eur. J. Biochem.* **76,** 251–261.
Gilman, A. G., and Nirenberg, M. (1971). *Nature (London)* **234,** 356–358.
Glick, M. C., Kimhi, Y., and Littauer, U. Z. (1973). *Proc. Natl. Acad. Sci. U.S.A.* **70,** 1682–1687.
Goodwin, C. D., and Margolis, S. (1978). *J. Lipid Res.* **19,** 747–756.
Greene, L. A., and Shooter, E. M. (1980). *Annu. Rev. Neurosci.* **3,** 353–402.
Hamprecht, B. (1977). *Int. Rev. Cytol.* **49,** 99–170.
Hamprecht, B., Jaffe, B. M., and Philpott, G. W. (1973). *FEBS Lett.* **36,** 193–198.
Hinnen, R., and Monard, D. (1980). *In* "Control Mechanisms in Animal Cells" (L. Jimenez de Asua, R. Levi-Montalcini, and R. Shields, eds.), pp. 315–323. Raven, New York.
Huff, K., End, D., and Guroff, G. (1981). *J. Cell Biol.* **88,** 189–198.
Inbar, M., and Shinitzky, M. (1975). *Eur. J. Immunol.* **5,** 166–170.
Ishii, D. N., and Maniatis, G. M. (1978). *Nature (London)* **274,** 372–374.
Johnson, J. D., Epel, D., and Paul, M. (1976). *Nature (London)* **262,** 661–664.

Jung, C., and Rothstein, A. (1967). *J. Gen. Physiol.* **50,** 917–931.
Kaplan, J. G. (1978). *Annu. Rev. Physiol.* **40,** 19–41.
Kates, J. R., Winterton, R., and Schlessinger, K. (1971). *Nature (London)* **299,** 345–347.
Kawasaki, Y., Wakayama, M., Koike, T., Kawai, M., and Amano, T. (1978). *Biochim. Biophys. Acta* **509,** 440–449.
Kimelberg, H. K. (1976). *Mol. Cell. Biochem.* **10,** 171–190.
Kimhi, Y., Palfrey, C., Spector, I., Barak, Y., and Littauer, U. Z. (1976). *Proc. Natl. Acad. Sci. U.S.A.* **73,** 462–466.
Klebe, R., and Ruddle, F. (1969). *J. Cell Biol.* **43,** 69A.
Knutton, S. (1976). *Exp. Cell Res.* **102,** 109–116.
Koch, K. S., and Leffert, H. L. (1979). *Cell* **18,** 153–163.
Koike, T. (1978). *Biochim. Biophys. Acta* **509,** 429–439.
Koike, T., and Pfeiffer, S. E. (1979). *Dev. Neurosci.* **2,** 177–182.
Kolber, A. R., Goldstein, M. H., and Moore, B. W. (1974). *Proc. Natl. Acad. Sci. U.S.A.* **71,** 4203–4207.
Koppel, D. E., Axelrod, D., Schlessinger, J., Elson, E. L., and Webb, W. W. (1976). *Biophys. J.* **16,** 1315–1329.
Lai, C.-S., Hopwood, L. E., and Swartz, H. M. (1980). *Biochim. Biophys. Acta* **602,** 117–126.
Lang, W. E., Jones, P. A., Nye, C. A., and Benedict, W. F. (1976). *Biochem. Biophys. Res. Commun.* **68,** 114–119.
Lanks, K. W., Turnbull, J. D., Aloyo, V. J., Darwin, J., and Papirmeister, B. (1975). *Exp. Cell Res.* **93,** 355–362.
Levi, A., Shechter, Y., Neufeld, E. J., and Schlessinger, J. (1980). *Proc. Natl. Acad. Sci. U.S.A.* **77,** 3469–3473.
Littauer, U. Z., Giovanni, M. Y., and Glick, M. C. (1980). *J. Biol. Chem.* **255,** 5448–5453.
Lubin, M. (1967). *Nature (London)* **213,** 451–453.
Maltese, W. A., and Volpe, J. J. (1980). *J. Neurochem.* **34,** 1522–1526.
Mandersloot, J. G., Roelofsen, B., and de Gier, J. (1978). *Biochim. Biophys. Acta* **508,** 478–485.
Martin, G. (1980). *Science* **209,** 768–776.
Mathews, R. G., Johnson, T. C., and Hudson, J. E. (1976). *Biochem. J.* **154,** 57–64.
Mendoza, S. A., Wigglesworth, N. H., Pohjanpelto, P., and Rozengurt, E. (1980). *J. Cell. Physiol.* **103,** 17–27.
Milenkovic, A. G., and Johnson, T. C. (1980). *Biochem. J.* **191,** 21–28.
Milenkovic, A. G., Rachmeler, M., and Johnson, T. C. (1978). *Biochem. J.* **176,** 695–704.
Miller, C. A., and Levine, E. M. (1972). *Science* **177,** 799–802.
Mills, B., and Tupper, J. T. (1975). *J. Membr. Biol.* **20,** 75–97.
Momoi, M., Kennett, R. H., and Glick, M. C. (1980). *J. Biol. Chem.* **255,** 11914–11921.
Monard, D., Solomon, F., Rentsch, M., and Gysin, R. (1973). *Proc. Natl. Acad. Sci. U.S.A.* **70,** 1894–1897.
Monard, D., Stöckel, K., Goodman, R., and Thoenen, H. (1975). *Nature (London)* **258,** 444–445.
Monard, D., Rentsch, M., Schuerch-Rathgeb, Y., and Lindsay, R. M. (1977). *Proc. Natl. Acad. Sci. U.S.A.* **74,** 3893–3897.
Moolenaar, W. H., de Laat, S. W., and van der Saag, P. T. (1979). *Nature (London)* **279,** 721–723.
Moolenaar, W. H., Boonstra, J., van der Saag, P. T., and de Laat, S. W. (1981a). *J. Biol. Chem.* (in press).
Moolenaar, W. H., Mummery, C. L., van der Saag, P. T., and de Laat, S. W. (1981b). *Cell* **23,** 789–798.
Morgan, J. I., and Seifert, W. (1979). *J. Supramol. Struct.* **10,** 111–124.

Morris, R. J., Gower, S., and Pfeiffer, S. E. (1980). *Brain Res.* **183,** 145–159.
Mummery, C. L., Boonstra, J., van der Saag, P. T., and de Laat, S. W. (1981a). *J. Cell. Physiol.* **107,** 1–9.
Mummery, C. L., Boonstra, J., van der Saag, P. T., and de Laat, S. W. (1981b). *J. Cell. Physiol.* (submitted).
Murphy, J. S., D'Alisa, R., Gershey, E. L., and Landsberger, F. R. (1978). *Proc. Natl. Acad. Sci. U.S.A.* **75,** 4404–4407.
Murphy, R. A., Pantazis, N. J., Arnason, B. G. W., and Young, M. (1975). *Proc. Natl. Acad. Sci. U.S.A.* **72,** 1895–1898.
Nicolson, G. L. (1976). *Biochim. Biophys. Acta* **457,** 57–108.
Obrénovitch, A., Sené, C., Nègre, M.-T., and Monsigny, M. (1978). *FEBS Lett.* **88,** 187–191.
Palfrey, C., Kimhi, Y., Littauer, U. Z., Reuben, R. C., and Marks, P. A. (1977). *Biochem. Biophys. Res. Commun.* **76,** 937–942.
Pall, M. L. (1977). *J. Biol. Chem.* **252,** 7146–7150.
Pardee, A. B. (1974). *Proc. Natl. Acad. Sci. U.S.A.* **71,** 1286–1290.
Pardee, A. B., Dubrow, R., Hamlin, J. L., and Kletzien, R. F. (1978). *Annu. Rev. Biochem.* **47,** 715–750.
Pardee, A. B., Shilo, B. Z., and Koch, A. L. (1979). *In* "Hormones and Cell Culture" G. H. Sato and R. Ross, eds.), pp. 373–392. Cold Spring Harbor Lab., Cold Spring Harbor, New York.
Perez-Polo, J. R., Werrbach-Perez, K., and Tiffany-Castiglioni, E. (1979). *Dev. Biol.* **71,** 341–345.
Pfenninger, K. H. (1979). *In* "The Neurosciences Fourth Study Program" (F. O. Schmitt and F. G. Worden, eds.), pp. 779–795. MIT Press, Cambridge, Massachusetts.
Pierce, G. B. (1967). *In* "Current Topics in Developmental Biology" (A. A. Moscona and A. Monroy, eds.), Vol. 2, pp. 223–246. Academic Press, New York.
Pierce, G. B., Shikes, R., and Fink, L. M. (1978). "Cancer: A Problem of Developmental Biology." Prentice-Hall, New York.
Pochedly, C. (1977). "Neuroblastoma." Arnold, London.
Porter, K., Prescott, D., and Frye, J. (1973). *J. Cell Biol.* **57,** 815–836.
Portier, M.-M., Eddé, B., Berthelot, F., Croizat, B., and Gros, F. (1980). *Biochem. Biophys. Res. Commun.* **96,** 1610–1618.
Prasad, K. N. (1971). *Nature (London)* **234,** 471–473.
Prasad, K. N. (1972). *Nature (London) New Biol.* **236,** 49–52.
Prasad, K. N. (1973). *Int. J. Cancer* **12,** 631–636.
Prasad, K. N. (1975). *Biol. Rev.* **50,** 129–165.
Prasad, K. N., and Hsie, A. W. (1971). *Nature (London) New Biol.* **233,** 141–143.
Prasad, K. N., and Sheppard, J. R. (1972). *Exp. Cell Res.* **73,** 436–440.
Prasad, K. N., and Sinha, P. K. (1976). *In Vitro* **12,** 125–132.
Prasad, K. N., and Sinha, P. K. (1978). *In* "Cell Differentiation and Neoplasia" (G. F. Saunders, ed.), pp. 111–141. Raven, New York.
Prasad K. N., and Vernadakis, A. (1972). *Exp. Cell Res.* **70,** 27–32.
Prasad, K. N., Gilmer, K., and Kumar, S. (1973a). *Proc. Soc. Exp. Biol. Med.* **143,** 1168–1171.
Prasad, K. N., Kumar, S., Gilmer, K., and Vernadakis, A. (1973b). *Biochem. Biophys. Res. Commun.* **50,** 973–977.
Prasad, K. N., Ramanujam, S., and Gaudreau, D. (1979). *Proc. Soc. Exp. Biol. Med.* **161,** 570–573.
Respess, J. G., Stubbs, J. D., and Chambers, D. A. (1978). *Biochim. Biophys. Acta* **529,** 38–43.
Revoltella, R. P., and Butler, R. H. (1980). *J. Cell. Physiol.* **104,** 27–33.
Revoltella, R., Bertolini, L., and Pediconi, M. (1974). *Exp. Cell Res.* **85,** 89–94.
Robinson, J. D., and Flashner, M. S. (1979). *Biochim. Biophys. Acta* **549,** 145–176.
Ross, J., Olmsted, J. B., and Rosenbaum, J. L. (1975). *Tissue Cell* **7,** 107–136.

Rossow, P. W., Riddle, V. G. H., and Pardee, A. B. (1979). *Proc. Natl. Acad. Sci. U.S.A.* **76,** 4446–4450.
Rozengurt, E., and Heppel, L. A. (1975). *Proc. Natl. Acad. Sci. U.S.A.* **72,** 4492–4495.
Rozengurt, E., and Mendoza, S. (1980). *Ann. N.Y. Acad. Sci.* **339,** 175–190.
Sachs, H. G., Stambrook, P. J., and Ebert, J. D. (1974). *Exp. Cell Res.* **83,** 362–366.
Sachs, L. (1980). *Proc. Natl. Acad. Sci. U.S.A.* **77,** 6152–6156.
Sandermann, H. (1978). *Biochim. Biophys. Acta* **515,** 209–237.
Sandquist, D., Williams, T. H., Sahu, S. K., and Kataoka, S. (1978). *Exp. Cell Res.* **113,** 375–381.
Sandquist, D., Black, A. C., Jr., Sahn, L., Williams, L., and Williams, T. H. (1979). *Exp. Cell Res.* **123,** 417–421.
Sandra, A., Paltzer, W. B., and Thomas, M. J. (1981). *Exp. Cell Res.* **132,** 473–477.
Schimke, R. N. (1980). *In* "Advances in Neuroblastoma Research" (A. E. Evans, ed.), pp. 13–24. Raven, New York.
Schubert, D. (1974). *Neurobiology* **4,** 376–387.
Schubert, D., and Jacob, F. (1970). *Proc. Natl. Acad. Sci. U.S.A.* **67,** 247–254.
Schubert, D., Humphreys, S., DeVitry, F., and Jacob, J. (1971). *Dev. Biol.* **25,** 514–546.
Schuerch-Rathgeb, Y., and Monard, D. (1978). *Nature (London)* **273,** 308–309.
Schwartz, J. P., and Costa, E. (1978). *Neuroscience* **3,** 473–480.
Scott, R. E., Carter, R. L., and Kidwell, W. R. (1971). *Nature (London) New Biol.* **233,** 219–220.
Seeds, N. W., Gilman, A. G., Amano, T., and Nirenberg, M. W. (1970). *Proc. Natl. Acad. Sci. U.S.A.* **66,** 160–167.
Shimuzu, T., Mizuno, N., Amano, T., and Hayashi, O. (1979). *Proc. Natl. Acad. Sci. U.S.A.* **76,** 6231–6234.
Shinitzky, M., and Barenholz, Y. (1978). *Biochim. Biophys. Acta* **515,** 367–394.
Singer, S. J., and Nicolson, G. L. (1972). *Science* **175,** 720–731.
Skaper, S. D., and Varon, J. (1980). *J. Neurochem.* **34,** 1654–1660.
Skaper, S. D., and Varon, S. (1981). *Exp. Cell Res.* **131,** 353–361.
Smith, G. L. (1977). *J. Cell Biol.* **73,** 761–767.
Smith, J. A., and Martin, L. (1973). *Proc. Natl. Acad. Sci. U.S.A.* **70,** 1263–1267.
Smith, J. B., and Rozengurt, E. (1978). *Proc. Natl. Acad. Sci. U.S.A.* **75,** 5560–5564.
Spector, I., Palfrey, C., and Littauer, U. Z. (1975). *Nature (London)* **254,** 121–124.
Strange, P. G., Birdsall, N. J. M., and Burgen, A. S. V. (1978). *Biochem. J.* **172,** 495–501.
Struve, W. G., Arneson, R. M., Chenevey, J. E., and Cartwright, C. K. (1977). *Exp. Cell Res.* **109,** 381–387.
Svetina, S. (1977). *Cell Tissue Kinet.* **10,** 575–581.
Tansik, R. L., and White, H. L. (1979). *Prostaglandins Med.* **2,** 225–234.
Torpier, G., Montaquier, L., Biquard, J. M., and Vigier, P. (1975). *Proc. Natl. Acad. Sci. U.S.A.* **72,** 1695–1698.
Truding, R., and Morell, P. (1977). *J. Biol. Chem.* **252,** 4850–4854.
Truding, R., Shelanski, M. L., Daniels, M. P., and Morell, P. (1974). *J. Biol. Chem.* **249,** 3973–3982.
Truding, R., Shelanski, M. L., and Morell, P. (1975). *J. Biol. Chem.* **250,** 9348–9354.
Tsiftsoglou, A. S., Bhargava, K. K., Rittmann, L. S., and Sartorelli, A. C. (1981). *J. Cell. Physiol.* **106,** 419–424.
Tupper, J. T., Zorgniotti, F., and Mills, B. (1977). *J. Cell. Physiol.* **91,** 429–440.
van der Saag, P. T., Feyen, A., Miltenburg-Vonk, W., and de Laat, S. W. (1981). *Exp. Cell Res.* **136,** 351–358.
van Wijk, R., and van de Poll, K. W. (1979). *Cell Tissue Kinet.* **12,** 659–663.
van Zoelen, E. J. J., van der Saag, P. T., and de Laat, S. W. (1981a). *Exp. Cell Res.* **131,** 395–406.
van Zoelen, E. J. J., van der Saag, P. T., and de Laat, S. W. (1981b). *J. Theor. Biol.* (in press).

van Zoelen, E. J. J., Tertoolen, L. G. J., Boonstra, J., van der Saag, P. T., and de Laat, S. W. (1981c). *Eur. J. Biochem.* (submitted).

Vinores, S., and Guroff, G. (1980). *Annu. Rev. Biophys. Bioengin.* **9,** 223–257.

Volpe, J. J., and Marasa, J. C. (1976). *Biochim. Biophys. Acta* **431,** 195–205.

Wachsman, J. T., and Biedler, J. L. (1974). *Exp. Cell Res.* **86,** 264–268.

Waris, T., Rechardt, L., and Waris, P. (1973). *Experientia* **24,** 1128–1129.

Warren, L., and Glick, M. (1969). *In* "Fundamental Techniques in Virology" (K. Habel and N. P. Salzman, eds.), pp. 66–71. Academic Press, New York.

Waymire, J. C., Gilmer-Waymire, K., and Haycock, J. W. (1978). *Nature (London)* **276,** 194–195.

Yanishevski, R. M., and Stein, G. H. (1981). *Int. Rev. Cytol.* **69,** 223–259.

Yoda, K., and Fujimura, S. (1979). *Biochem. Biophys. Res. Commun.* **87,** 128–134.

Zagyansky, Y., Benda, P., and Bisconte, J. C. (1977). *FEBS Lett.* **77,** 206–208.

Zisapfel, N., and Littauer, U. Z. (1979). *Eur. J. Biochem.* **95,** 51–59.

INTERNATIONAL REVIEW OF CYTOLOGY, VOL. 74

Mechanisms That Regulate the Structural and Functional Architecture of Cell Surfaces

JANET M. OLIVER AND RICHARD D. BERLIN

Department of Physiology, University of Connecticut Health Center, Farmington, Connecticut

I. Introduction

The mechanisms controlling cell surface molecular and functional topography have been debated for over a decade. Nevertheless, few general principles governing the organization of mammalian membranes are generally recognized.

In this article we will emphasize several advances that may simplify and focus the search for general principles governing membrane properties. First, we will show that asymmetric surface topography is a predictable correlate of asymmetric shape and microfilament distribution, and that it is independent of how asymmetry is induced and which cell type is studied. Thus precisely analogous membrane heterogeneities will be demonstrated in capped, phagocytizing, oriented, and dividing cells and general relationships drawn by comparison between these systems. Second, we will establish that receptors can be segregated out of as well as into regions of microfilament accumulation. This contradicts the persistent hypothesis that microfilament–receptor connections are sufficient to determine surface topography and

ISBN 0-12-364474-7

prepares the way for a more general hypothesis. Third, we will show that topographical heterogeneity is not restricted to membrane molecular determinants but extends to a range of endocytic functions as well as to a macromolecular assembly, the coated pit. This broadens both the scope and significance of research into the control of cell surface topography. Fourth, we will document the remarkable arrest of dynamic surface events during mitosis.

A set of general principles of surface control will be derived from these experimental data. These principles form the basis of a new model presented

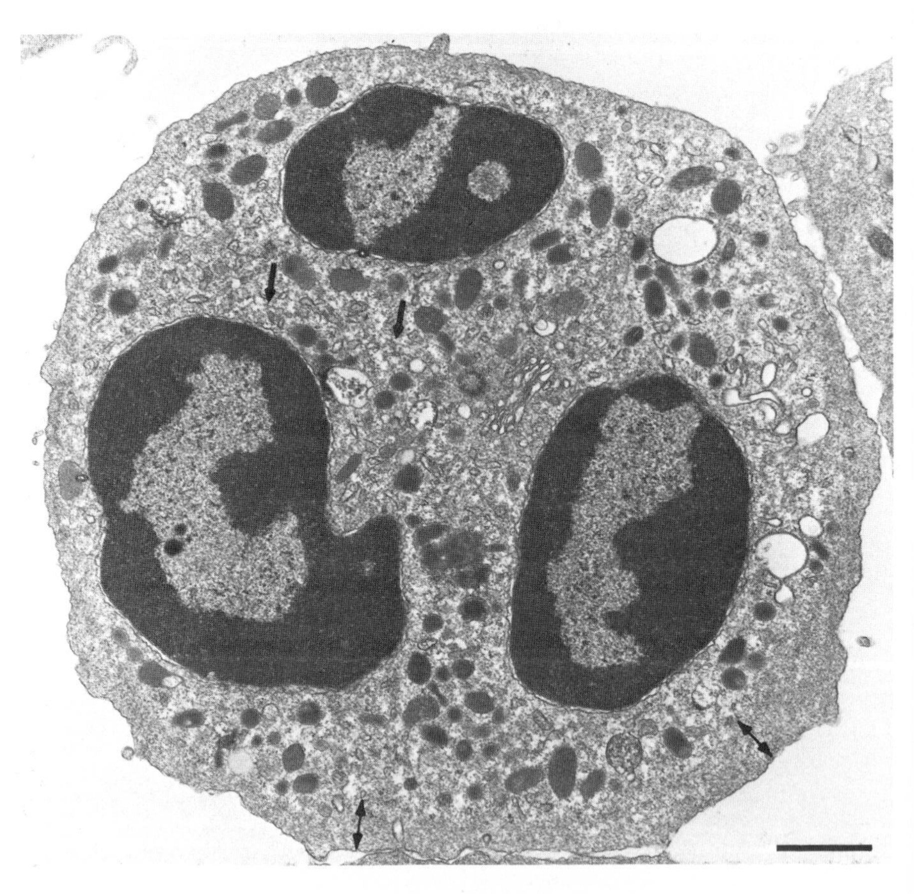

FIG. 1. A human peripheral blood polymorphonuclear leukocyte incubated for 5 minutes at 37°C with 50 μg/ml Con A and fixed in suspension. This symmetrical cell displays a rather uniform, submembranous meshwork of microfilaments (double arrows). Microtubules, barely visible at this magnification (arrows), originate from the centriole and radiate toward the cell periphery. Bar = 1 μm.

here to explain and predict the development of membrane topographical patterns.

II. Routes to Cell Shape and Cytoskeletal Asymmetry

In our experience membrane molecular and functional asymmetry in mammalian cells is always associated with the development of asymmetries in cell shape and cytoskeletal organization. We therefore begin this article by describing four routes to major cell shape changes that are particularly accessible to experimental analysis. These descriptions emphasize leukocytes and macrophages because they exhibit all of the relevant shape changes. However, the same relationships between cell shape and cytoskeletal properties can be derived in other cultured cells.

Around 90% of resting leukocytes [polymorphonuclear leukocytes (PMN), lymphocytes, monocytes, macrophages], like most mammalian cells, maintain a more or less rounded shape with a uniformly ruffled membrane when suspended in simple media (phosphate-buffered saline or Hanks' medium with 1% bovine serum albumin) (Fig. 1). In these cells, microfilaments are typically arranged in a rather uniform, submembranous meshwork. Microtubules originate from centrioles and radiate toward the cell periphery, approaching, but rarely associating directly with the plasma membrane. Intermediate filaments are typically seen near the nucleus and in the centriolar region of resting leukocytes.

A small proportion of "resting" leukocytes exhibits a characteristically different morphology. In simple buffers, between 2 and 15% of cells develop a bulge or protuberance, often bordered by highly plicated membrane. This protuberant region is supported by a meshwork of microfilaments visible by immunofluorescence using antiactin antibody (Oliver *et al.*, 1977) and by electron microscopy (Albertini *et al.*, 1977: see Figs. 2 and 3). Intermediate filaments frequently occupy the core of the protuberance (Berlin and Oliver, 1978). A thin band of submembranous microfilaments persists away from the protuberance and is particularly evident at the nuclear pole of the macrophage depicted in Fig. 3. Two manipulations may be used to increase the proportion of cells with this morphology. Around 25% of human peripheral blood lymphocytes and T lymphoblasts become protuberant following transfer from simple buffers into RPMI medium with 10% fetal calf serum. In these cells, cytoplasmic microtubules run toward the protuberance but stop short of the microvillous region (Fig. 2). In addition, the majority (around 80%) of PMN, monocytes, macrophages, and lymphocytes develop a protuberance following microtubule disassembly by colchicine, nocodozole, diamide, and other antimicrotubule agents (Fig. 3; Albertini *et al.*, 1977).

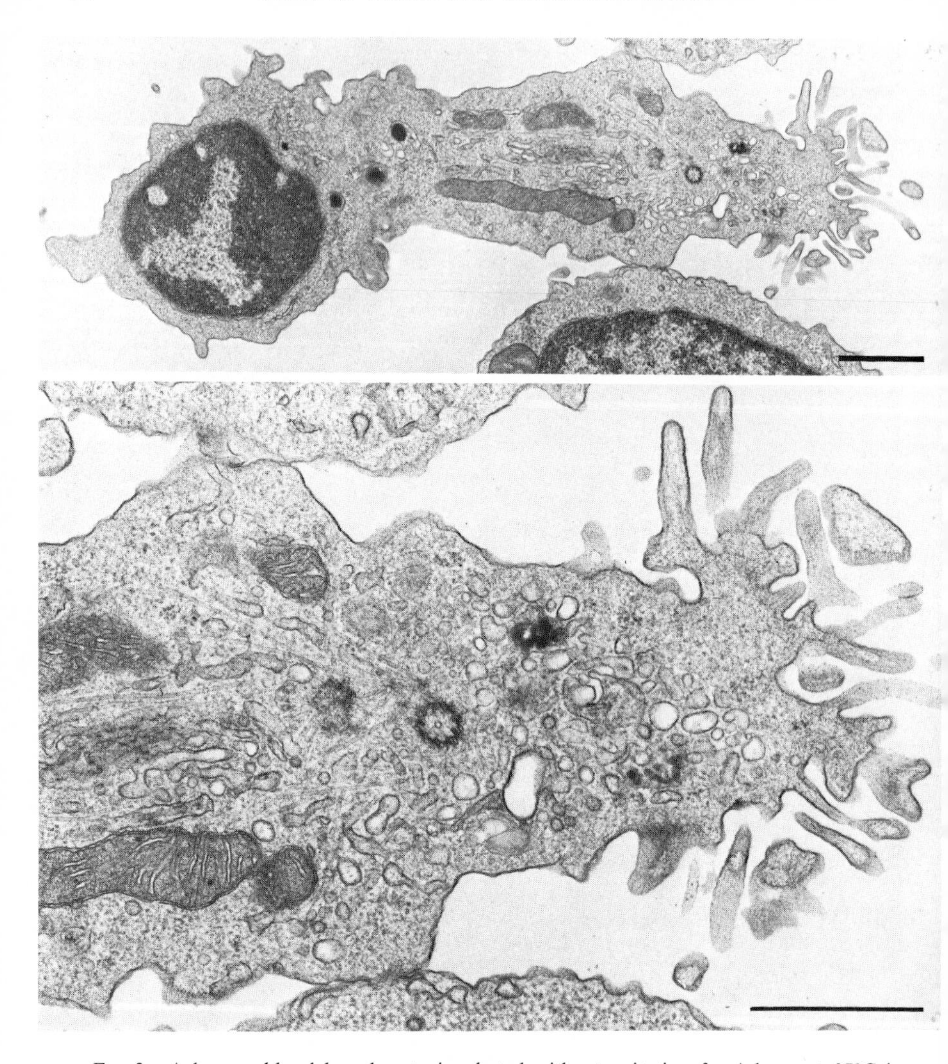

FIG. 2. A human blood lymphocyte incubated without agitation for 1 hour at 37°C in RPMI medium with 10% fetal calf serum, then fixed *in situ* by addition of 1 volume of 4% glutataldehyde in 0.1 *M* sodium cadodylate pH 7.4. The elongated shape and ruffled protuberance of this cell is typical of motile lymphocytes. The tip of the protuberance (uropod) excludes organelles and is packed with microfilaments that extend to the tips of the microvilli. Microtubules originate from two anterior centrioles and stretch toward the cell nucleus and lamellipodium. Bar = 1 μm.

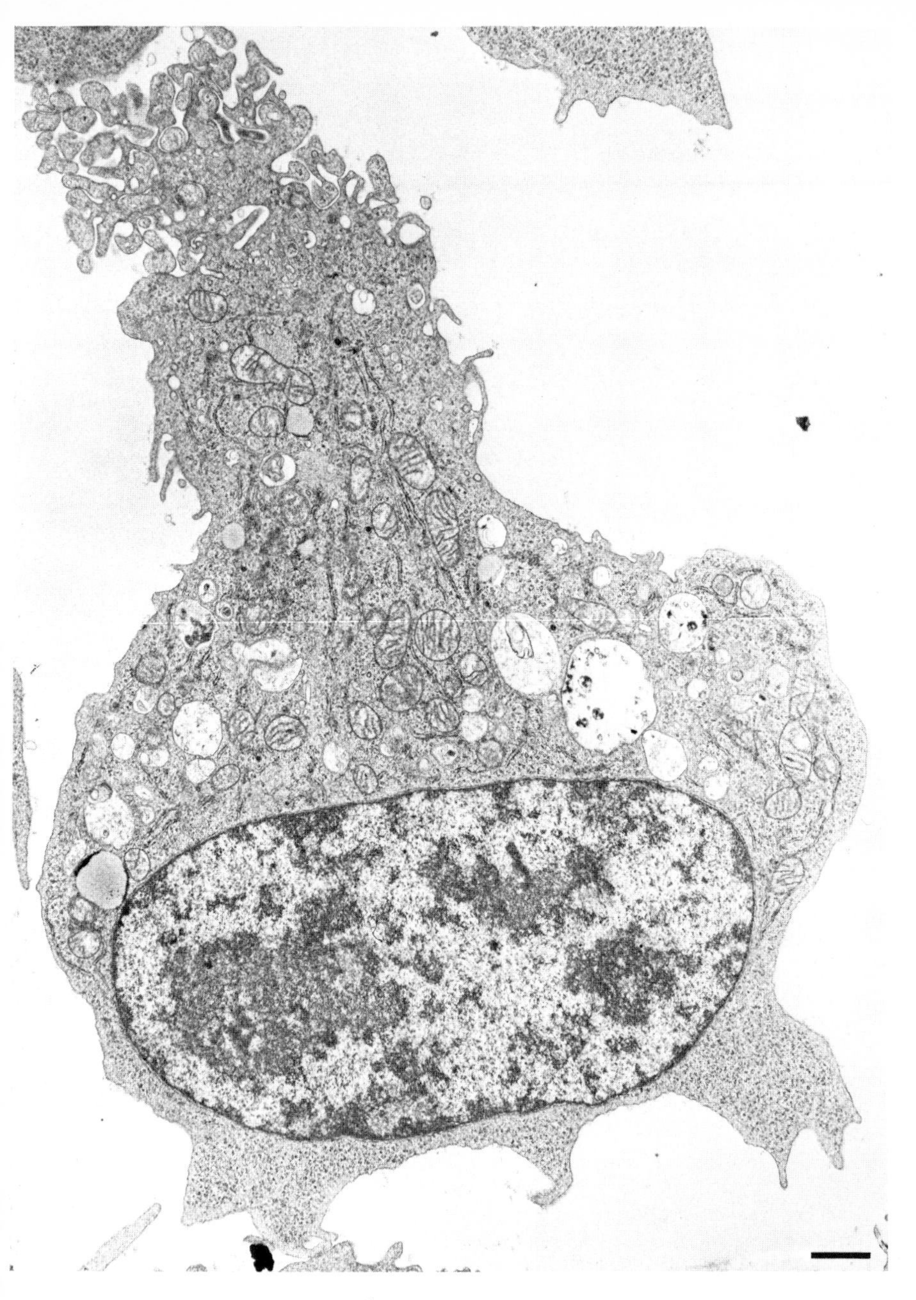

FIG. 3. A colchicine-treated (10^{-6} M, 1 hour) J774.2 macrophage, showing a ruffled protuberance that excludes organelles and is enriched for microfilaments. Submembranous microfilaments are also present away from the protuberance, particularly as a thin band at the nuclear pole of the cell as suggested by exclusion of cell organelles. Bar = 1 μm.

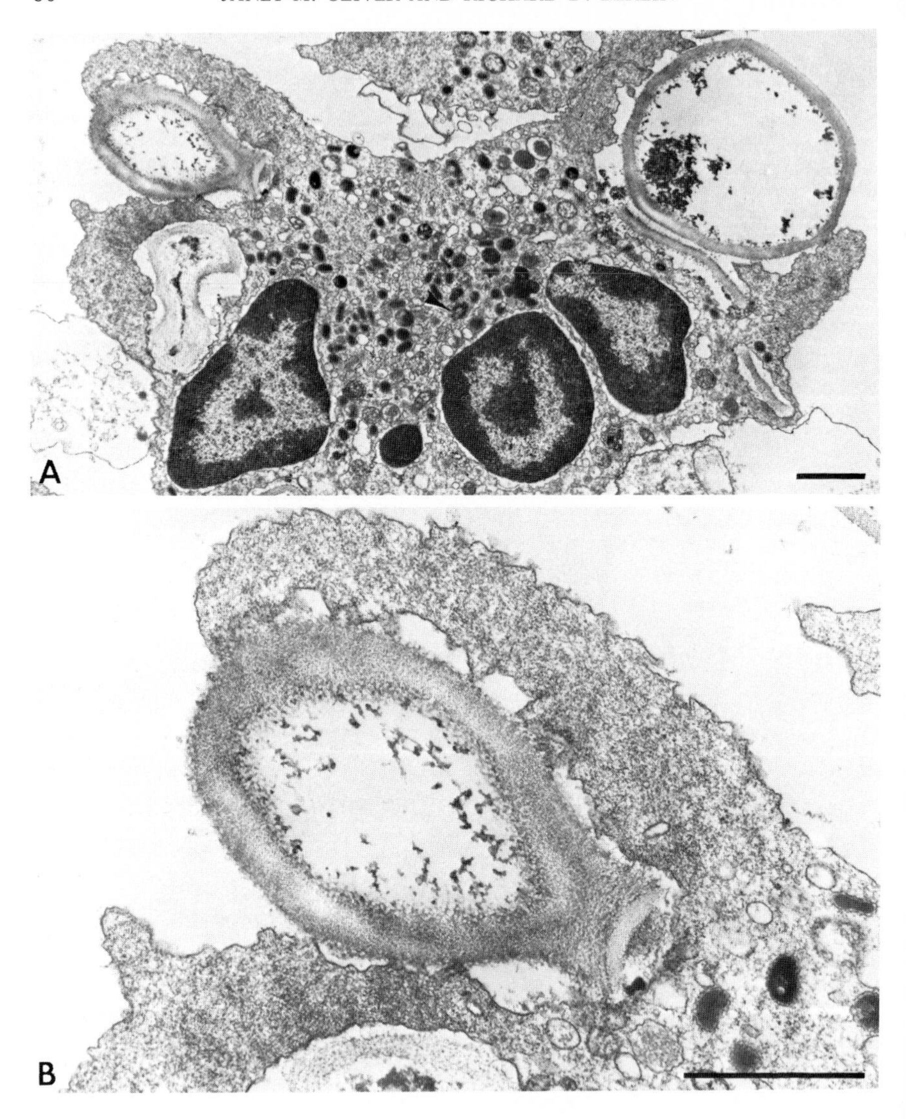

FIG. 4. A human peripheral blood PMN fixed after 30 seconds exposure to complement-opsonized zymosan. Panel A demonstrates the formation of pseudopodia that enclose zymosan particles at points around the whole cell surface. Panel B emphasizes the remarkable recruitment of microfilaments in meshworks associated with pseudopod formation. These microfilaments have disassembled at the base of the developing phagocytic vesicle and lysosomal degranulation has already begun. Microtubules (not visible at this magnification) original from the centriole (arrow) but do not enter the pseudopodia. From Berlin and Oliver (1978). Bar = 1 μm.

This dramatic stimulation of protuberance formation can occur within 5 minutes after addition of the glutathione oxidant, diamide, to human leukocytes, and can be precisely correlated with microtubule disassembly (Oliver *et al.*, 1980). It occurs both in simple buffers and in culture media. A decrease in cell volume of around 20% accompanies, and may be related to, the change in cell shape (Melmed *et al.*, 1980).

Unanue considers protuberant cells to be motile forms whose protuberance is properly described as a uropod. This reflects his observation that protuberent lymphocytes in contact with a substratum may show directed movement over short distances with the protuberance acting as a tail (probably the case for Fig. 2). However the protuberant morphology is also readily observed in suspension where no directed cell movement can occur (e.g., Fig. 3). Thus "protuberant" leukocytes is a more comprehensive description than the terms "motile" or "uropod-bearing" cells.

Cell shape and cytoskeletal organization also become asymmetric during the process of phagocytosis (Berlin and Oliver, 1978). Phagocytic leukocytes can engulf suitable particles by their enclosure in microfilament-packed pseudopodia (e.g., Fig. 4A and B). The pseudopodia eventually fuse, sealing particles within intracellular vacuoles. Pseudopod formation is extremely rapid: the cell in Fig. 4 was fixed 30 seconds after addition of zymosan particles. Furthermore, the recruitment of microfilaments into pseudopodia is readily reversed, with the bases of forming phagocytic vacuoles often losing their microfilament "coats" prior to completion of pseudopod fusion.

A third example of shape and microfilament asymmetry in mammalian cells occurs during the process of cytokinesis (Schroeder, 1976). As cells move from metaphase into anaphase and telophase, a cleavage furrow forms that is underlain by a contractile ring of microfilaments (Fig. 5A and B) and that encloses residual interpolar microtubules. In Fig. 5B, intermediate filaments lie immediately below the microfilamentous layer. The ring supports the rapid invagination and eventual separation of membrane between daughter cells.

Finally, we have studied the polarization of shape and cytoskeleton in glass-adherent PMN during orientation in a gradient of the chemotactic peptide F-Met-Leu-Phe (Oliver *et al.*, 1978; Davis *et al.*, 1982). A dense meshwork of microfilaments occupies the anterior lamellipodium of oriented cells (Fig. 6). Microfilaments are also prominent in the uropod region, often surrounding a central core of intermediate (100 Å) filaments. In such cells, abundant cytoplasmic microtubules originate from more or less centrally located centrioles and run toward, but not into, the microfilament-rich poles. This oriented morphology may develop whether or not chemotaxis follows. Thus Marasco *et al.* (1980) found ionic conditions that allowed orientation but not movement of PMN in a gradient of peptide. We have

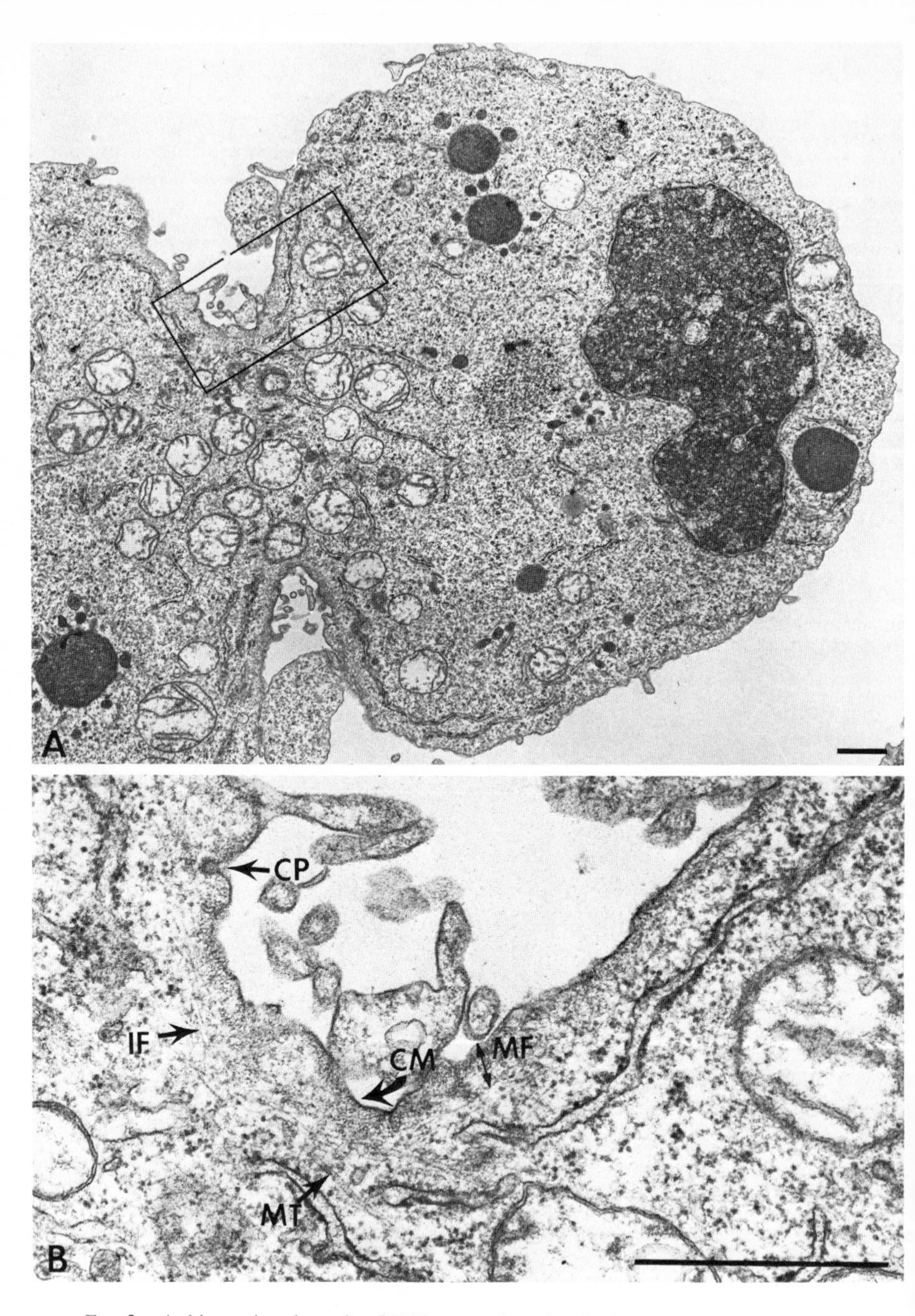

FIG. 5. A thin section through a J774.2 macrophage in telophase reveals a typical cleavage furrow subtended by microfilaments (MF). Intermediate filaments (IF) lie below the microfilaments and residual microtubules (MT) are visible within the furrow region. The arrows indicate the presence of coated pits (CP) and stretches of coated membrane (CM). Bar = 1 μm.

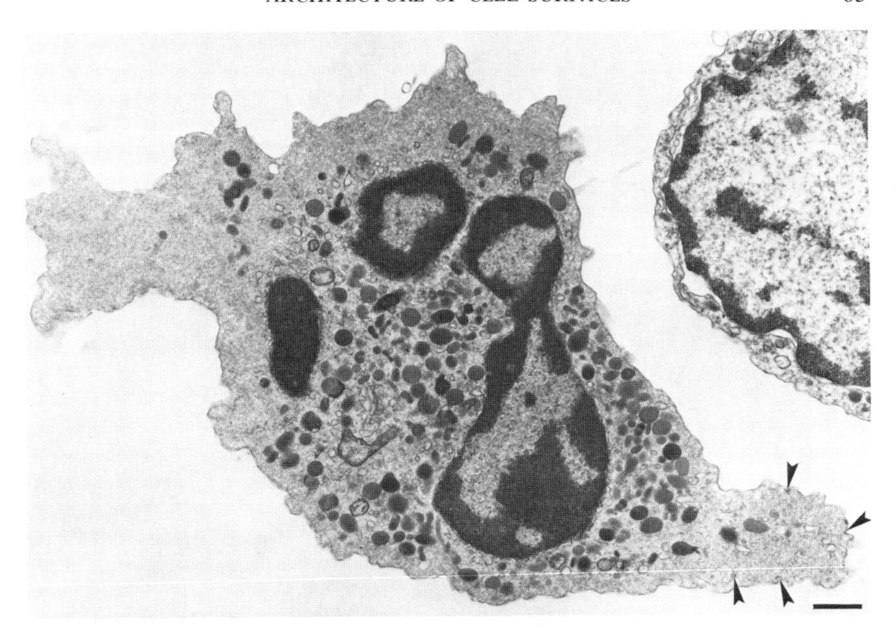

FIG. 6. Ultrastructure of a human peripheral blood PMN oriented in a gradient of the chemotactic peptide *N*-formyl-methionyl-leucyl-phenylalanine (F-Met-Leu-Phe) according to the procedure of Zigmond (1977). Both the anterior lamellipodium and the posterior uropod contain microfilaments. The lamellipodial microfilaments are typically organized as a loose meshwork. The posterior microfilaments are less numerous, more orderly, and may be organized about a central "core" of intermediate filaments (not visible at this magnification). Microtubules originate from centrioles and radiate between, but not into, the microfilamentous poles. Cytoplasmic vesicles (coated and smooth) and coated pits (arrows) are restricted to the uropod and are almost never observed in the lamellipodium. Bar = 1 μm.

also noted that PMN can assume the same oriented morphology simply by incubation in suspension with F-Met-Leu-Phe: migration, which depends on substrate adherence and a gradient of chemoattractant, is obviously precluded under these conditions.

We emphasize that present data do not permit selection of one set of cytoskeletal structures over another as the primary regulator of cell shape. Increased cortical microfilament density is most obviously associated with regions of cell deformation. However microtubules appear to influence cell shape, at least partly through their effect on microfilament distribution. This is apparent in the enormous increase in protuberance formation following microtubule disassembly. It is also apparent in subtler ways: for example, an intact microtubule system is essential for PMN to ingest phagocytic particles at points of contact over the entire cell surface (Berlin and Oliver, 1978), for PMN to orient and move directly up a gradient of chemotactic factors (Allan and Wilkinson, 1978), and for the proper positioning of the contractile ring in dividing cells (Rappaport, 1975; Oliver and Berlin, 1979; Berlin *et al.*, 1980). In addition, we point out that the rapid redistribution of intermediate filaments into cytoplasm subtending the protuberance, uropod, and cleavage furrow is consistent with a role for these fibers in cell shape control.

III. Functional and Molecular Topography of Asymmetric Cells

A. The Distribution of Concanavalin A–Receptor Complexes

The data above establish two common features of protuberance, uropod, pseudopod, and cleavage furrow membranes. They are all regions of marked surface deformation. They all overlie regions of microfilament accumulation. These common features are emphasized in Fig. 7. Analysis of the fate of

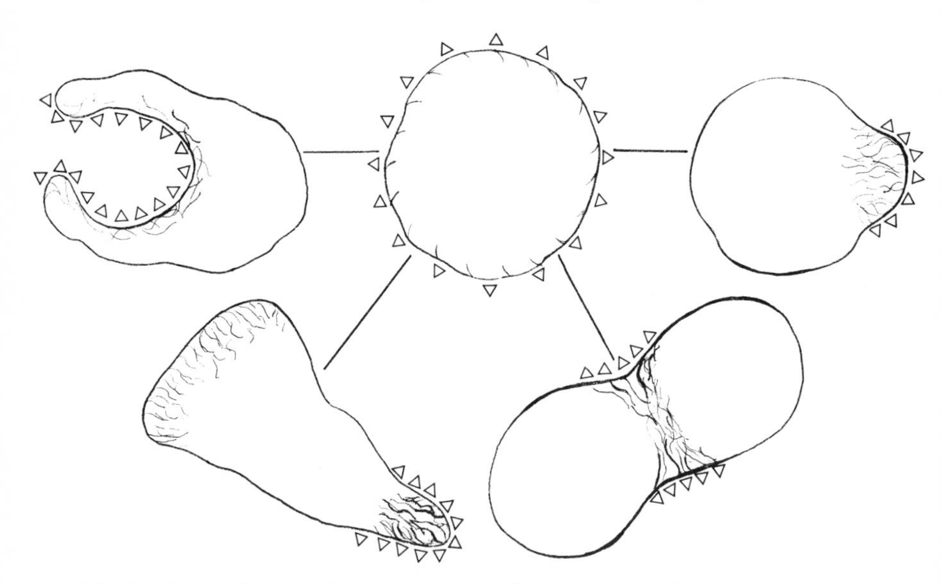

Fig. 7. Schematic view of the shapes, microfilament asymmetries, and foci of Con A–receptor complex accumulation on cells undergoing various dynamic processes. The resting cell (center) is approximately symmetrical with respect to shape, microfilament distribution, and Con A–receptor complex distribution. In contrast cells undergoing the dynamic processes of (L-R) phagocytosis, chemotaxis, cytokinesis, and capping all accumulate Con A at regions of microfilament accumulation and membrane deformation. This redistribution of Con A receptors requires ligand binding and brief (5–10 minute) warming of labeled cells. Although microfilaments are indicated only at regions of their concentration, it should be noted that a thin band of submembranous microfilaments is usually present under the entire membrane of polarized cells.

concanavalin A (Con A)–receptor complexes on a variety of cells demonstrates that the similarity between these specialized membrane regions extends to their molecular and functional properties.

Figure 8A illustrates the distribution of fluorescein–Con A on a resting human PMN following incubation for 5 minutes at 37°. The lectin maintains a uniform surface distribution during brief labeling periods. With longer incubation, it is internalized by pinocytosis from points along the entire cell periphery (Fig. 9A). Lateral movement into aggregates and patches may precede this internalization. In contrast, Con A bound to a protuberant (microtubule-depleted) cell is accumulated on the protuberance forming a cap (Fig. 8B). The cap, too, is eventually internalized (Fig. 9B).

Con A also accumulates in the pseudopodia of phagocytizing PMN (Berlin and Oliver, 1978, Fig. 8C); it accumulates in the uropod of PMN oriented in a chemotactic gradient (Davis *et al.*, 1982; Fig. 8D); and it accumulates in the cleavage furrow of cells ranging from sea urchin eggs to rat kangaroo cells (but best studied in J774.2 mouse macrophages, Fig. 8E, Fig. 10; Berlin *et al.*, 1978). That is, a predictable, preferential accumulation of Con A occurs in regions of maximum surface deformation and microfilament accumulation. This accumulation is independent of the event causing shape and microfilament asymmetry, which can range from drug-induced microtubeule disassembly to the physiological process of cell division.

Several additional properties of Con A receptor topography deserve mention. First, the inherent distribution of Con A receptors, determined by labeling at 4°C or after fixation, is largely uniform on both rounded and polarized cells (e.g., Fig. 11A and B). A proportion of Con A receptors was shown in biochemical (Oliver *et al.*, 1974) and fluorescence (Berlin and Oliver, 1978) studies to concentrate in pseudopod membrane in the absence of ligand. Similarly, dividing cells labeled at 4°C may show a hint of Con A receptor accumulation at the cleavage furrow. Nevertheless, the quantitative redistribution of Con A receptors requires ligand binding followed by incubation at 37°C. When the latter conditions apply, ligand–receptor complexes move into regions of membrane deformation and microfilament accumulation whether these already exist (as in cells that are protuberant or oriented prior to Con A labeling) or develop after ligand binding (as in cells labeled with Con A prior to cytokinesis: see Fig. 12 below).

Second, although all these foci of Con A accumulation represent membrane overlying microfilament-rich cytoplasm, factors other than the mere presence of microfilaments are required for the specific entrapment of Con A. As noted above, asymmetric cells in general show a persistent band of submembranous microfilaments away from regions of maximum shape change and maximum microfilament accumulation. This band is particularly

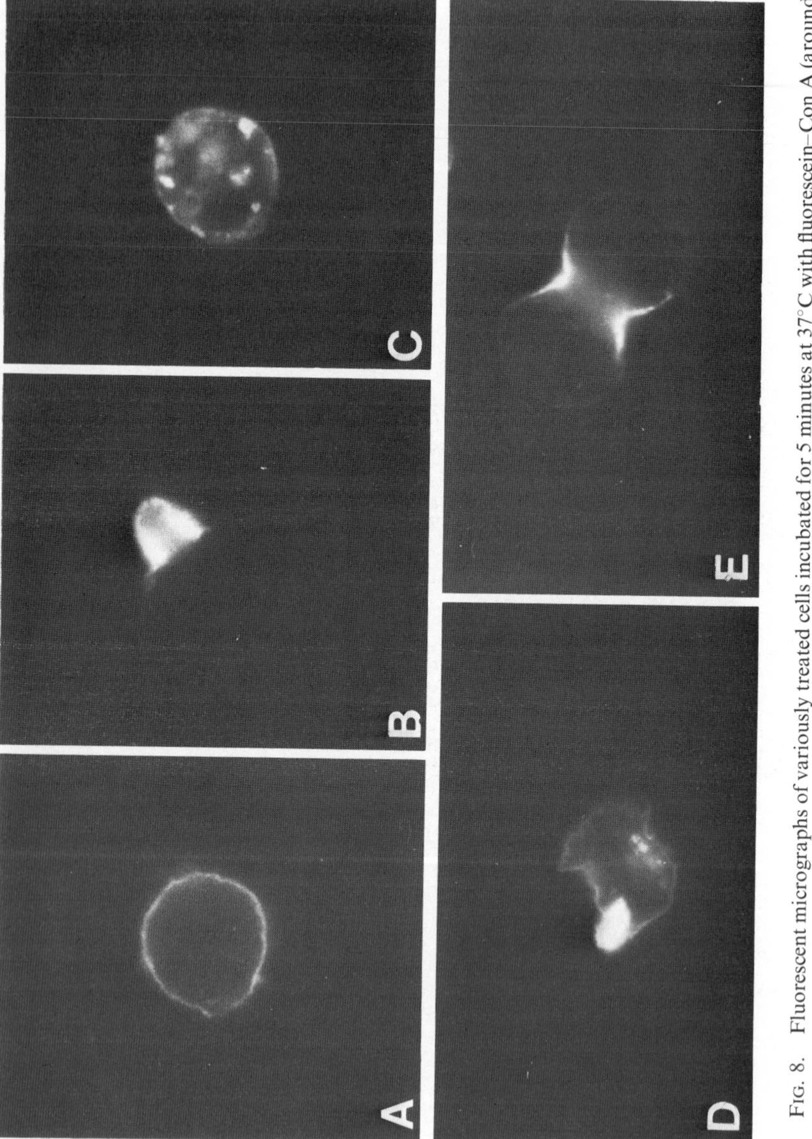

FIG. 8. Fluorescent micrographs of variously treated cells incubated for 5 minutes at 37°C with fluorescein–Con A (around 10 μg/ml). The cells are, respectively, (A) untreated; (B) microtubule-depleted (protuberant); (C) phagocytizing (carboxylated polystyrene particles that do not bind Con A); (D) oriented (by incubation for 20 minutes in F-Met-Leu-Phe), and (E) dividing. All cells are human PMN except for the mitotic, which is a Chinese hamster ovary cell. In each case Con A is accumulated at regions of membrane deformation and microfilament accumulation.

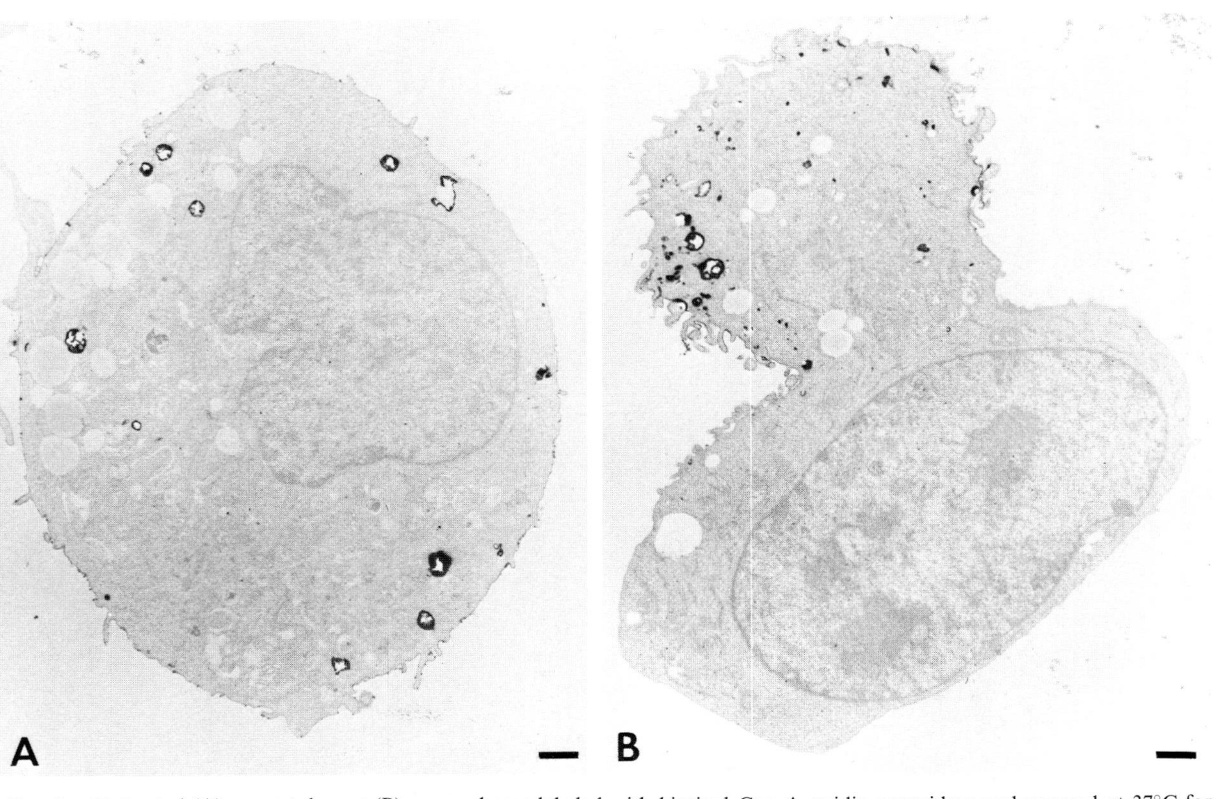

FIG. 9. Untreated (A) or protuberant (B) macrophages labeled with biotinyl Con A–avidin peroxidase and warmed at 37°C for 10 minutes before fixation. The rounded cell has removed a large proportion of Con A–receptor complexes by adsorptive pinocytosis over the whole cell surface. Con A–receptor complexes were capped on the protuberant cell prior to internalization. Bar = 1 μm.

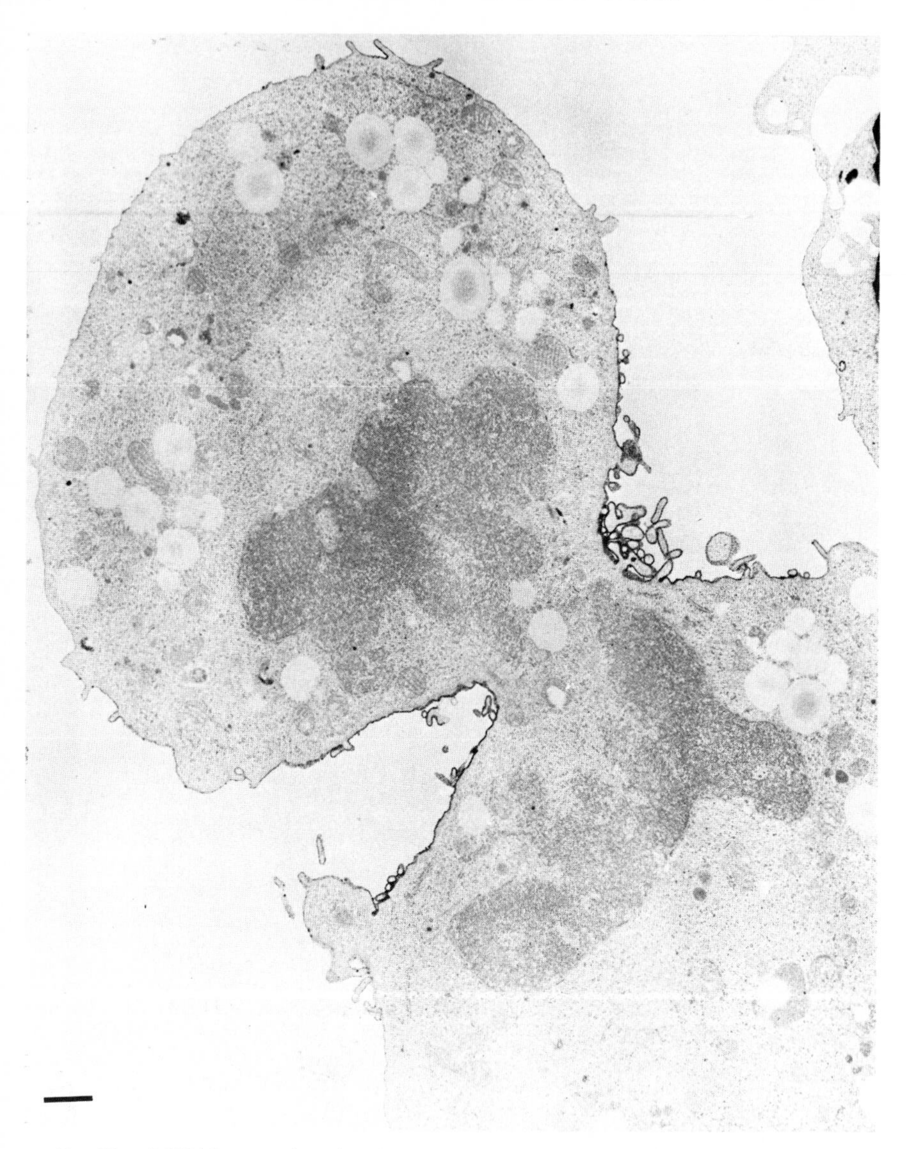

FIG. 10. A J774.2 macrophage incubated for 5 minutes at 37°C with biotinyl Con A–avidin peroxidase during cytokinesis. Con A–receptor complexes accumulate in a region corresponding to the full extent of the contractile region supporting the cleavage furrow. Bar = 1 μm.

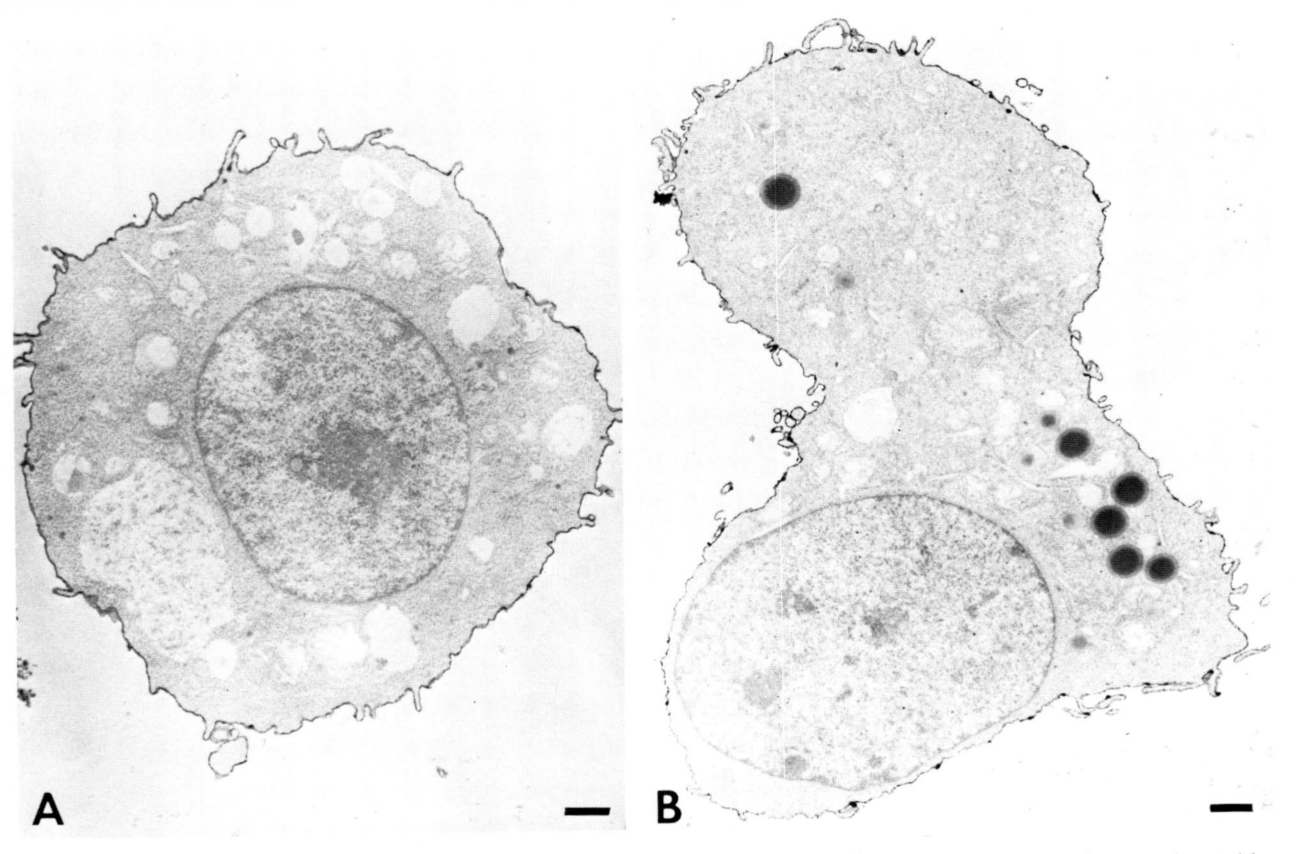

FIG. 11. The uniform distribution of Con A receptors on an untreated (A) and protuberant (B) J774.2 macrophage as observed by labeling at 4°C with biotinyl Con A and avidin peroxidase and immediate fixation. Bar = 1 μm.

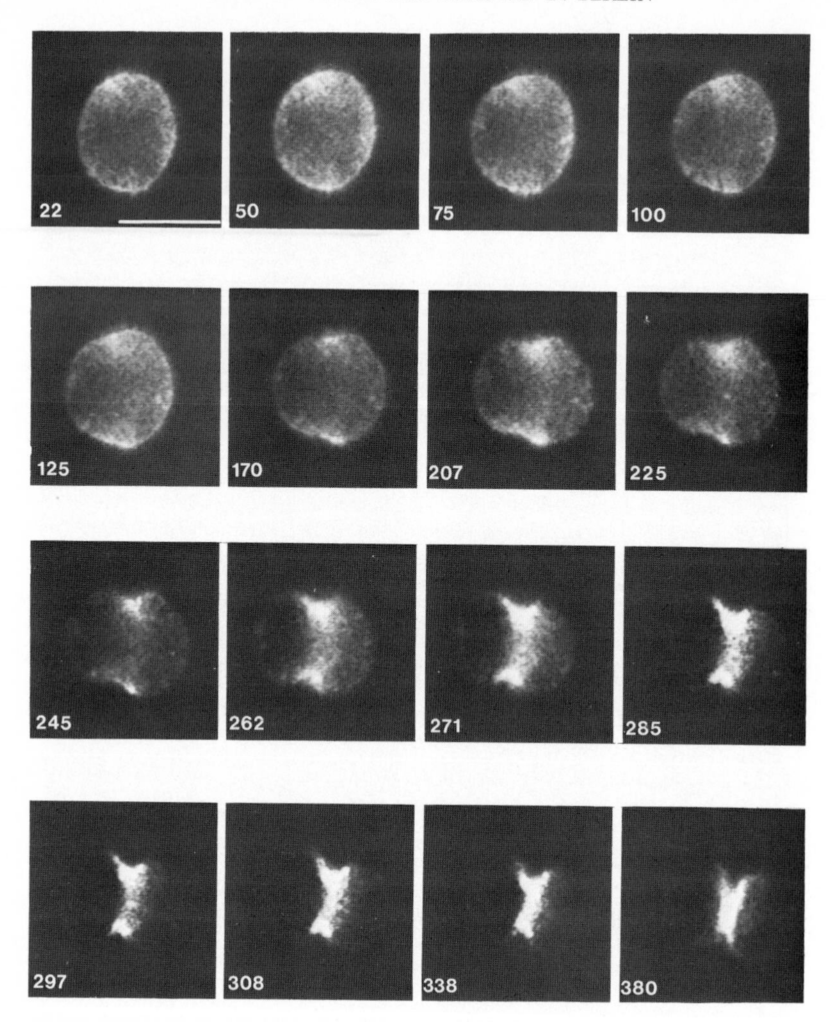

FIG. 12. Frames from a video recording of a J774.2 macrophage labeled for 1 minute at 37°C with fluorescein-conjugated succinyl–Con A (Vector), rinsed, mounted in culture medium on the stage of a Zeiss Universal microscope, and viewed through an image intensification TV camera. The cell was selected for observation at metaphase. A rapid and complete redistribution of Con A from a uniform distribution into the cleavage furrow occurred during the transition from metaphase to anaphase and telophase. Time is given in seconds. Bar = 20 μm.

evident at the nuclear (nonprotuberant) pole of protuberant J774.2 macrophages (Walter *et al.*, 1980b; Fig. 3). In addition, oriented PMN show microfilament aggregates at both the posterior and anterior poles (Fig. 6) but only the uropod region and not the lamellipodium accumulates Con A.

Third, the ligand-induced movement of Con A–receptor complexes is clearly not dependent on extensive cross-linking by multivalent ligand. Capping and other redistributions of Con A occur readily with succinyl-Con A (e.g., see Fig. 12). Furthermore, the long range Con A–receptor redistributions described here are not preceded by extensive formation of surface patches visible by either fluorescence or electron microscopy (Albertini *et al.*, 1977; Berlin and Oliver, 1978; Walter *et al.*, 1980b; Oliver *et al.*, 1980).

Fourth, the movement of Con A is remarkably rapid. Figure 12 is a series of photographs of a video record of a macrophage labeled with succinyl–Con A at metaphase and filmed during cytokinesis at 32°C (Koppel *et al.*, 1982). Receptors are uniformly distributed until the onset of cytokinesis (at about 75 seconds). Their subsequent quantitative redistribution into a cleavage furrow occurs in a period hardly exceeding 3 minutes. If we assume that the receptors diffuse freely until they are trapped at the furrow, then it may be calculated that on a cell whose diameter ranges from 20 μm at metaphase (first frame) to 23 μm at late telophase/G_1 (last frame) a diffusional rate constant on the order of 10^{-9} cm^2/second would be required. A similar analysis can be applied to the kinetics of cap formation. In PMN, Con A caps can develop within 90 seconds, again corresponding to a diffusion constant of 10^{-9} cm^2/second. These rates exceed by an order of magnitude the usual estimates of diffusion of the same ligand on mammalian cells (Elson and Reidler, 1979).

Fifth, the movement of Con A may be unidirectional and anisotropic. Unidirectional movement was demonstrated by repetitive quantitative laser scans over a succinyl–Con A-labeled cell that was subjected to laser bleaching of the fluorescent lectin at one pole as it entered cytokinesis. No fluorescence recovery followed photobleaching. Rather fluorescence increased over the cleavage furrow while decreasing over both the bleached and opposite pole (Koppel *et al.*, 1982). Related studies by Smith *et al.* (1979) have demonstrated that Con A–receptor complexes can undergo anisotropic movement. Their measurements of fluorescence recovery after photobleaching a grid pattern onto the surface of fibroblasts yielded diffusion rates that were at least five-fold greater parallel than perpendicular to the long axis of the attached cells.

B. Polarization of other Receptors

We emphasize that the segregation of membrane receptors on asymmetric cells is not restricted to Con A receptors nor is it necessarily dependent on the presence of ligand. For example, a short list of receptors that may segregate spontaneously (that is, independently of ligand binding) to membrane at regions of cell shape and microfilament asymmetry can be developed from morphological data obtained in several laboratories. These include receptors for wheat germ agglutinin [observed by Lustig *et al.* (1980) to accumulate in cleavage furrow membrane] and receptors for antithymocyte serum, H_2 antigens, Thy-1.2 antigen, and immunoglobulins (all claimed to cap spontaneously on polarized mouse lymphocytes: Braun *et al.*, 1978; Yahara and Kakimoto-Sameshima, 1979).

In addition, many membrane determinants accumulate in caps after ligand binding. A list of at least 15 distinct determinants that can be capped after lectin or antibody binding on lymphocytes alone was provided by DePetris (1977). The list can be enormously expanded by reference to other cell types. Detailed studies of these capping processes are not always available. In some cases, particularly when several antibodies are bound sequentially to the same receptor, capping may depend most importantly on chemical cross-linking to form increasingly large aggregates and, eventually, caps. This type of capping is typically slow, even at 37°C, and has been claimed to occur on cells without obvious cell shape or cytoskeletal asymmetries (e.g., Braun *et al.*, 1978). However, we think it likely from our own data that capping that follows binding of a *single* antibody to a membrane receptor, like Con A capping, may frequently occur by a segregative process that is probably not dependent on progressive crosslinking of patches of receptors by multivalent ligand. For example, we have found that fluorescent anti-IgM (Cappel) caps to protuberance and cleavage furrow membranes of human B cells and B cell lines during 5 minutes incubation at 37°C (Fig. 13D). No evidence for multiple patching *en route* to cap formation was obtained, although large patches were typically observed on cells that maintained a rounded shape (Fig. 13C). Antibody–antigen complexes appeared as small, uniformly distributed patches on both rounded and protuberant cells when labeling was at 4°C (Fig. 13A and B): thus the "spontaneous capping" of Ig receptors reported on mouse splenic lymphocytes does not extend to human peripheral blood B cells and their derived cell lines.

Membrane proteins may also segregate out of regions of cell shape and microfilament asymmetry. This is apparently the case for several classes of transport carriers (which are integral membrane proteins). Tsan and Berlin (1971) reported that the transport of purine bases and amino acids occurs by carrier-mediated, saturable processes in alveolar macrophages. The initial rate of adenine and lysine transport is unaffected by removal of up to 40% of the cell surface by phagocytosis. Studies with the impermeable transport inhibitor *p*-chloromercuribenzoate sulfonate showed that no insertions of carriers occurred during phagocytosis. Hence it was concluded that carriers are preserved on the cell surface during phagocytosis. This experiment was in fact the first demonstration of cell surface heterogeneity in nonepithelial cells.

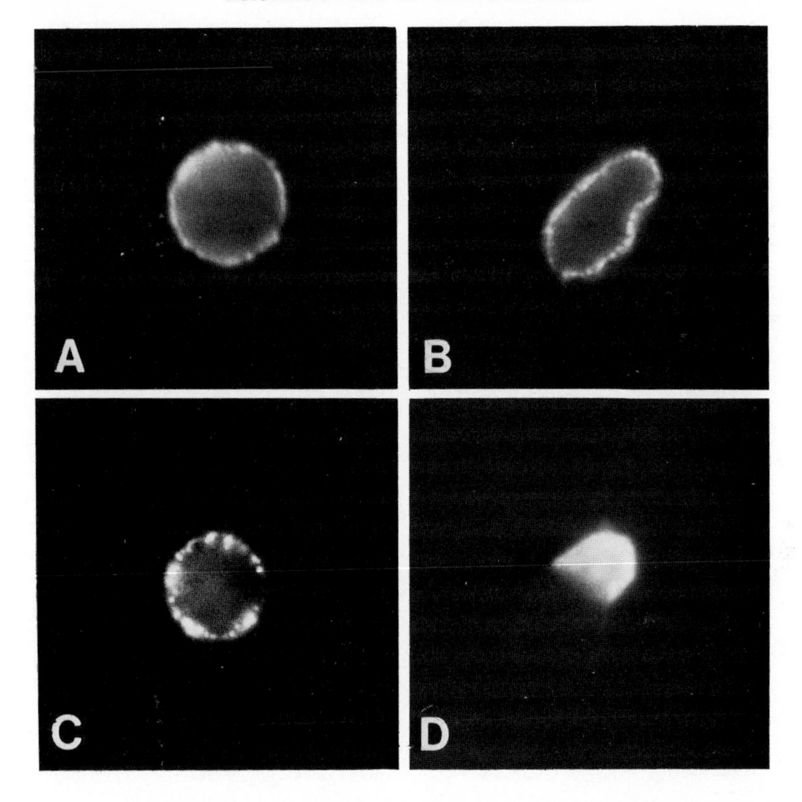

FIG. 13. The distribution of surface IgM on human peripheral blood B cells. Lymphocytes were incubated for 5 minutes with fluorescein-antihuman IgM (Cappel). At 4°C, IgM–anti-IgM complexes were distributed as random clusters over both rounded and protuberant cells (A,B). At 37°C, complexes were in large patches on rounded cells and in polar caps on protuberant cells (C,D).

More recent studies (Walter *et al.*, 1980a,b) suggest that two biologically important membrane markers, the Fc and C3b receptors that mediate phagocytosis in macrophages and PMN, may also segregate away from regions of cell shape and microfilament asymmetry. Colchicine-treated (protuberant) J774.2 cells and oriented PMN were labeled at 4°C for 5 minutes or at 37°C for no more than 30 seconds with two Fc probes (IgG-opsonized erythrocytes and Ig aggregates) and with two probes for C3b receptors (complement-opsonized erythrocytes and zymosan). These conditions, which were chosen to minimize ligand-induced receptor redistribution, revealed the majority of surface labeling to be *away* from the protuberance or uropod (Fig. 14a).

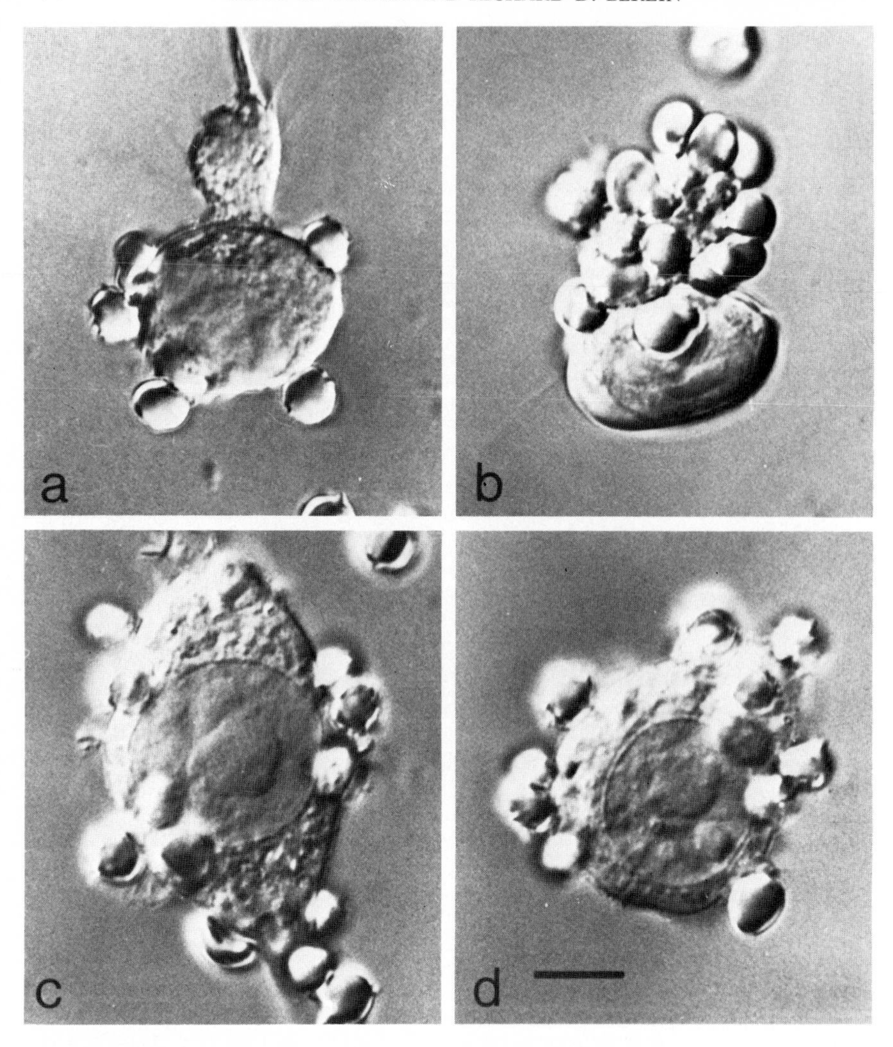

FIG. 14. The distribution of IgG-opsonized erythrocytes on J774.2 macrophages. In colchicine-treated cells, erythrocytes are concentrated over the cell body after incubation for 5 seconds at 37° (a) but move into the protuberance when incubation is continued (after rinsing) for 5 minutes at 37°C (b). Untreated cells maintain a uniform distribution of particles whether incubation is for 5 seconds (c) or 5 minutes (d). From Walter *et al.* (1980b). Bar = 10 μm.

In subsequent analyses, the cells were rinsed to remove excess particles and then warmed to 37°C for 5 or 10 minutes. Upon warming, particle–receptor complexes that formed away from the protuberance of colchicine-treated J774.2 macrophages rapidly redistributed into the protuberance (Fig. 14b). Particle–Fc receptors but not particle–C3b receptor complexes were then ingested by the macrophages. Harris (1973) and Wilkinson *et al.* (1980) have described a similar posterior accumulation at 37°C of IgG-opsonized red cells on oriented PMN. Thus occupied receptors can assume different, asymmetric distributions from unoccupied receptors.

An equivalent distribution of Fc and complement receptors is probably maintained during phagocytosis. Petty *et al.* (1980) recently reported the preservation of C3b receptor activity on the surface of macrophages depleted of Fc receptors by phagocytosis of lipid vesicles opsonized with antibody. This exclusion of unoccupied C3b receptors from internalization is the predicted result based on the analogy drawn here between protuberance and pseudopod membranes.

These observations establish that topographical asymmetry may characterize a wide range of receptors. They also indicate that the regulation of receptor topography may be a good deal more complex than suggested from studies of Con A alone. In particular, unoccupied as well as occupied receptors may assume asymmetric distributions. Furthermore, receptors can segregate out of as well as into regions of shape and microfilament asymmetry depending on the particular receptor and the presence or absence of ligand.

C. Polarization of Membrane Lipids

Cells with shape and microfilament asymmetries may develop heterogeneities not only in membrane protein topography but also in lipid topography. We have determined "microviscosity" before and after phagocytosis by the method of fluorescence depolarization of diphenylhexatriene in isolated plasma membranes of PMN (Berlin and Fera, 1977). Surface membrane isolated after removal of pseudopod membrane by its internalization shows a substantially increased depolarization (decreased "microviscosity") as compared to control membrane. This difference persists in liposomes derived from these membranes, suggesting (but not proving) the selective removal of less fluid lipids into the membrane of phagocytic vacuoles. By analogy, we predict the enrichment of less fluid lipid in proturberance, uropod, and cleavage furrow membranes.

Not surprisingly, this polarization does not include all classes of lipids. We have observed the topography of a surface ganglioside (GM_1) from the distribution of fluorescein-conjugated chlorea toxin on protuberant and dividing macrophages. An absolutely uniform distribution of toxin–GM1 complex is maintained even over long periods of incubation at 37°C.

D. Polarization of Membrane Functions

Topographical asymmetry is not limited to molecular determinants. We have found that a range of endocytic processes is restricted to specific regions of the cell membrane. The restriction of phagocytosis to the protuberance of cells was described above. However, this could be explained by receptor asymmetries. On the other hand, fluid pinocytosis, by definition, occurs without receptor involvement. We have shown that the uptake by fluid pinocytosis of fluorescein-conjugated dextran, which normally occurs over the entire cell surface, is restricted to the protuberance of microtubule-depleted cells (Walter *et al.*, 1980b; Fig. 15) and to the uropod of cells oriented in a chemotactic gradient (Davis *et al.*, 1982). Furthermore, the resumption of pinocytosis after cell division is first observed at the region of cell–cell contact, which represents the former site of the cleavage furrow (Berlin and Oliver, unpublished). This topographical restriction of fluid pinocytosis is particularly interesting in light of the persistence of a small rim of subcortical microfilaments in nonprotuberant regions of colchicine-treated cells (Fig. 3)

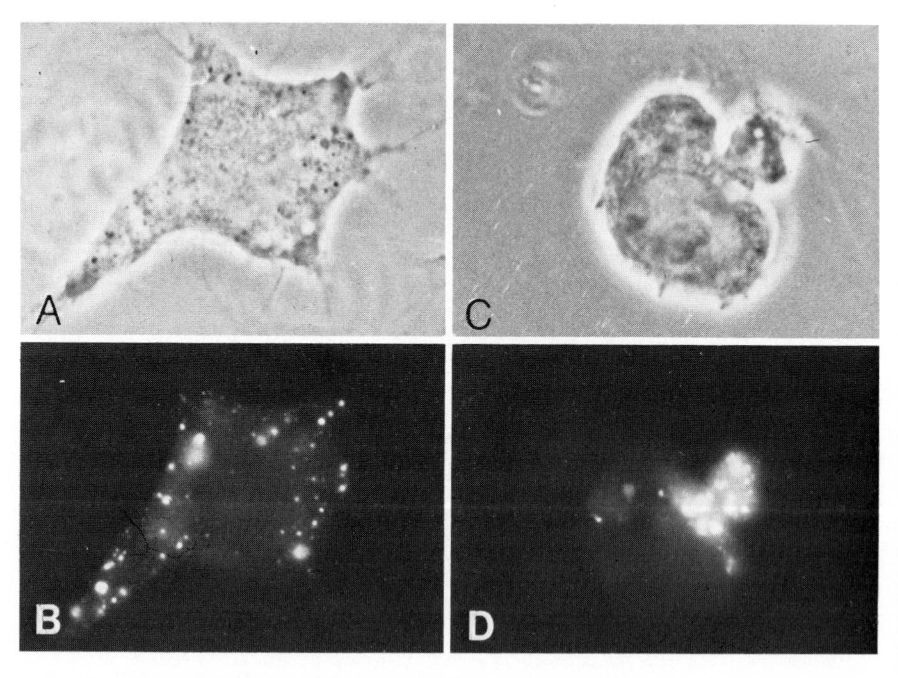

FIG. 15. The polarization of fluid pinocytosis in protuberant cells. Control and microtubule-depleted (protuberant) cell monolayers were incubated for 5 minutes at 37°C with 10 mg/ml rhodamine-dextran. The control cell (A) is well spread and displays a uniform distribution of cytoplasmic vesicles containing the marker for fluid pinocytosis (B). The protuberant cell (C) shows rhodamine-dextran primarily located in the protuberance (D), which thus appears to be the region of initial uptake.

and especially the presence of a dense network of microfilaments in the lamellipodium of oriented cells (Fig. 6).

A directly related example of topographical asymmetry associated with asymmetry of cell shape was revealed in experiments designed to map the distribution of a major membrane macromolecular assembly, the coated pit (Pfeiffer *et al.*, 1980). Coated pits are specialized regions of membranes recognized by their characteristic bristle coat composed primarily of clathrin (Fig. 16). They mediate the uptake by adsorptive pinocytosis of a wide range of hormones and growth factors (Goldstein *et al.*, 1979). On resting cells, we found that these structures are distributed more or less randomly: analysis in symmetrical J774.2 macrophages revealed their distribution as a random display of small clusters. However coated pits are highly polarized on asymmetrical cells. Thus careful mapping on protuberant cells from colchicine-treated macrophage suspensions indicated their quantitative removal into the membrane of the protuberance (Pfeiffer *et al.*, 1980; Fig. 17). Similarly, we have shown that coated pits are restricted to the uropod of oriented cells (Davis *et al.*, 1982; Fig. 6), while Aggeler and Heuser (1980) have demonstrated the specific accumulation of clathrin coats in pseudopod membrane. In preliminary studies, we have also observed significant stretches of thickened membrane at the cleavage furrow of J774.2 macrophages with coated pits typically forming at the outer edges of the furrow membrane (see Fig. 5). This asymmetry of coated pit topography on polarized cells has important implications for the regulation of action of various growth substances and hormones.

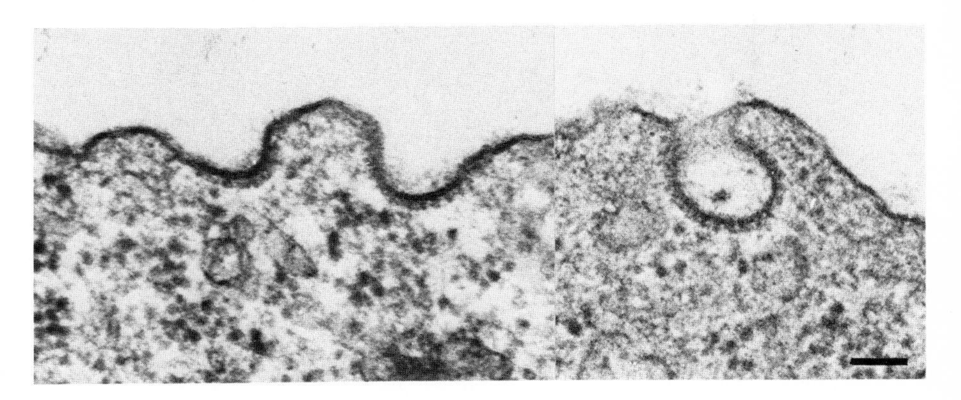

FIG. 16. Typical appearance of coated pits, showing a thickened membrane, bristle coat on the cytoplasmic surface, and variable degrees of membrane deformation. From Pfeiffer *et al.* (1980). Bar = 0.1 μm.

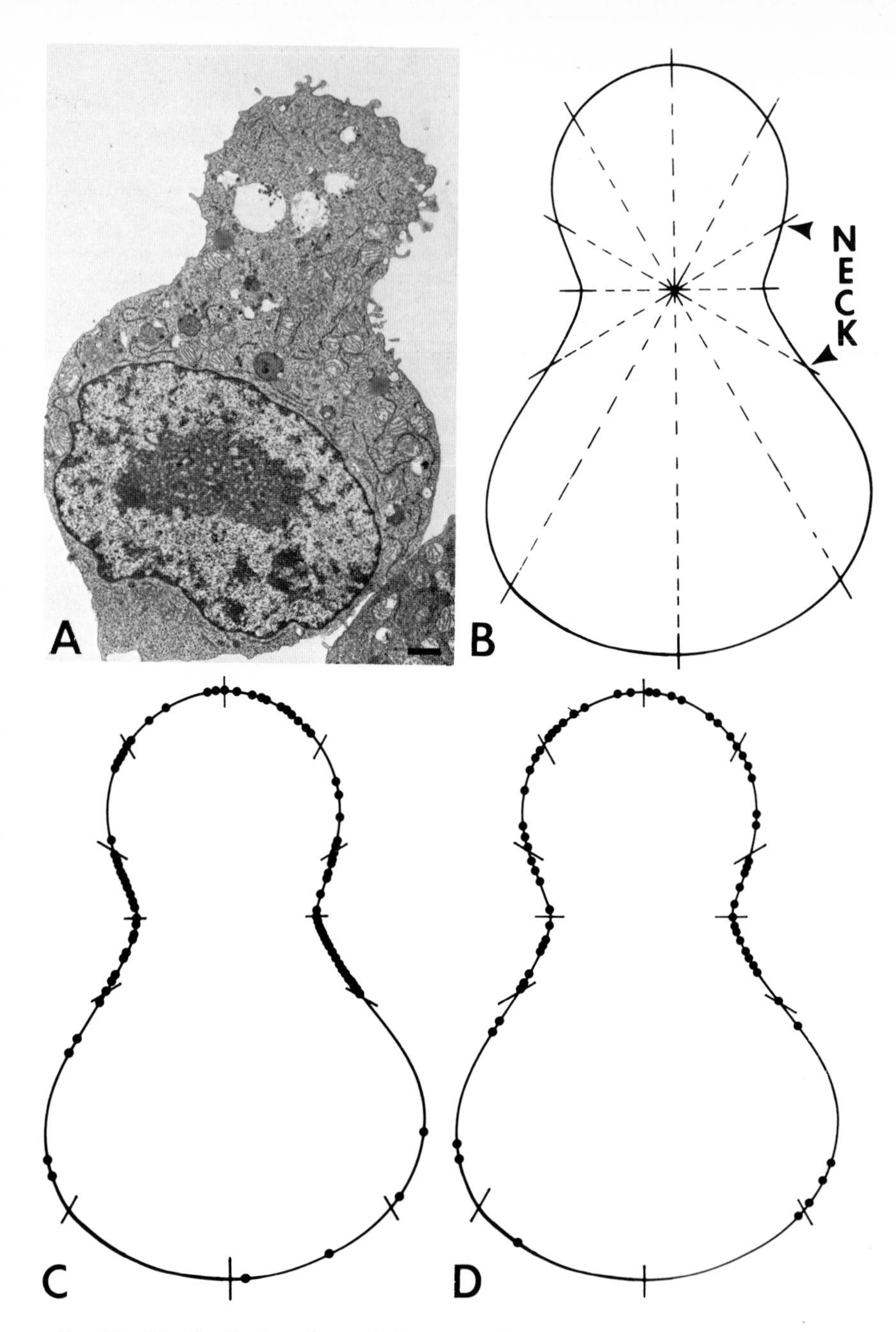

FIG. 17. The distribution of coated pits on protuberant cells. A cumulative map of coated pit topography was built by translocating pits from individual cells (A) onto equivalent quadrats of a diagrammatic cell (B) whose dimensions were proportional to those of the original cell. The results of two typical experiments, each analyzing 12 randomly chosen protuberant J774.2 macrophages from colchicine-treated (10^{-5} *M*, 60 minutes) populations, are given in (C) and (D). They demonstrate the concentration of pits in the "neck" and "head" regions of protuberant cells. Modified from Pfeiffer *et al.* (1980).

IV. Membrane Topography during Mitosis

In striking contrast to the asymmetries described above (and summarized in Table I), the membranes of mitotic cells maintain an extraordinary molecular symmetry and absence of dynamic function.

During prophase, J774.2 macrophages, like other cell lines, disassemble cytoplasmic microtubules, assemble a mitotic spindle, and show a variable degree of rounding. A rounded shape is maintained through metaphase, followed by cleavage furrow formation and cytokinesis to form two daughter cells. At least through metaphase, microfilaments are seen as a thin submembranous band. The recruitment of microfilaments at anaphase to form a contractile ring was illustrated above.

TABLE I
PATTERNS OF MEMBRANE ASYMMETRY ON POLARIZED CELLS[a]

A. Components and functions that maintain symmetrical distributions on polarized cells
Con A receptors
Gangliosides
B. Components and functions excluded from protuberance, uropod, pseudopod, and/or cleavage furrow membranes
Transport carriers
Fc receptors
C3b receptors
C. Components and functions concentrated in protuberance, uropod, pseudopod, and/or cleavage furrow membranes
Con A–receptor complexes
Fc–receptor complexes
C3b–receptor complexes
Fluid pinocytosis
Adsorptive pinocytosis
Phagocytosis
Coated pits
Lipids contributing to elevated membrane "microviscosity"

[a] This table lists membrane probes studied in our laboratory. In most cases the same asymmetry is produced whether analysis is performed with protuberant, oriented, phagocytizing, or dividing cells. However not all probes have been studied under all four conditions. List C, in particular, can be greatly expanded by reference to the literature on membrane capping.

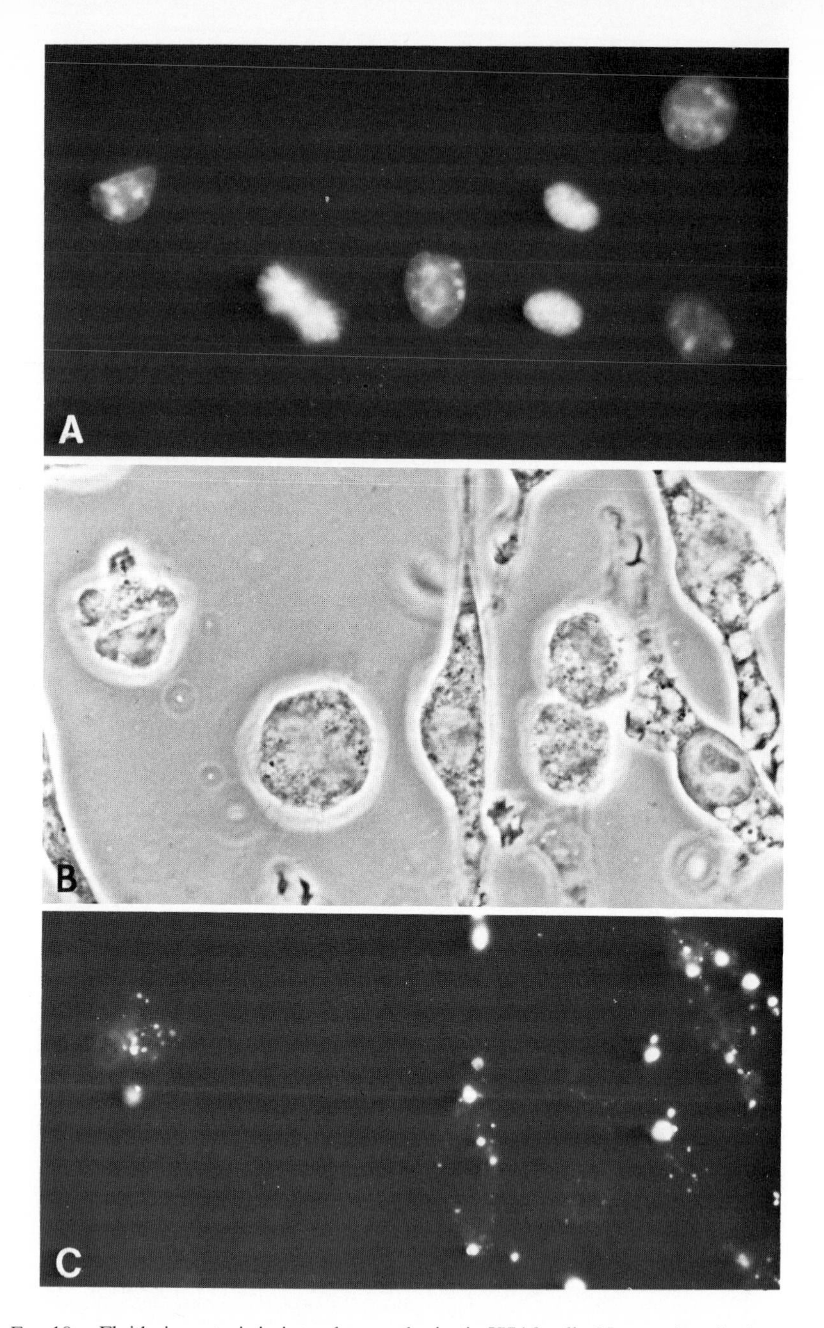

FIG. 18. Fluid pinocytosis in interphase and mitotic J774.2 cells. Nonsynchronized macrophages (B) were incubated with 10 mg/ml fluorescein dextran for 2 minutes at 37°C (C), then fixed and their nuclei stained with Hoechst 33258 (A). Interphase cells show multiple pinocytic vesicles. Mitotic (metaphase and telophase) cells show no evidence of pinocytosis. From Berlin and Oliver (1980).

Analyses of surface structure and function of mitotic J774.2 cells have revealed that the onset of mitosis is accompanied by remarkable changes in membrane properties (Berlin *et al.*, 1978; Berlin and Oliver, 1980; Berlin *et al.*, 1980). For example Con A–receptor complexes maintain a uniform surface distribution and essentially no ligand-induced patching or capping is seen from prophase to anaphase (when Con A redistribution occurs during cleavage furrow formation; see Figs. 10 and 12). The uptake of fluorescein-dextran and horseradish peroxidase by fluid pinocytosis stops within 30 seconds at the G_2–prophase border (Fig. 18) and resumes only at the telophase–G_1 transition. Ig aggregates and Ig-opsonized particles bind to uniformly distributed Fc receptors but no ingestion follows binding to mitotic cells (Fig. 19). Complement-opsonized particles completely fail to bind even to C3b receptors during mitosis, perhaps reflecting a cell cycle-dependent expression of receptors or the inability to form sufficient particle–receptor links when receptor mobility is severly restricted (Walter, Oliver, and Berlin, unpublished).

Similar changes in membrane properties have been measured in a range of other cultured cells (3T3 fibroblasts, PTK2 epitheloid cells, CHO cells, etc.; Berlin *et al.*, 1978; Berlin and Oliver, unpublished). Furthermore DeLaat *et al.* (1980) have determined by fluorescence recovery after photobleaching that both protein and lipid mobilities are lowest during mitosis in neuroblastoma cells.

The mechanism of this rapid and complete arrest of membrane function is an unanswered question of major importance. Clearly mitotic cells do not behave merely as cells depleted of cytoplasmic microtubules. Rather they behave as cells whose microfilament system is frozen and/or whose membrane properties are so changed that appropriate signals no longer elicit

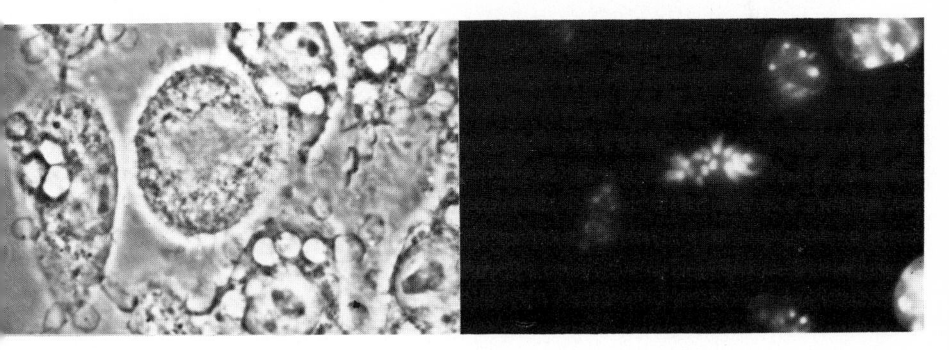

FIG. 19. Uptake of IgG-opsonized erythrocytes by J774.2 macrophages. The mitotic cell (center) shows no phagocytosis whereas surrounding interphase cells, including a G_1 pair (identified by its residual midbody), are filled with endocytized erythrocytes. Modified from Berlin *et al.* (1978).

typical responses. The net result of such inhibition may be to isolate cells from their environment during this critical stage of the cell cycle. In essence, mitosis may be an island in time where internal events override external pressures.

V. The Control of Cell Surface Topography

A. The Range of Published Models

Numerous models to explain asymmetries in membrane molecular topography have been advanced over the past decade. Most deal only with caping processes and fail to recognize that analogous asymmetries exist under other conditions.

The most widely quoted models represent variations in themes set by Edelman in 1973 when he proposed that microtubules may anchor receptors for ligands such as IgG and Con A, while microfilaments may attach to receptors to cause their redistribution (Edelman *et al.*, 1973; Edelman, 1976; Schreiner and Unanue, 1976; DePetris, 1977). Until the precise molecular associations of receptors and their intramembranous environment and the cytoskeleton are understood, this hypothesis cannot be fully analyzed. However, it is clear that this model is, at best, an oversimplified view of cytoskeleton involvement in surface events. For example, Edelman's basic observation was the inhibition of lymphocyte surface Ig patch and cap formation by Con A. From this, he suggested that lectin binding may stimulate cytoskeleton–membrane interactions that in turn immobilize surface components. Consistent with Edelman's original interpretation, Henis and Elson (1981) recently studied Ig mobility by the technique of fluorescence recovery after photobleaching and confirmed that Con A binding indeed decreases Ig mobility. However, they also reported that Ig mobility could be increased by cytochalasin B treatment and further increased by cytochalasin B plus colchicine. Since these latter treatments *inhibit* capping, Henis and Elson's data can in fact be used to dissociate local receptor mobility from long-range receptor redistribution.

The temporal separation of membrane and cytoskeletal rearrangements is difficult to reconcile with the hypothesis of comigration of microfilaments with receptors. We have shown that ligand–receptor complexes may redistribute into regions of *existing* microfilament accumulation. Furthermore we (Albertini *et al.*, 1977) and others have established that addition of a ligand for a second species of receptor can lead to its capping over membrane already capped by a different ligand–receptor complex. The process can be repeated many times. If microfilament–receptor linkages were directly involved in propelling the receptors one must suppose a most extraordinary traffic in which microfilaments can be recruited indefinitely in response to surface binding events and can subsequently move past each other within a protuberance. Because this seems excessively complex, we prefer the view that microfilaments accumulated beneath the membrane do not directly move receptors. However these microfilaments may signal the presence of of a "trap" which spatially fixes ligand–receptor complexes *after* their movement.

An essential role for direct microfilament–receptor linkages in determining inherent receptor topography is also not supported by our data. As noted above, Braun *et al.* (1978) and Yahara and Kakimoto-Sameshima (1979) have claimed that several lymphocyte antigens, like coated pits on macrophages, may redistribute spontaneously into membrane overlying the microfilament-rich uropod. However, we determined that Fc and C3b receptor activities as well as membrane transport carriers segregate away from such regions on macrophages. Furthermore in our hands unoccupied receptors for Con A and anti-IgM show little or no inherent polarization on asymmetric cells. Thus individual receptor species can assume different and characteristic inherent distributions that cannot be determined by the same sets of microfilaments.

Finally, our studies of phagocytic and oriented cells do not support a role microtubules as anchors. In both cases, microtubule assembly is stimulated yet Con A–receptor complexes move from membrane overlying microtubules into pseudopod or uropod membrane that specifically excludes microtubules. Conversely in mitotic cells, where cytoplasmic microtubules are considered to be generally depleted, there is no formation of caps.

The most recent versions of this capping model are particularly concerned with elements that may link microfilaments to crosslinked surface receptors [the substance X of Bourguignon and Singer (1977)]. In particular the accumulation of clathrin under patches and caps of anti IgM–IgM complexes on B lymphoid cells has been taken as evidence that clathrin may link aggregated surface receptors to cytoskeletal components (Salisbury *et al.*, 1980; McKeon *et al.*, 1980). However we have demonstrated that coated pits are clustered on rounded cells and "cap" to regions of membrane deformation and microfilament aggregation whether or not ligand–receptor complexes are present at the cell surface (Pfeiffer *et al* 1980). Hence the similar topographical responses of clathrin and antibody complexes may reflect their parallel but independent responses to changes in cell shape and cytoskeletal organization and their ultimate "trapping" in regions overlying microfilaments.

The apparent increase in both protuberance and cap formation that occurs after binding of specific antibodies and ligands to receptors on some cells, particularly lymphocytes, has been taken as evidence for comigration of receptors and microfilaments. However another explanation is possible. We have seen that lymphocytes, like PMN and macrophages, constantly produce and retract uropod-like (protuberant) structures. Consequently a small proportion of cells is always asymmetric in shape and microfilament distribution. It is possible that capping of certain complexes to these microfilament-enriched regions prevents retraction so that protuberances can now be resolved only by the slower process of endocytosis. As a result, the proportion of cells with coincident microfilament, shape, and receptor asymmetries should increase in the presence of ligand.

A radically different scheme for the regulation of surface topography, again based on capping, was proposed by Bretscher (1976) and modified by Harris (1976). In their models membrane lipids (Bretscher) or whole membrane (Harris) were postulated to flow continuously from the front of the cell to the rear where lipids were removed and recycled via endocytosis to the front. This flow caused the movement of ligand–receptor complexes whereas unoccupied receptors would escape flow by back-diffusion. A molecular filter excluded ligand–receptor complexes from internalization and recycling. The net result was cap formation at the site of filtration. This model relieved cytoplasmic microfilaments of an obligatory direct role in topographical control. However, the filter was not defined and the exclusion of ligand–receptor complexes from the cytoplasmic milieu seemed inconsistent with the rapid internalization of complexes that usually follows capping of multivalent ligands.

Recently Hewitt (1979) has proposed a model for capping that avoids several difficulties of the Bretscher and Harris models. In Hewitt's proposal, membrane-bound particles and ligand–receptor complexes are entrained by waves that move continuously over the cell surface. Entrainment results from positive interaction of the complexes with a segment of the moving wave front. Small molecules, for example unoccupied receptors, do not interact and thus do not move. Although the wave motion is initiated by the contraction of submembranous microfilaments, the geometrical interaction of wave shape and cross-linked complex, not microfilament–membrane interaction, ultimately determines which membrane components will be selected for movement. This model avoids the requirement for constant internalization of membrane. However mechanisms to orient the waves and determine the focus of the accumulation of complexes are not considered and capping of ligands that are not extensively cross-linked is not readily explained.

Finally Unanue's view (Braun *et al.*, 1978) that certain membrane determinants can cap by an entirely passive process deserves further mention. In their hands, lymphocyte antigens such as H2 and Thy. 1 can be capped only after crosslinking in patches with two antibodies and prolonged (30–60 minutes) incubation at 37°C. These caps appear to be formed by the random and progressive crosslinking of surface components into increasingly large patches. They do not overlie increased concentrations of myosin. Similar capping processes have been reported to occur over long periods of time on lipsomes where, obviously, no cytoskeletal involvement is possible.

Other investigators disagree with this completely passive route to cap formation on lymphocytes: for example Bourguignon and Singer (1977) have argued for the redistribution of actin and myosin associated with capping of the same ligand–receptor complexes as studied by Braun and co-workers, while Yahara and Kakimoto-Sameshima (1977) have reported that these receptors cap spontaneously (that is, in the absence of ligand) on lymphocytes incubated in hypertonic media. We are unable to explain these discrepancies. However, we emphasize that the kind of capping process described by Unanue is distinct from the processes under discussion here in which topographical changes occur actively and rapidly.

B. A New Model for the Control of Membrane Molecular and Functional Topography

We have developed a model that can explain the segregation of membrane components and functions both into and out of regions of cell shape asymmetry. The model applies not only to membrane asymmetry due to capping but to the analogous asymmetries that accompany phagocytosis, chemotaxis and cytokinesis.

Our model draws from Hewitt's concept that mechanical surface waves may be involved in the movement of membrane components. In addition, it rests on a series of four general conclusions drawn from our experimental data reviewed above. These conclusions are as follows:

1. Mammalian membranes are not organized at random. Polarization of surface receptors and functions are the rule.
2. Asymmetric membrane molecular (protein and lipid), macromolecular (coated pits), and functional (endocytosis) topography is associated with asymmetries in cell shape and cytoskeletal organization that arise from diverse causes.
3. Membrane determinants can segregate out of as well as into regions of shape and microfilament asymmetry depending on the particular determinant and the presence or absence of ligand.
4. Ligand-induced receptor movement on asymmetric cells can occur by a directional, anisotropic process whose rates exceed those predicted for simple diffusion of proteins in membranes (see below). This movement is not

dependent on obligatory connections between receptors and cytoskeletal structures. It does not require extensive receptor crosslinking by multivalent ligand.

The essence of our model is illustrated in Figs. 20 and 21. It may be described as follows:

The first event leading to the formation of a uropod, protuberance, pseudopod, or cleavage furrow is the recruitment of microfilaments. These organize into bundles and interact with the membrane to generate tension. This interaction has several consequences (Fig. 20). First, it produces a characteristic change in cell shape. Second, this asymmetric tension exerted inward and along the cell surface initiates wave motion over the cell surface. This wave is directed toward the region of tension. In the case of protuberant, oriented, and dividing cells, wave generation may be amplified by extension of lamellipodia at the pole opposite to the region of tension. The thin band of microfilaments underlying the bulk membrane may play a role in the transmission of tension over the entire cell surface. We suppose that intermittent application of tension leads to repeated wave generation.

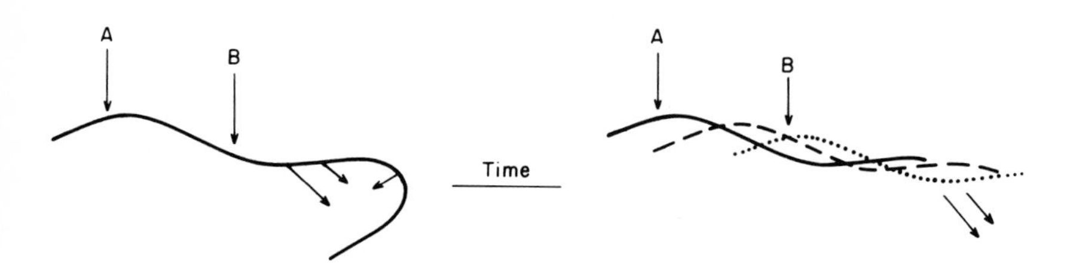

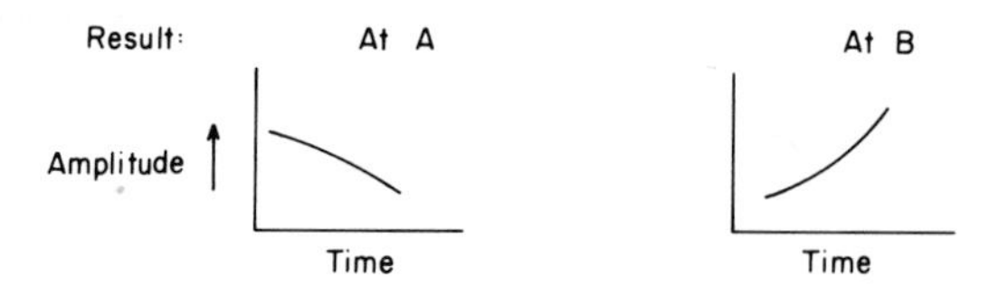

FIG. 20. A model for the generation of oriented waves in cell surfaces. Tension acting inward and along the membrane (arrows) provides force for the progressive displacement of the membrane. The resulting oscillation of amplitude with time at any chosen point on the membrane constitutes a wave oriented toward the region of force generation.

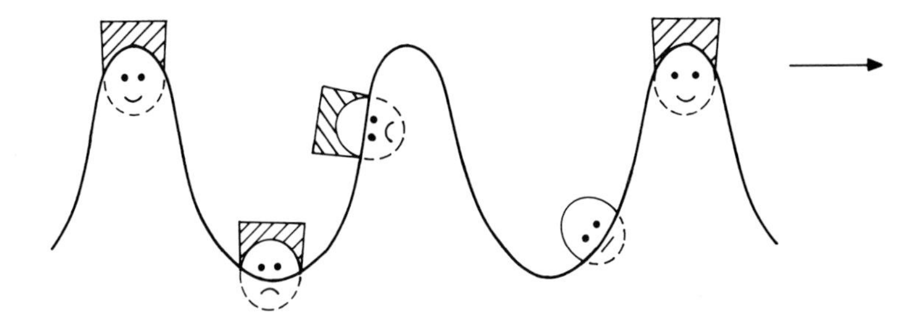

FIG. 21. Response of membrane determinants to the passage of waves. Certain receptors sense a change in orientation but no large change in environment as the waves progress. These occupy random positions on the cell surface. Other receptors and ligand–receptor complexes are preferentially associated with particular regions of the waves (the crests in this diagram) and retain this distribution. These move with the waves to the region of force generation.

Once oriented wave motion is initiated, membrane determinants may respond in one of two ways (Fig. 21). Some components behave essentially as corks on water: as the wave passes they may undergo changes in orientation but little net movement occurs. Other components "ride the waves like surfboards. Essentially these components associate with waves such that they attain a maximum interaction at some point on the surface of the wave. Then they move along with the wave in order to preserve this maximal interaction, i.e., any displacement from this point requires work. They are "beached" by the termination of the wave at the protuberance, uropod, pseudopod, cleavage furrow, or equivalent region.

This model seems able to account for all of the experimental data generated so far on membrane topographic patterns. Thus, it may be predicted from the model that all regions of membrane that overly microfilamentous aggregates and are engaged in pinocytic activity will provide foci for ligand–receptor accumulation. The concentration of Con A–receptor complexes in protuberance, uropod, pseudopod, and cleavage furrow membrane is thus anticipated. The failure of Con A redistribution into lamellipodial membrane is also anticipated: although lamellipodia are underlain with microfilaments, microfilament–membrane forces act outward in this region, thus favoring the generation of waves directed backward toward the uropod.

The anistropy and rapid rates of ligand–receptor redistribution are consistent with this model. Once a membrane determinant is entrained in a wave its motion is no longer determined by its intrinsic diffusion but is primarily a function of the speed and orientation of the wave.

The model predicts that in general, because of geometrical and other considerations discussed below, relatively larger membrane determinants will move with the waves, while the distribution of many smaller components may be relatively unaffected. Consistent with this, all ligand–receptor complexes studied so far, as well as coated pits, which constitute macromolecular aggregates, assume asymmetric distributions on polarized cells. In contrast, most unoccupied receptors (i.e., those for Con A and anti IgM as well as Fc and C3b receptors) are not accumulated and may in fact show a preferential distribution away from regions of shape and microfilament assymetry.

According to this model, membrane–microfilament interactions are important for translocation of ligand–receptor complexes but not because microfilaments directly engage and move with the complexes. Rather, contractions set up a wave motion which in turn leads to receptor movement into protuberance, pseudopod, etc., membrane by a "surf-board" mechanism. This is completely consistent with our evidence that membrane determinants can redistribute with respect to existing regions of microfilament accumulation. It retains the microfilament dependence (and hence cytochalasin sensitivity) of capping and similar processes without the cumbersome requirement for direct links between microfilaments and every surface receptor during receptor movement. Furthermore, it accommodates the ability of antibodies to stimulate capping of a range of endogenous and introduced surface determinants that specifically should not interact with submembranous microfilaments. These include Forssman antigen, a membrane glycolipid (Stern and Bretscher, 1979), *N*-2,4,6-trininophenyl 1-acyl-2-(*N*-4-nitrobenzo-2-oxa-1,3-diazole) aminocaproyl phosphatidylethanolamine, a fluorescent phospholipid analog (Schroit and Pagano, 1981), and stearoyl-dextran, an impermeable fatty acid analog (Wolf *et al.*, 1980).

We emphasize that the proposed wave motion does not involve the bulk movement of a section of membrane from one region to another, say as one might pull a blanket up and over a bed. Rather the waves involve vertical movements of the surface with the peak or crest of a wave displaced laterally with time. This means that an entrained membrane constituent moving through the surface must overcome drag due to the viscosity of the lipid bilayer, to interactions with other membrane components, and to interaction of cytoskeletal elements with the membrane bilayer structure. That is, entrainment will depend not only on the relationship of receptor–complex to wave geometry (Hewitt's major concern) but also on the presence of factors that modulate drag.

One modulating factor might simply be the variable surface tension at different points along the surface of a sharply peaked wave. Since the radius of curvature becomes very small at the crest of the wave, for a stationary wave (the simplest case) of uniform composition, the surface tension at the crest must be far greater than at the sides of the wave. However this may be offset by a local reorganization of membrane lipids and/or protein which could relieve inequalities at surface tension. Positive interaction of a particular receptor or ligand–receptor complex with a locally reconstructed membrane could promote its entrainment and propagation with the wave.

Local decreases in cytoskeleton–membrane interaction as a wave passes would also facilitate entrainment and wave movement of surface components. This contradicts the common view that protein redistribution is mediated, not impaired, by links to cytoskeletal components, primarily microfilament. Nevertheless the hypothesis that cytoskeleton–membrane interaction may block the movement of membrane proteins is elegantly supported by recent studies of membrane diffusion in normal and spherocytic mouse erythrocytes (Sheetz *et al.*, 1980; Koppel *et al.*, 1981). Spherocytes lack the principal components of the erthrocyte cytoskeletal matrix (normally consisting of spectrin, actin, and bands 4.1 and 4.9). Lipid diffusional rates are closely similar between normal and mutant red cells. In contrast, the diffusion coefficient of the major membrane proteins measured by the technique of fluorescence recovery after photobleaching in normal erythrocytes ($4.5 \pm 0.8 \times 10^{-11}$ cm^2/seconds) is two orders of magnitude lower than that measured in spherocytic cells ($2.5 \pm 0.6 \times 10^{-9}$ cm^2/second. This is consistent with an arrest of protein movement due to cytoskeleton–membrane interaction in normal cells. Extrapolating these data to more dynamic cells, one would indeed predict membrane protein *immobilization* at regions of increased interaction of cytoskeletal components with membrane and, conversely, increased protein movement at sites of decreased cytoskeletal density.

Changes in the density of intrinsic membrane proteins could also influence drag. In another recent study, this time using a reconstituted multibilayer, Schindler and co-workers (1980) determined that the packing density of intrinsic membrane proteins can modulate diffusion of other membrane components. If density is affected by wave motion or surface tension, then the integral membrane proteins and wave motion may be mutually dependent: the proteins affecting drag; the wave motion, protein density. In this way, wave motion could account indirectly for anisotropic lateral diffusion.

This view that entrainment will depend on the balance of several forces in addition to the size and degree of crosslinking of membrane components provides a basis for analysis of membrane determinants that segregate in the absence of ligand as well as those whose lateral redistribution after ligand binding appears to be independent of intermediate patch formation. In addition, the concept that drag due to membrane viscosity or cytoskeleton–membrane interaction can antagonize entrainment may explain the uniform or "antiprotuberant" topography of certain receptors under conditions where other membrane components are accumulated at regions of cell shape change.

The physical existence of oriented wavelike motion on asymmetric leukocytes has already been suggested. Ramsey (1972), Senda *et al.* (1975, 1979), and Englander (1980) have all described a constricting ring that flows backward in a wavelike motion on migrating neutrophils. This ring is readily observed by light and scanning electron microscopy. It is difficult to discern by transmission electron microscopy, apparently because no accumulation of cytoplasmic microfilaments or other obvious cytoplasmic event accompanies the membrane change (as would be predicted if as suggested the wave is initiated by tension generated at a protuberance or uropod). Less well characterized, but nevertheless clearly evident, membrane movement, independent of net forward cell movement, is readily observed on protuberant leukocytes in suspension. In particular Yahara and Kakimoto-Sameshima (1979) have recently emphasized the broad, sharply peaked, parallel lamellae that characteristically occur on the cell bodies of protuberant lymphocytes observed in suspension. These lamellae are distinct from the multiple, fine microvilli that occupy the protuberance. They may very well represent waves of membrane activity. Finally Abercrombie, Harris, and others have described in detail the flow of membrane and attached particles (beads, Con A–receptor complexes) away from the lamellipodia of moving (adherent) fibroblasts (reviewed in Harris, 1973). These observations could be explained in terms of a wave model of membrane movement.

Fluid pinocytosis, whose function has remained obscure, is given a new significance in our model. It is seen as the basic mechanism by which membrane asymmetries created by wave motion are relieved. "Excess" membrane constituents swept into the protuberance, for example, are removed by internalization. This hypothesis implies that the composition of endocytic vesicle membrane differs from that of bulk membrane, a prediction that has received considerable support in our studies of phagocytosis (for review, see Oliver and Berlin, 1978) but has been denied for pinocytosis (Mellman *et al.*, 1980).

Of course a second route to relieve these membrane asymmetries is also available. Membrane components can be randomized and/or redistributed into new, asymmetric patterns by the relaxation of microfilament-induced tension and its establishment elsewhere on the membrane. This release and redevelopment of asymmetry occurs constantly during chemotaxis, when cells undergo cycles of anterior elongation followed by retraction of uropods toward the cell body. Similarly phagocytosis and cytokinesis are transient processes, while protuberant cells observed by time lapse cinematography constantly extend, remodel, and retract their projections. Presumably membrane architecture is remodeled during all these cell shape changes.

Finally, it is useful to consider the resting cell, shown above to maintain a rather symmetrical shape and to endocytize from points over the entire membrane. We think it likely that the same processes of microfilament recruitment and wave generation occur continuously over the whole membrane of such cells. This results in the constant production of multiple, randomly oriented ripples over the cell surface. These ripples may contribute to surface patching of numerous ligand–receptor complexes on rounded cells. They may also explain the characteristic distribution of coated pits in random clusters on rounded cells. However interference between interacting waves prevents their support of long range movements.

We suppose that the remarkable topographical symmetry on mitotic cells reflects the absence of even such short-range wave motion.

VI. Summary

In this article we have emphasized several advances that may simplify and focus the search for mechanisms controlling cell surface molecular and functional topography. Most importantly, we have recognized that precisely analogous membrane heterogeneities may be studied in protuberant, phagocytizing, oriented and dividing cells. Thus asymmetric surface topography is a predictable correlate of membrane deformation and microfilament asymmetry, independent of how asymmetry is induced and which cell type is studied. In addition, we have demonstrated that receptors can be segregated out of as well as into regions of microfilament accumulation. This contradicts the hypothesis that microfilament–receptor connections are sufficient to determine surface topography. We have also shown that topographical heterogeneity is not restricted to membrane molecular determinants but extends to a range of endocytic functions as well as to a macromolecular assembly, the coated pit. Finally, we have documented the remarkable arrest of dynamic surface events during mitosis.

Our working model to explain the topographical data depends on initial recruitment of microfilaments to incipient protuberance, uropod, pseudopod, or cleavage furrow membranes. The resulting microfilament–membrane interaction acting inward and along the cell surface leads to the generation of tension which is propagated over the cell surface as a wave. Certain receptors "sense" a change in orientation but no large change in environment as the wave progresses. Like corks bobbing as waves pass, these remain more or less uniformly over the membrane. Other receptors and the majority of ligand–receptor complexes tend to occupy regions of maximal interactional energy on the wave; these move by a surfboarding process to the region where tension is generated.

This model is consistent with the various topographical properties of all the membrane molecular and functional determinants analyzed to date. It reinforces and also simplifies the analogy drawn previously between protuberance, pseudopod, uropod, and cleavage furrow membranes: these all become sites of increased tension where cell shape is deformed, and wave motion is generated. Predictable membrane topographical asymmetries follow. In each instance, microfilament polarization is required to establish and maintain membrane structural asymmetry and by endocytosis to eventually relieve the accumulation of membrane components.

We suppose that the process of microfilament recruitment and generation of tension occurs at random over uniformly shaped cells producing asynchronous ripples that may gather surface determinants in patches but cannot support long-range movement due to interference between opposing waves. Wave generation may be suspended altogether during mitosis.

We have little doubt that the concepts developed here will be refined and altered by future work. However we hope that our model offers new orientation and focus to the search for mechanisms that regulate the structural and functional architecture of cell surfaces.

Acknowledgments

Supported in part by NIH grants CA-15544 and ES-01106 and grant BC-179 from the American Cancer Society. Janet M. Oliver holds an ACS Faculty Research Award.

References

Aggeler, J., and Heuser, J. G. (1980). *J. Cell Biol.* **87,** 93a.

Albertini, D. F., Berlin, R. D., and Oliver, J. M. (1977). *J. Cell Sci.* **26,** 57–75.

Allan, R. B., and Wilkinson, P. C. (1978). *Exp. Cell Res.* **111,** 191–203.

Berlin, R. D., and Fera, J. P. (1977). *Proc. Natl. Acad. Sci. U.S.A.* **74,** 1072–1076.

Berlin, R. D., and Oliver, J. M. (1978). *J. Cell Biol.* **77,** 789–804.

Berlin, R. D., and Oliver, J. M. (1980). *J. Cell Biol.* **85,** 660.

Berlin, R. D., Oliver, J. M., Walter, R. J. (1978). *Cell* **15,** 327–341.

Berlin, R. D., Pfeiffer, J. R., Walter, R. J., and Oliver, J. M. (1980). *In* "Mononuclear Phagocytes" (R. Van Furth, ed.). Blackwell, Oxford.

Bourguignon, L. Y. W., and Singer, S. J. (1977). *Proc. Natl. Acad. Sci. U.S.A.* **74,** 5031–5035.

Braun, J., Fujiwara, K., Pollard, T. D., and Unanue, E. R. (1978). *J. Cell Biol.* **79,** 409–418.

Bretscher, M. S. (1976). *Nature (London)* **260,** 21–23.

Davis, B. H., Walter, R. J., Pearson, C. B., Becker, E. L., and Oliver, J. M. (1982). *J. Cell Biol.* , in press.

DeLaat, S. W., Van Der Saag, P. T., Elson, E. L., and Schlessinger, J. (1980). *Proc. Natl. Acad. Sci. U.S.A.* **77,** 1526–1528.

DePetris, S. (1977). *In* "Dynamic Aspects of Cell Surface Organization" (G. Poste and G. L. Nicolson, eds.), pp. 644–728. Elsevier, Amsterdam.
Edelman, G. M. (1976). *Science* **192,** 218–226.
Edelman, G. M., Yahara, I., and Wang, J. L. (1973). *Proc. Natl. Acad. Sci. U.S.A.* **70,** 1442–1446.
Elson, E. L., and Reidler, J. A. (1979). *J. Supramol. Struct.* **12,** 481–489.
Englander, L. L. (1980). *J. Cell Biol.* **87,** 89a.
Goldstein, J. L., Anderson, R. G. W., and Brown, M. S. (1979). *Nature (London)* **279,** 679–684.
Harris, A. K. (1973). *Ciba Found. Symp.* **14,** 3–26.
Harris, A. K. (1976). *Nature (London)* **263,** 781–783.
Henis, Y. I., and Elson, E. L. (1981). *Proc. Natl. Acad. Sci. U.S.A.* **78,** 1072–1076.
Hewitt, J. A. (1979). *J. Theor. Biol.* **80,** 115–127.
Koppel, D. E., Sheetz, M. P., and Schindler, M. (1981). *Proc. Natl. Acad. Sci. U.S.A.* **78,** 3576–3580.
Koppel, D. E., Oliver, J. M., and Berlin, R. D. (1982). *J. Cell Biol.*, in press.
Lustig, S., Fishman, P., Djaldetti, M., and Plusnik, D. H. (1980). *Exp. Cell Res.* **129,** 321–328.
McKeon, F. D., Reichart, L. F., and Heuser, J. Z. (1980). *J. Cell Biol.* **87,** 93a.
Marasco, W. A., Becker, E. L., and Oliver, J. M. (1980). *Am. J. Pathol.* **98,** 749–767.
Mellman, I. S., Steinman, R. M., Unkeless, J. C., and Cohn, Z. A. (1980). *J. Cell Biol.* **86,** 712–722.
Melmed, R. N., Karanian, P., and Berlin. R. D. (1980). *In* "Microtubules and Microtubule Inhibitors" (M. DeBrabander and J. DeMay eds.), pp. 85–90. Elsevier, Amsterdam.
Oliver, J. M. (1978). *Am. J. Pathol.* **93,** 221–270.
Oliver, J. M., and Berlin, R. D. (1979). *Symp. Soc. Exp. Biol.* **33,** 277–298.
Oliver, J. M., Yin, H. H., and Berlin, R. D. (1976). *In* "Leukocyte Membrane Determinants Regulating Immune Reactivity" (V. P. Eijsvoogel, D. Roos, and W. P. Zeijlemaker, eds.), pp. 3–17. Academic Press, New York.
Oliver, J. M., Lalchandani, R., and Becker, E. L. (1977). *J. Reticuloendothee. Soc.* **21,** 357–364.
Oliver, J. M., Krawiec, J. A., and Becker, E. L. (1978). The distribution of actin during chemotaxis in rabbit neutrophils. *J. Reticuloendothel. Soc.* **24,** 697–704.
Oliver, J. M., Gelfand, E. W., Pearson, C. B., Pfeiffer, J. R., and Disch, H.-M. (1980). *Proc. Natl. Acad. Sci. U.S.A.* **77,** 3499–3503.
Petty, H. R., Hafeman, D. G., and McConnell, H. M. (1980). *J. Immunol.* **125,** 2391–2396.
Pfeiffer, J. R., Oliver, J. M., and Berlin, R. D. (1980). *Nature (London)* **286,** 727–729.
Poo, M. M., and Cone, R. A. (1974). *Nature (London)* **247,** 438–441.
Ramsey, W. S. (1972). *Exp. Cell Res.* **72,** 489–501.
Rappaport, R. (1975). *In* "Molecules and Cell Movement" (S. Inoue and R. E. Stephens, eds.), pp. 287–304. Raven, New York.
Salisbury, J. L., Condeelis, J. S., and Satir, P. (1980). *J. Cell Biol.* **87,** 132–141.
Schindler, M., Osborne, M. J., and Koppel, D. E. (1980). *Nature (London)* **283,** 346–350.
Schreiner, G. F., and Unanue, E. R. (1976). *Adv. Immunol.* **24,** 38–165.
Schroeder, T. E. (1976). *In* "Molecules and Cell Movement" (S. Inoue and R. E. Stephens, eds.), pp. 305–334. Raven, New York.
Schroit, A. J., and Pagano, R. E. (1981). *Cell* **23,** 105–112.
Senda, N., Tamura, H., Shibata, N., Yoshitake, J., Kondo, K., and Tanaka, K. (1975). *Exp. Cell Res.* **91,** 393–407.
Senda, N., Shibata, N., Tamura, H., and Yoshitake, J. (1979). *Methods Achiev. Exp. Pathol.* **9,** 169–186.
Sheetz, M. P., Schindler, M., and Koppel, D. E. (1980). *Nature (London)* **285,** 510–512.
Silverstein, S. C., Steinman, R. M., and Cohn, Z. A. (1977). *Annu. Rev. Biochem.* **46,** 669–722.
Smith, B. A., Clarke, W. R., and McConnell, H. M. (1979). *Proc. Natl. Acad. Sci. U.S.A.* **76,** 5641–5644.

Stern, P. L., and Bretscher, M. S. (1979). *J. Cell Biol.* **82,** 829–833.
Tsan, M. F., and Berlin, R. D. (1971) *J. Exp. Med.* **134,** 1016–1035.
Walter, R. J., Berlin, R. D., and Oliver, J. M. (1980a). *Nature (London)* **286,** 724.
Walter, R. J., Berlin, R. D., Pfeiffer, J. R., and Oliver, J. M. (1980b). *J. Cell Biol.* **86,** 199–211.
Wilkinson, P. C., Michl, J., and Silverstein, S. C. (1980). *Cell Biol. Int. Rep.* **4,** 736.
Wolf, D. E., Henkart, P., and Webb, W. W. (1980). *Biochemistry* **19,** 3893–3904.
Yahara, I., and Kakimoto-Sameshima, F. (1977). *Proc. Natl. Acad. Sci. U.S.A.* **74,** 4511–4515.
Yahara, I., and Kakimoto-Sameshima, F. (1979). *Cell Struct. Funct.* **4,** 143–152.
Zigmond, S. H. (1977). *J. Cell Biol.* **77,** 606–616.

INTERNATIONAL REVIEW OF CYTOLOGY, VOL. 74

Genome Activity and Gene Expression in Avian Erythroid Cells

KARLEN G. GASARYAN

Laboratory of Molecular Basis of Differentiation, Institute of Molecular Genetics, 123182, USSR Academy of Sciences, Moscow, USSR

I. The Cell System

Avian erythropoiesis is one of the best models for the investigation of processes underlying the inactivation of genes during terminal differentiation, the expression of individual tissue-specific genes and the accumulation of their product in the cytoplasm, and the relationship between the processes ensuring division and differentiation etc. (Scherrer *et al.*, 1966; Ingram, 1972; Bruns and Ingram, 1973; Ringertz and Bolund, 1974; Brown and Ingram, 1974; Sinclair and Brasch, 1975; Weintraub, 1975; Gasaryan, 1977).

A characteristic feature of the erythroid system is its high degree of phenotypic variability, both in the "vertical" (the existence of several versions of erythropoiesis replacing each other in ontogenesis) and the "horizontal" (a multiplicity of stages in the cell's development) perspectives.

Avian embryogenesis involves at least three generations of erythroid cells (Bruns and Ingram, 1973; Weintraub, 1975; Tobin *et al.*, 1976): (1) the cells

ISBN 0-12-364474-7

of primitive erythropoiesis occurring in the yolk sac, (2) the cells of definitive erythropoiesis in the yolk sac, and (3) the cells of definitive erythropoiesis in the bone marrow.

Erythroid cells of different generations differ in morphology, dimensions, the size of the nucleus, the type of hemoglobin, etc. According to early works (Bruns and Ingram, 1973; Brown and Ingram, 1974) primitive erythrocytes contain embryonic hemoglobins only (E, P, M, P′, P″), whereas definitive erythrocytes of the yolk sac contain only adult hemoglobins (A, D) and a fetal hemoglobin H. This suggests that the functions of the globin genes coding for embryonic and adult globins are switched over at appropriate stages of development by starting a new cell generation rather than intracellularly through modulation of gene activity (repression of the genes for embryonic hemoglobins and activation of the genes for definitive hemoglobins). This concept was supported by some of the later studies (Shimizu, 1976; Mahoney *et al.*, 1977). Recently, however, somewhat different results have been obtained. Thus according to the results of Chapman and Tobin (1979) the first embryonic definitive erythrocytes contain both adult and embryonic hemoglobins. Evidence was also presented showing that the primitive erythroid population can synthesize adult hemoglobins (Chui *et al.*, 1979; Zanjani *et al.*, 1979). These authors believe, therefore, that the switching is gradual rather than of the "all-or-nothing" kind. Further research is required to clarify the matter.

Erythroid development is characterized by a complex of morphological, biochemical, and physiological changes which spring from changes in genome activity and gene expression. These phenotypic changes have been meticulously analyzed (Ringertz and Bolund, 1974; Sinclair and Brasch, 1975; Tobin *et al.*, 1976). The researchers concentrate on the mechanisms which condition the genome changes involving inactivation of thousands of genes while the expression of the globin genes (as well as some other genes) is allowed to continue at a high level. Another extremely interesting question is how globin mRNAs are selectively accumulated in the cytoplasm of intermediate-stage cells and what the state of the active genes is in the nuclei of mature erythrocytes.

The studies in this field were confined mostly to gross analysis of the genome organization and the synthesis of major RNA classes in populations more or less enriched with either early or intermediate erythroid forms or mature erythrocytes. Recently these overall analyses have been complemented with data concerning the organization and expression of individual genes, especially the genes for globin polypeptides.

The profusion of phenotypic manifestations makes the erythroid model into a challenging object to study, because the cell populations used for biochemical studies may be different even within one research project.

As to the primitive erythropoiesis, the development of an entire generation is triggered off simultaneously and all subsequent stages are passed synchronously, so that one may have homogeneous cellular material for biochemical research into cytodifferentiation (see Ingram, 1972; Bruns and Ingram, 1973; Ringertz and Bolund, 1974; Sinclair and Brasch, 1975; Weintraub, 1975; Tobin *et al.*, 1976). By contrast, definitive erythropoiesis is a continuous steady-state process and here the blood (or the bone marrow) contains cells at different stages of development at the same time. As a result homogeneous populations of cells at various stages of development are very hard to obtain. Besides, erythroid cells constitute only 50% of the bone marrow in adult animals (Gasaryan *et al.*, 1971; Gasaryan, 1977). Different erythroid forms can be separated with respect to their density in Ficoll gradients or bovine serum albumin (see, e.g., Williams, 1972). Anemization is another approach. As Fig. 1 shows the erythroid forms of the bone marrow increase in quantity up to 80% in anemic pigeons, the proportion of the most immature cells (basophilic erythroblasts) growing from 10 to 65%. This shift in the ratio of early to late erythroblasts is reversible and reaches a narrow peak after which the proportion of basophilic forms falls again and that of polychromatic and orthochromatic erythroblasts increases. Cells obtained from the bone marrow at the peak of basophilic erythroblasts (see Fig. 1) represent a suitable material for analysis of the processes that occur at early stages of erythroid differentiation (Gasaryan *et al.*, 1971; Gasaryan, 1977).

The immature forms which appear in the peripheral blood of anemic birds, referred to as "erythroblasts" or "reticulocytes," are widely used for biochemical research into the early to intermediate-stage erythropoietic processes. However, a problem exists concerning the cytodifferentiational

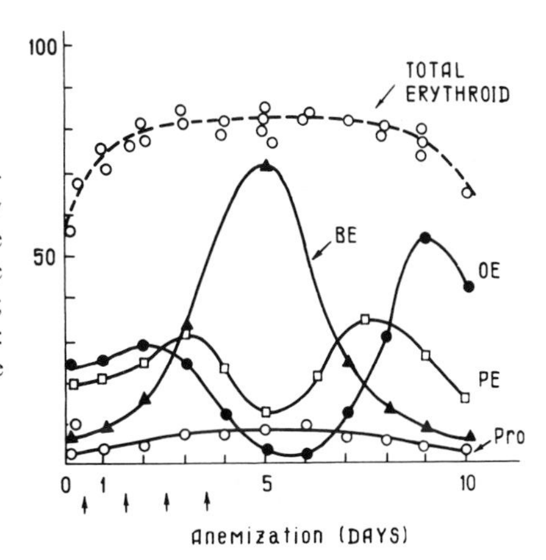

FIG. 1. Change in the proportion of different erythroid forms in pigeon bone marrow induced by anemization with phenylhydrazine (arrows). Pro, Proerythroblast; BE, basophilic erythroblast; PE, polychromatic erythroblast; OE, orthochromatic erythroblast. Ordinate: proportion of each erythroid form, percentage of the total. (From Gasaryan *et al.*, 1971.)

nature of the immature erythroid cells of anemic animals and in particular the significance of their premature appearance in the circulation. In anemic mammals circulating immature forms are obviously atypic and most probably do not lead to normal erythrocytes. In birds, however, no atypic forms like those observed in mammals were found: immature erythroid forms appearing in the blood during acute anemia of ducks (Scherrer *et al.*, 1966), hens (Williams, 1972), and pigeons (Gasaryan, 1977) look quite normal in their cytological and developmental characteristics. They include cells with basophilic, polychromatic, and orthochromatic cytoplasms (Scherrer and Marcaud, 1965; Scherrer *et al.*, 1966; Gasaryan *et al.*, 1971; Gasaryan, 1977); some of these forms incorporate [^{3}H]thymidine (Modak *et al.*, 1975; Gasaryan, 1977). Many of these cells are slightly (1.5 times) larger than the corresponding forms (erythroblasts, reticulocytes) of the normal definitive erythropoiesis (Williams, 1972; Kulminskaya *et al.*, 1978) but not as large as the macrocytes in mammals (see, e.g., Papies *et al.*, 1973). Most of the circulating basophilic forms and all the cells with a polychromatic or orthochromatic cytoplasm have an oblong shape (see, e.g., Scherrer *et al.*, 1966; Gasaryan, 1977) which is characteristic of mature forms rather than erythroblasts. Though it is certain that these immature forms in anemized birds subsequently grow into erythrocytes the question is whether the morphological and biochemical processes occurring in these immature forms in the anemic situation reflect the normal definitive erythropoiesis of adult birds or another line of erythroid cells (Bruns and Ingram, 1973; Gasaryan and Kulminskaya, 1975; Gasaryan, 1977).

The routes of the development of avian erythroid cells have been thoroughly investigated in anemic birds (Scherrer *et al.*, 1966; Gasaryan *et al.*, 1971; Williams, 1972; Gasaryan and Kulminskaya, 1975; Gasaryan, 1977). Studies on pigeons have suggested that most of the forms occurring in the anemic state represent a new line of erythroid cells ("reserve erythropoiesis," Gasaryan and Kulminskaya, 1975). The analysis of cell population kinetics (Gasaryan and Kulminskaya, 1975; Gasaryan, 1977) and cytophotometry of DNA and hemoglobin content (Kulminskaya *et al.*, 1978; Korvin-Pavlovskaya *et al.*, 1978, 1980) suggested the following scheme for "reserve erythropoiesis."

1. When cells of this line reach the stage of basophilic erythroblasts they are arrested at the G_2 phase (or at the end of the S_1 phase). This results in a transient accumulation of a population of basophilic erythroblasts (see Fig. 1) with subtetraploid or tetraploid DNA content (Kulminskaya *et al.*, 1978; Korvin-Pavlovskaya *et al.*, 1978).

2. These cells enter the peripheral blood; some of them (apparently those at the end of the S_1 phase) continue synthesizing DNA at a low level (see

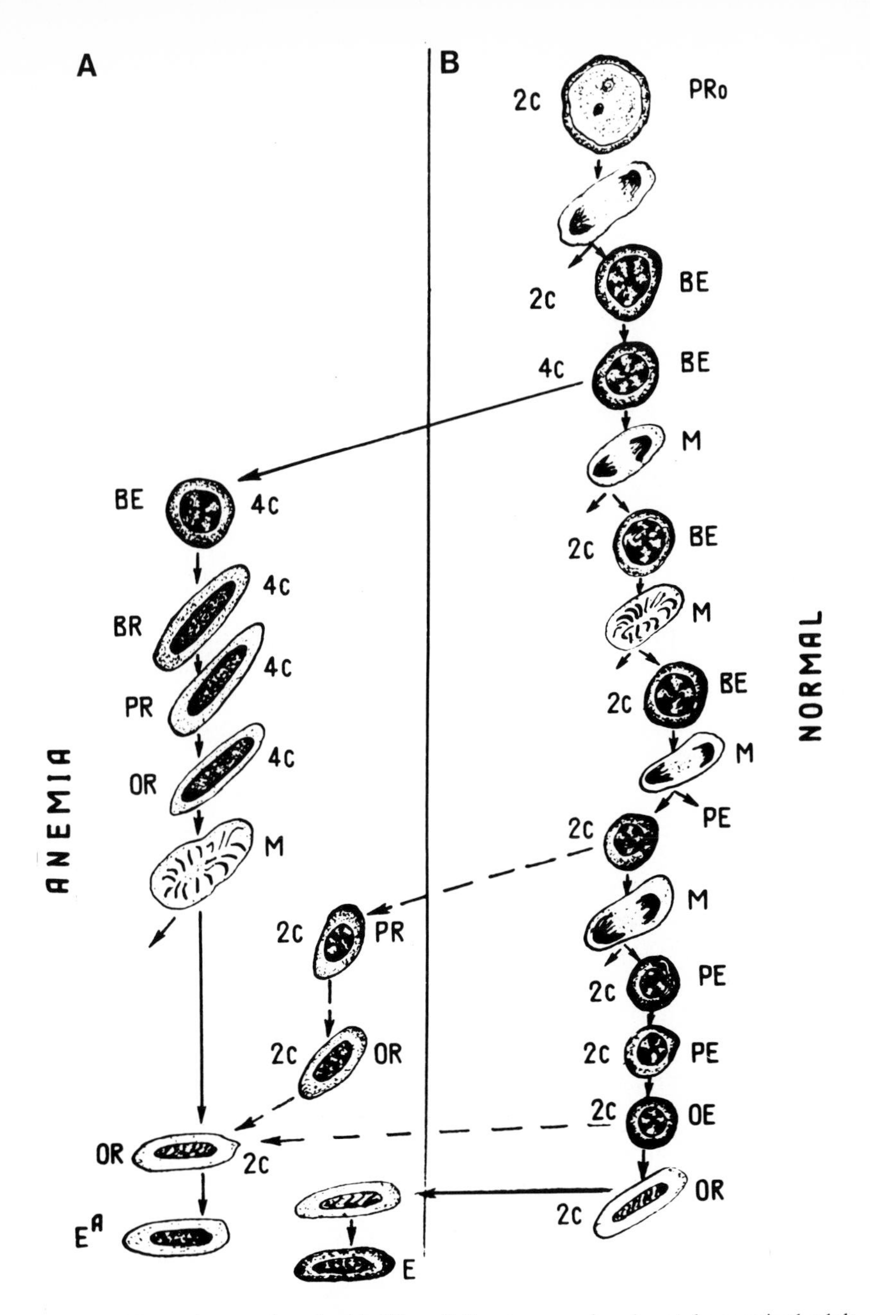

FIG. 2. The pathways of erythroid differentiation in normal and acutely anemized adult pigeons. (A) Peripheral blood; (B) bone marrow. PRo, Proerythroblast; BE, basophilic erythroblast; PE, polychromatic erythroblast; OE, orthochromatic erythroblast; BR, basophilic reticulocyte; PR, polychromatic reticulocyte; OR, orthochromatic reticulocyte; E, erythrocyte (normal); E^A, erythrocyte formed in anemic condition; M, mitosis; C, ploidy (DNA equivalent). (Summarized from Gasaryan and Kulminskaya, 1975; Kulminskaya *et al.*, 1978; Korvin-Pavlovskaya *et al.*, 1978.)

Modak *et al.*, 1975; Gasaryan, 1977). At first circulating basophilic tetraploid (or subtetraploid) forms are round-shaped (erythroblast-like) but soon they transform into oblong (reticulocyte-like) cell types—"basophilic reticulocytes."

3. The population of "basophilic reticulocytes" with a DNA content of about 4C is transformed into cells with a polychromatic cytoplasm.
4. These cells divide forming cells with a diploid DNA content, which develop into young erythrocytes.

Over 50% of the young erythrocytes formed in acutely anemized pigeon develop in this way (Fig. 2). Though otherwise quite normal these erythrocytes have a 25–30% deficit of hemoglobin (Korvin-Pavlovskaya *et al.*, 1980).

It is interesting that the specialization of these cells, including their morphological and biochemical differentiation, proceeds in the G_2 phase (or at the end of the S_1 phase) and not in the G_1 as in normal erythropoiesis (Yataganas *et al.*, 1970). Another interesting point is that the cell divisions which were suppressed at early stages (in basophilic erythroblasts) occur later, after the cells have accumulated a large amount of hemoglobin and lost most of their basophilia (Kulminskaya *et al.*, 1978). There is only one division wave which lasts about 24 hours (during the seventh to eighth days after the onset of anemia).

Since the erythrocytes which develop in this way (1) are blocked in the G_2 phase (or at the end of the S phase) while they accumulate hemoglobin and (2) have a hemoglobin deficit on maturation, the functioning of their genes may be somewhat different from the gene activity in normal erythropoiesis. These differences may be important for interpreting the results of research into genome activity and gene expression involving immature forms of anemized animals along with erythroid cells of normal animals.

II. Structural Organization of the Genome and the Globin Genes in Avian Erythroid Cells

A. The Genome

A mature chicken erythrocyte has been reported to contain 1.7–3.8 pg of DNA (for a revue see Zentgraf *et al.*, 1975; see also Epplen *et al.*, 1979). The reason for the discrepancies most probably lies in the different methods of analysis. The results of quantitative determination (Table I) give no grounds to believe that avian erythrocytes have any losses or underreplication of any DNA fraction. Though the available methods cannot detect a 3–5% DNA deficiency and so we cannot draw a final conclusion, we have no reason to

TABLE I

SEQUENCE ORGANIZATION OF AVIAN GENOME

Object	G[a] (pg)	Kinetic components of DNA: Fraction of genome	Kinetic components of DNA: Repetition frequency (average)	Length of interspersed sequences (kb): Repeated	Length of interspersed sequences (kb): Unique	References
Chicken (liver)	1.6	87	1	2–4	4.5	Eden and Hendrick (1978)
		5.5	15			
		7.5	1500			
Duck (erythrocytes)	1.75	73	1	0.8	?	Epplen *et al.* (1979)
		13	400			
		14	?			
Ostrich (erythrocytes)	1.9	87	1	?	?	Eden *et al.* (1978)
		7.2	20			
		3.2	4000			
Pigeon (reticulocytes)	1.9	69.5	1	2	3.7	Tarantul *et al.* (1973)
		6.0	50			Gasaryan *et al.* (1982)
		3.5	3000–5000			
		16.0	Foldback			

[a] Amount of DNA per haploid genome.

assume that the genome of a mature erythrocyte is deficient either quantitatively or sequence-wise.

The avian genome contains the same major classes of nucleotide sequences identified through reassociation of 300–500 base-long fragments, as all the other higher animals (Table I). Pigeon and chicken DNAs have been analyzed in detail (Tarantul *et al.*, 1973; Jemenez *et al.*, 1974; Gasaryan *et al.* 1977b, 1982; Arthur and Straus, 1978; Epplen *et al.*, 1978). About 70% of pigeon erythroid DNA are single-copy sequences, 10% are moderately repeated sequences, and 15–20% are very rapidly reassociating sequences (Tarantul *et al.*, 1973; Gasaryan *et al.*, 1977b, 1982). Birds seem to have few satellite DNAs (Cortadas *et al.*, 1979).

The fraction of middle repetitive DNA of pigeon, comprising a heterogeneous set of sequences with different frequencies of reiteration, has been shown to include two major classes (Gasaryan *et al.*, 1977b): the sequences repeated in average ~50 times ("rare" repeats) or about 2000 times (the "moderate" repeats) per haploid genome. "Rare" repeats have been purified and thoroughly investigated (Gasaryan *et al.*, 1977b, 1982). These repeats are transcriptionally extremely active: they account for only 6% of the genome, but their transcripts make at least one-third of hnRNA (Gasaryan *et al.*, 1977b).

The organization of repeated and single-copy sequences in pigeon and chicken DNAs has been analyzed (Arthur and Straus, 1978; Epplen *et al.*, 1978; Gasaryan *et al.*, 1982). Most of the avian repeated sequences have sizes ranging from 1.0 to 4 kb and are interspersed with 4-kb-long single-copy sequences (Table I). The majority of single-copy sequences are much longer (from 10 kb upward). Thus the sequence organization in avian genome is different from the two known patterns—those of *Drosophila* and *Xenopus* (Davidson *et al.*, 1975) constituting an intermediate type.

B. Globin Genes

Table II shows the globin polypeptides of chicken which are known so far. Although these polypeptides are accumulated in very large amounts in erythrocytes, the dosage of globin genes is the same in the liver and in erythroid cells, i.e., these genes are not amplified (Bishop *et al.*, 1972; Packman *et al.*, 1972). Two slightly different variants of the α and β globin genes are found in birds (Bishop *et al.*, 1972; Packman *et al.*, 1972; Cummings *et al.*, 1978; Dodgson *et al.*, 1979). The coding sequences of avian α and β globin genes are similar to the coding sequences of mammalian α and β globin genes (Engel and Dodgson, 1978; Richards *et al.*, 1979).

Chicken α and β globin genes are located in different chromosomes (Hughes *et al.*, 1979). The β and β' genes have been shown to be in the same

TABLE II
CHICKEN HEMOGLOBIN POLYPEPTIDES[a]

Primitive cell (embryonic globin)					Definitive cell (adult globin)				
Hb	α type	%[b]	β type	%	Hb	α type	%	β type	%
	π[c]								
P	π'	35	ρ	35	A	α^A	37.5	β	37.5
E	α^A	10	ε	10	D	α^D	12.5	β	12.5
M	α^D	5	ε'[d]	5	H	(Present at 17–22 days)			

[a] From Engel and Dodgson (1978); based on the data of Brown and Ingram (1974).
[b] Percentage of substituent polypeptide chain in cell type.
[c] Globins π and π' are assumed to be allelic but may be identical.
[d] Globin ε may be identical with globin ε'.

chromosome and very close to each other: the distance between them is only 3 kb (Dodgson *et al.*, 1979; Ginder *et al.*, 1979) and this raises a question about the mechanism of their autonomous expression. The sequence organization of the β globin gene has been more thoroughly investigated. It was shown that the chicken β globin gene is similar to the mammalian β globin gene in that it contains two introns: a small one (about 100 base pairs) and a large one (about 800 base pairs) (Dodgson *et al.*, 1979; Ginder *et al.*, 1979). The introns are similarly located in the chicken and in mammals (for a revue see Dawid and Wahli, 1979).

One might expect (Kabat, 1972) the organization of globin genes to change in the course of ontogenesis like that of immunoglobulin genes (for a revue see Dawid and Wahli, 1979). It seems, however (Engel and Dodgson, 1978), that the structure of globin genes remains constant in the process of development, apart from some changes in the degree of base methylation (McGhee and Ginder, 1979; Van der Ploeg and Flavell, 1980).

Very little is known about the nature and structure of the other genes operating in avian erythroid cells.

III. Organization of the Genome and the Globin Genes in the Chromatin of Avian Erythroid Cells

A. THE GENOME

The very first studies that reported the nucleosomal organization of the chromatin showed that the chromatin of avian erythroid cells, including mature erythrocytes, were basically organized in the same way as the chromatin of other cells (see Kornberg, 1977; Felsenfeld, 1978). The nucleosome in erythroid, as well as other, cells comprises typical DNA repeat in which a 140-bp-long section winds around an octameric protein globule with a diameter of 76–80 Å and a molecular weight of 100,000 (Shaw *et al.*, 1976). The fact that the chromatin maintains its typical structure in mature erythrocytes, in spite of their virtual inactivity, is another indication that their chromosomes suffer no degradation. Serine-rich (H5) histone is one of the main peculiarities of the chromatin structure of avian erythroid cells.

The structure and properties of the H5 histone are of great interest also because it represents another (after hemoglobin) highly tissue-specific protein which gets accumulated in erythroid cells at about the same stages of terminal differentiation as hemoglobin. H5 histone appears in the bone marrow (Appels *et al.*, 1972; Billett and Hindley, 1972; Gasaryan and Andreeva, 1972; Gasaryan *et al.*, 1976a, 1978b). H5 histone forming in these early erythroid cells is strongly phosphorylated (Adams *et al.*, 1970; Seligy and Neelin, 1970; Tobin and Seligy, 1975; Sung, 1977). No other histone has as many phosphorylation sites as H5 (Sung and Freedlender, 1978). Late erythroid cells have a lower degree of phosphorylation (so far it is not clear whether this is due to dephosphorylation or to the synthesis of less phosphorylated histone molecules) but are capable to accept phosphates *in vitro* when chromatin is incubated with even heterologous enzyme (Kurochkin *et al.*, 1977). H5 synthesis goes on in intermediate erythroid forms after the synthesis of all other histones, as well as DNA synthesis, has stopped, i.e., the regulation of H5 synthesis is quite different from the other histones, it is not linked to the cell cycle (Ruiz-Carrillo *et al.*, 1976).

In recent years investigations of the primary structure and conformation of H5 histone have made progress: the primary structure of H5 has been completely decoded in chicken (Briand *et al.*, 1980) and goose (Yaguchi *et al.*, 1979). However the nature of its association with the chromatin and its relation to H1 are still obscure. In the early studies H5 was reported to replace H1 (Dick and Johns, 1969). According to other data, it only partially replaces H1, whereas Weintraub (1978) has recently reported that H1 is fully retained in the erythrocyte. The matter is important because H5, as well as H1 (Whitlock and Simpson, 1976), is believed to be located in linker DNA and not in the nucleosome octameric particles (Shaw *et al.*, 1976; Noll and Kornberg, 1977; Allan *et al.*, 1980).

B. Globin Genes

Organization of active genes in the chromatin has been shown to possess a somewhat different conformation than the genes which are not transcribed in the tissue concerned. The first indications of the specific state of the globin genes in the chromatin of erythroid cells were obtained in experiments involving transcription *in vitro* by bacterial RNA polymerases. It has been demonstrated that bacterial RNA polymerases, which are unable to recognize the initiation sites of the globin genes (Axel *et al.*, 1973; Gadski and Chi-Bom Chae, 1978) and transcribe them less effectively by two orders of magnitude than animal polymerases (Fodor and Doty, 1977), preferentially transcribe globin genes and possible other genes which are active *in vivo* in these cells (Axel *et al.*, 1973; Barrett *et al.*, 1974; Steggles *et al.*, 1974; Gadski and Chi-Bom Chae, 1978). This suggests that genes in the chromatin of erythroid cells have structural peculiarities which cause the polymerase to give them preference.

The conclusion that the globin genes must have some structural peculiarities also results from the data on a limited digestion of the chromatin by nucleases. The results have shown that though the genes which are actively transcribed in the tissue are organized in the nucleosomal structure as the total chromatin (Bellard *et al.*, 1977) they are preferentially digestible in the chromatin treated with nucleases. Weintraub and Groudine (1976) were the first to demonstrate that the globin gene is preferentially hydrolyzed when chromatin of erythroid cells is digested with DNase 1. This observation has been confirmed by other authors (Chambon, 1977; Bellard *et al.*, 1977; 1980; Garel *et al.*, 1977; Levy and Dixon, 1977; Felsenfeld, 1978; Zasloff and Camerini-Otero, 1980) who found not only the globin genes in the chromatin of erythroid cells but genes actively transcribed in other tissues, e.g., the ovalbumin gene in the oviduct to possess a specific conformation making them more nuclease-sensitive than the inactive genes. The ovalbumin gene in erythroid cells and the globin gene in the oviduct do not possess these properties.

Interesting results were obtained when this approach was applied to embryonic or definitive globin genes in primitive erythroid cells and in immature erythroid cells of adult birds. It was found that (1) not only active embryonic globin gene but also yet inactive definitive globin gene is preferentially hydrolyzed in primitive embryonic erythroid cells (Stalder *et al.*, 1980; Zasloff and Camerini-Otero, 1980); (2) in mature erythrocytes, where the globin genes are believed (though not yet proved) to be inactive, both the definitive gene and the embryonic one are preferentially hydrolyzed, although the latter is somewhat more resistant (Stalder *et al.*, 1980). The data suggest that at the early stages of embryonic development not only the embryonic globin genes but their definitive analogs in primitive cells assume the active conformation. This premature change of conformation in a gene that is as yet inactive but is going to operate in this tissue later on is referred to as the gene's conversion to the "preactivated" state (Stalder *et al.*, 1980). In nonerythroid cells all these genes remain in the inactive conformation.

What is the reason for the peculiar conformation of the globin genes in erythroid cells? Since they contain nucleosomal core particles (Bellard *et al.*, 1977) it might be expected that the globin (and other active) genes are preferentially hydrolyzed because they do not possess the higher level of condensation brought about by H1 or H5. This possibility seems to be rejected by the experiments in which the chromatin was deprived of H1 and H5 then digested with DNase 1: the globin genes were still preferentially hydrolyzed (Villeponteaux *et al.*, 1978).

The specific conformation of the active genes which is manifested by enhanced sensitivity to nucleases could most probably be due to specific nonhistone proteins which are known to abound in the active chromatin regions (Cadski and Chi-Bom Chae, 1978; Berkowitz and Doty, 1975).

IV. Transcriptional Activity

A. RNA Types and Amount in the Avian Erythroid Cells

The erythroid system is a consecutive series of specializing cells with a gradually decreasing steady-state mass of the nuclear and cytoplasmic RNA (see Schweiger, 1962; Sinclair and Brasch, 1975; Zentgraf *et al.*, 1975).

1. *Morphological Evidence*

The decrease in the cytoplasmic RNA amount is morphologically manifested in diminished basophilia and in falling concentrations of polysomes and ribosomes in electron micrographs (Grasso and Woodard, 1966; Tooze and Davies, 1967; Small and Davies, 1972). Early erythroblasts (proerythroblasts) have a high RNA content according to both parameters. Basophilic erythroblasts have an appreciably lower concentration of polysomes, while polychromatic erythroblasts and the corresponding immature forms in the peripheral blood of anemic animals, whose major translation product is hemoglobin, have still less polysomes. The cytoplasm of mature erythrocytes shows no polysomes at all in electron micrographs (Fig. 3).

The nuclei of young bone marrow erythroblasts show large areas of interchromatin granules and multiple perichromatin granules and fibrils. In intermediate forms (late erythroblasts, reticulocytes) the perichromatin granules and fibrils are rare but there are still many interchromatin granules (Fig. 3). Mature erythrocytes show no perichromatin RNA structures. As to the interchromatin granules, the interior of a mature erythrocyte's condensed nucleus shows areas of RNA-containing structures (Small and Davies, 1972; Zentgraf *et al.*, 1975; Smetana and Likovsky, 1978; Stvolinskaya and Gasaryan, 1980a; Stvolinskaya *et al.*, 1981). When assayed by Bernhard's technique revealing specifically RNP structures (Bernhard, 1969) the central

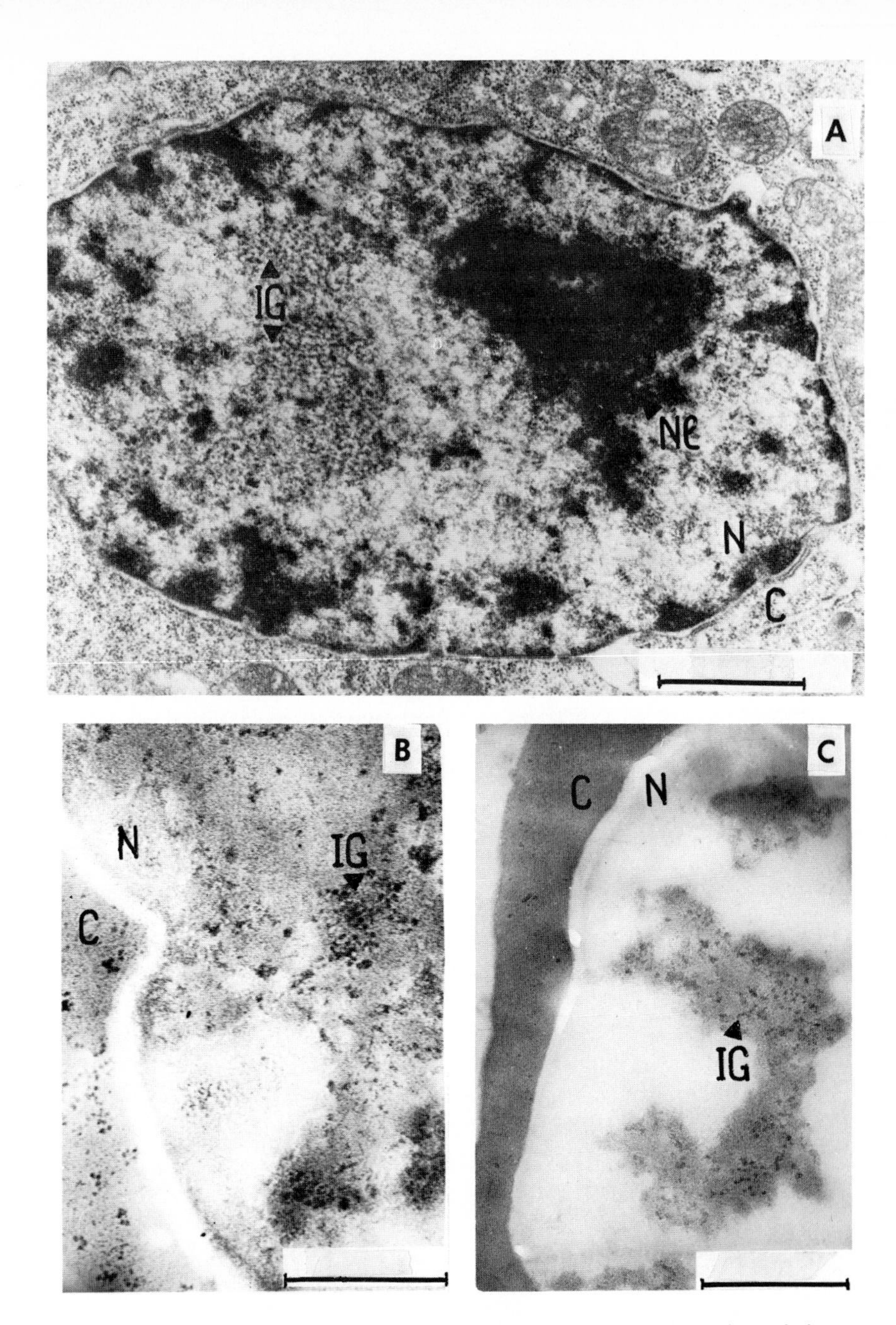

FIG. 3. Electron microscopic visualization of the ribonucleoproteins in sections of pigeon erythroid cells at different stages of terminal differentiation. (A) Early erythroblast. Fixation glutaraldehyde–OsO_4, embedding medium: Epon 812. (B) Reticulocyte; (C) erythrocyte. EDTA method for preferential staining of RNA-containing structures (Bernhard, 1969). N, Nucleus; C, cytoplasm; Nl, nucleolus; IG, interchromatin granules. Bars = 1 μm. (From Stvolinskaya *et al.*, 1981.)

part of the nucleus shows, against the background of blanched chromatin clusters of interchromatin granules and a netlike structure similar to the RNP net described by Puvion and Bernhard (1975) on thick sections of liver cells. In erythrocyte nuclei this net structure can be observed on ultrathin sections (Stvolinskaya *et al.*, 1981), whereas it was undetectable on thin liver sections (Puvion and Bernhard, 1975). Possibly this region is highly compactized in the erythrocyte chromatin and becomes distinguishable on thin sections (see Fig. 3).

Highly characteristic changes occur in the nucleoli (Small and Davies, 1972; Smetana and Likovsky, 1978). Smetana and his associate (Likovsky and Smetana, 1975; Smetana and Likovsky, 1978) have given a detailed account of the morphological rearrangements that occur in the nucleoli of differentiating erythroid cells in chicken and pigeon. Four morphologically different kinds of nucleoli which reflect a gradual decrease in ribosome production at consecutive stages of erythrocyte development were described by the authors: (1) compact nucleoli, (2) nucleoli with nucleolonemata, (3) ring-shaped nucleoli, and (4) micronucleoli. Compact nucleoli are characteristic of the youngest erythroblasts whereas reticulocytes and mature erythrocytes have nothing but micronucleoli which seem to be functionally inactive. In pigeons no nucleoli have been detected by the available morphological methods in 30% of the reticulocytes and 70% of the mature erythrocytes (for comparison, all the erythrocytes in chick and in hen contain micronucleoli).

2. *Biochemical Analysis*

Biochemical analysis of bone marrow cells of anemized pigeons has shown that the cells contain the normal amount of RNA [~3 pg, on average, per cell (Table III)]. About 40% of RNA is extracted from the nuclei and include all the known major classes: hnRNA, rRNA, and 4 to 7 S fraction. About 2% of the cytoplasmic RNA in these cells contain long poly(A) sequences and globin coding sequences amount to ~4% of the polyadenylated RNA (Gasaryan *et al.*, 1980).

TABLE III
RNA CONTENT IN PIGEON ERYTHROID CELLS[a]

Cell type	RNA per cell (pg)	RNA per nucleus (pg)
Erythroblasts	2.8 ± 0.247	1.35 ± 0.108
Reticulocytes	1.28 ± 0.048	0.33 ± 0.024
Erythrocytes	0.13 ± 0.008	0.084 ± 0.001

[a] From Dubovaya and Gasaryan (unpublished).

In immature forms of anemized birds there is 2–2.5 times less RNA per cell than in the bone marrow (Table III). The proportion of RNA in the nuclei of these forms is also smaller (15–25%).

Lasky *et al.* (1978) have analyzed the complexity of the poly(A) fraction of hnRNA in chicken's immature blood cells and found that it amounted to 5×10^6 nucleotides (1–1.5 orders of magnitude less than in the L cells). According to their data, 50% of hnRNA in the immature blood cells is represented by ~4000 kinds of sequences (~2000 nucleotides long), 0.5 copies per nucleus on an average. The remaining 50% of hnRNA is represented by 25 kinds of sequences, 80 copies of each per nucleus. There are discrepant data as to the number of globin mRNA sequences in the nuclei of avian immature erythroid blood cells (see Scherrer *et al.*, 1979; Tobin, 1979; Reynaud *et al.*, 1980). The amount of poly(A)$^+$ RNA in the cytoplasm of immature blood cells is less (1–1.5%) than in the cytoplasm of the bone marrow.

According to Tobin's group (Tobin *et al.*, 1978; Tobin, 1979) 25–50% of poly(A)$^+$ RNA is in polysomes, the rest is in nonpolysomal particles. Of the polysomal poly(A)$^+$ RNA 90% are, according to the same study, represented by globin RNAs with a total complexity of about 1700 nucleotides. This complexity is believed to be due to three kinds of mRNA, each 600 nucleotides long and making a total of 4500 copies. The remaining 10% of the polysomal poly(A)$^+$ RNA is made up by 100 kinds of nonglobin mRNA, 7 copies of each per cell, i.e., the concentration of globin mRNAs is, according to this study, 200 times higher than the concentration of each nonglobin mRNA (Table IV). Maundrel *et al.* (1979) have analyzed the poly(A)$^+$ RNA of polysomal and nonpolysomal RNP in the blood cells of anemized ducks. According to their data, polysomal RNP contain 240 kinds of poly(A)$^+$ mRNA, of which 50 to 70% are represented by three globin mRNAs. As to the nonpolysomal mRNAs, 30% are globin mRNAs and the remaining 70% are represented by 2000 kinds of nonglobin mRNA.

Practically nothing is known so far about the poly(A)$^-$ mRNA of erythroid cells. Knöchel (1975) has reported that mRNA coding for the H5 histone can be revealed among poly(A)$^-$ mRNAs. However, recently it has been shown that mRNA for this histone is polyadenylated (Molgaard *et al.*, 1980).

The amount of RNA decreases as the cells of intermediate stage develop into mature erythrocytes (Table III). Many authors reported a total lack of RNA in mature erythrocytes. However a more careful analysis reveals some RNA located predominantly in the nuclei. Zentgraf *et al.* (1975) have shown the nucleus of a mature chicken erythrocyte to contain a minimal amount of 0.02–0.04 pg of RNA. According to our analysis pigeon erythrocyte nucleus contains ~0.084 pg RNA (see Table III). The results of biochemical analysis accord with electron microscopic observations (see above). While

TABLE IV
AMOUNT AND THE SPECTRUM OF mRNA IN ERYTHROID AND NONERYTHROID CELLS[a]

	L cells			Friend cells (uninduced)			Immature blood cells (chicken)	
Total mRNA per cells (pg)	0.18			0.26			0.001	
Average size of mRNA (NT)	2000			1900			Globin 600	Nonglobin 1300
Number of average-sized mRNA molecules per cell	170,000			260,000			5000	
Classes of mRNA[b]	A	I	R	A	I	R	Globin	Nonglobin
Complexity of each class (number of average-sized mRNAs represented)	3	300	7600	3	1700	8700	3	87
Fraction of total mRNA in each concentration–class	0.045	0.4	0.45	0.13	0.77	0.1	0.9	0.1
Concentration of individual mRNA molecules (molecules per cell)	2250	300	9	12,000	120	3	1500	7

[a] From Tobin (1979).
[b] A, Abundant; I, intermediate; R, rare.

comparing electron microscopic and biochemical evidence one may suggest that nuclei of mature erythrocytes contain only the RNA of the interchromatin granules and possibly the RNA of nuclear RNP net.

B. RNA METABOLISM

It was reported in the early studies of RNA metabolism in avian erythroid cells that (1) the level of total RNA synthesis fell with advancing specialization and (2) RNA synthesis stopped altogether in a mature erythrocyte (Cameron and Prescott, 1963; Kolodny and Rosenthal, 1973; Ringerts and Bolund, 1974). Other research has shown that RNA synthesis does not completely cease in mature erythrocytes; it goes on at a lower level, 20–100 times below that of early erythroblasts according to different reports (Scherrer *et al.*, 1966; Attardi *et al.*, 1970; Gasaryan *et al.*, 1971; Zentgraf *et al.*, 1975).

The level of RNA synthesis in pigeon's erythroid cells has been autoradiographically shown to fall starting from basophilic erythroblasts and then continue falling throughout the process of maturation (Gasaryan *et al.*, 1971). Each morphologically identifiable erythroid form is not homogeneous either, i.e., it comprises cells with different levels of transcription. Hence the morphological criteria fail to distinguish cells with slight but measurable differences in the rate of RNA synthesis. Since there are a number of mor-

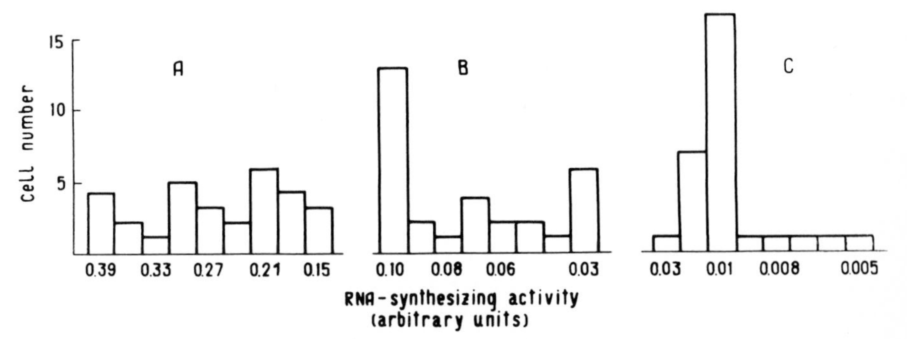

FIG. 4. Histograms showing differences in the RNA-synthesizing activity of the cells in morphologically distinguishable erythroid populations (see the text). (A) Basophilic erythroblasts; (B) polychromatic erythroblasts; (C) orthochromatic erythroblasts. (From Gasaryan *et al.*, 1971.)

phologically different stages each comprising cells with different levels of RNA synthesis, the inactivation of transcription in erythroid cells seems to be a many-step, practically a continuous process (Fig. 4).

Some biochemical studies of the RNA metabolism have been carried out using bone marrow cells of anemized birds enriched by basophilic erythroblasts (Gasaryan *et al.*, 1970). Pulse-labeled RNA extracted from those cells was shown to have a typical polydisperse distribution in sucrose gradient with a maximum in the 35–45 S region; about one-third of the newly formed molecules with sedimentation constant over 7 S are represented by pre-rRNA, the rest by hnRNA. Selective inhibition of rRNA synthesis with small doses of actinomycin has shown that the high-molecular fraction of newly formed RNA contains nonribosomal molecules with sedimentation constants up to 60 S. The newly formed RNA from the immature peripheral blood cells is characterized by a very low pre-rRNA content (Attardi *et al.*, 1966; Scherrer *et al.*, 1966). Physicochemical properties of the hnRNA forming in the nuclei of these intermediate-stage erythroid cells in the duck, pigeon, and chicken have been thoroughly characterized (Attardi *et al.*, 1966; Scherrer *et al.*, 1966; Gasaryan *et al.*, 1970; Spohr *et al.*, 1974; Williamson and Tobin, 1977). The early assessments of the size of hnRNA molecules were made without the use of denaturing agents to prevent aggregation. Though one must be cautious in interpreting these early data it is noteworthy that the concept of "giant" hnRNA molecules based on the analysis of newly formed RNA in the avian erythroid cells without using the denaturing agents has been *in principle* confirmed by recent experiments carried out in completely denaturing conditions. Imaizumi *et al.* (1973) and Williamson and Tobin (1977) have shown the pulse-labeled hnRNA of immature blood cells from duck and chicken respectively to contain molecules sized up to 50 S in the presence of 99%

DMSO. The most reliable conditions of denaturation and disaggregation of RNA molecules have recently been used in Scherrer's group: electrophoresis in absolute formamide at 45–50°C (Spohr *et al.*, 1976; Scherrer *et al.*, 1979; Reynaud *et al.*, 1980). In this situation the authors found a sizable fraction of large (>28 S) primary transcripts, the largest hnRNA molecules (from duck erythroid cells) having lengths of about the same order as were observed under nondenaturing conditions. However there were significantly fewer giant molecules under denaturing conditions. The reasons for this may be not only the exclusion of artificial aggregates but the dissociation of true "giant" molecules which had single breaks due to nonspecific nuclease effects or the onset of natural processing (Scherrer *et al.*, 1979).

In erythroid and other cells hnRNA molecules are transcribed from genome regions containing both repeated and single-copy sequences (Gasaryan *et al.*, 1974, 1976b, 1977c). Judging by the maximum attainable levels of hybridization between pulse-labeled hnRNA and a large excess of DNA, about one-third of hnRNA is transcribed from repeats and the rest from single-copy sequences (Table V).

The proportion of transcription from repeated and single-copy sequences has been compared in three populations of pigeon erythroid cells differing in the degree of maturity. The transcription from repeats and palindromes proved to decrease with advancing maturation (Gasaryan *et al.*, 1976b). A

TABLE V
HYBRIDIZATION OF PIGEON BONE MARROW PULSE-LABELED hnRNA AND STABLE NUCLEAR RNAs (snRNA) WITH THE EXCESS OF DNA FRACTIONS[a]

	Hybridization (%)		
RNA	Crude preparation of repeats	Purified unique sequences	Purified "rare repeats"[b]
45 S hnRNA[c]	21.5	40.0	15.5
28 S snRNA[d]	26.5	21.8	26.7
18 S snRNA[d]	—	9.4	13.2
28 S rRNA[d]	24.2	0.5	0.3
18 S rRNA[d]	13.0	0.2	0.2

[a] See Gasaryan *et al.* (1976b, 1979b).

[b] A kinetically homogeneous fraction of intermediate repeats consisted of ~50 copies per haploid genome (Gasaryan *et al.*, 1977b).

[c] Labeled 1 hour *in vitro*.

[d] Labeled *in vivo* by [^{3}H]uridine injected to pigeons 30 hours before sacrificing the animals.

similar situation (diminished transcription from repeats) was observed in differentiating myoblasts (Man and Cole, 1974).

hnRNA is characterized by a high metabolic activity as has been demonstrated for many kind of cells (see Lewin, 1975) including avian erythroid cells. According to early estimations the average half-life of hnRNA in duck immature blood cells is about 20–25 minutes (Attardi *et al.*, 1966; Scherrer *et al.*, 1966). Recently however it was reported that the hnRNA of avian erythroid cells includes a class of slowly metabolizing molecules. Analysis of the metabolic properties of three fractions of pulse-labeled RNA of duck's immature blood cells has shown (Spohr *et al.*, 1974; Scherrer *et al.*, 1979) the highest molecular-weight fraction (molecules sized $5-20 \times 10^6$) to have a half-life of 20–30 minutes, the fraction of molecules sized $1-5 \times 10^6$ to have a half-life of 3 hours, and the fraction of still smaller molecules to have a half-life of ~15 hours. Highly stable species of hnRNA (with a half-life ~15 hours) were detected also in the bone marrow of anemized pigeons (Tarantul *et al.*, 1974; Gasaryan *et al.*, 1977a, 1979a,b). The stable nonribosomal RNA species were detectable not only in preparations of total deproteinized nuclear RNA but in RNP particles with a floating density of 1.4 gm/cm^3 extracted from chromatin of erythroid cells (Gasaryan *et al.*, 1979b; Gasaryan and Stvolinskaya, 1979). Sedimentation analysis of prelabeled stable RNA from the nuclei of bone marrow cells (or from preparations of chromatin RNP particles) has revealed, apart from the fraction which is similar in size to cytoplasmic mRNA (16–18 S), a population of a stable high-molecular RNA sized 26–30 S. This high-molecular (~28 S) fraction has some interesting properties: (1) Nuclear ~28 S RNA which remain labeled 30 hours after the injection of [^{3}H]uridine to pigeons (i.e., at least 20 hours after practically no radioactivity is left in the nucleotide pool; see Tarantul *et al.*, 1974) is nonpolyadenylated (Gasaryan *et al.*, 1977a); (2) up to 50% of this RNA form RNase-resistant hybrids with purified single copy DNA and with highly transcribable fraction of the middle repetitive DNA (the "rare" repeats; Gasaryan *et al.*, 1977b), i.e., consists of nonribosomal RNA species (Table V). (3) Stable nuclear ~28 S RNA contains an appreciable amount of sequences homologous to pulse-labeled poly(A)$^-$ ~45 S hnRNA but has practically no homology to poly(A)$^+$ ~45 S hnRNA. Based on this result it was suggested that at least part of the stable ~28 S molecules originate from longer nonpolyadenylated hnRNA sized ~45 S and are maintained in the nucleus (Gasaryan *et al.*, 1978a, 1979a). (4) Stable prelabeled (see above) nuclear ~28 S RNA shows some homology to cytoplasmic RNA (Fig. 5). As nuclear ~28 S fraction of these cells includes a very small (0.01%) amount of globin-specific sequences (Gasaryan *et al.*, 1980) the above evidence suggested that in the erythroid cells nuclear ~28 S fraction contains stable nonpolyadenylated and non-globin pre-mRNA species (Gasaryan *et al.*, 1979a).

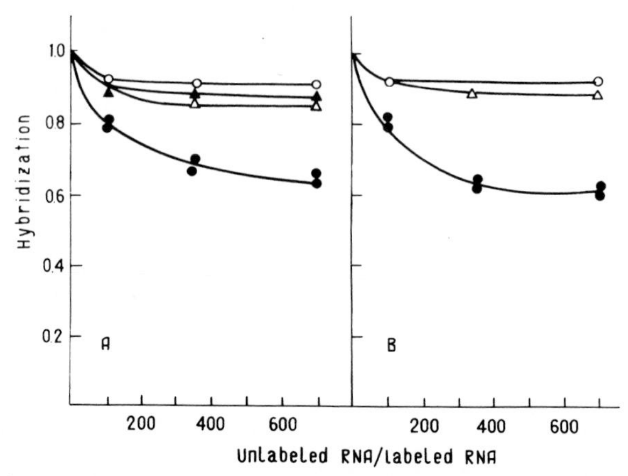

FIG. 5. Existence of a homology between nuclear stable 26–30 S (~28 S) RNA and 8–18 S fraction of the cytoplasmic RNA of pigeon bone marrow cells. ~28 S [3H]RNA fraction was obtained from nuclei 30 hours after injection of [^{3}H]uridine to anemic pigeons, RNA was purified in a sucrose gradient containing 85% formamide at 16°C and hybridized to the single-copy (B) or middle repetitive (A) DNA fractions. Competitor RNAs added: *E. coli* RNA (○); 28 S ribosomal RNA from pigeon reticulocytes (▲); 28 S fraction of pigeon bone marrow cytoplasmic RNA (△); 8–18 S fraction from the same cytoplasmic RNA (●). (From Gasaryan *et al.*, 1979a.)

It seems that the nuclei of differentiating erythroid cells contain RNA of even a higher degree of stability. This suggestion is based on electron microscopic indications that mature erythrocyte nuclei contain stable interchromatin RNP granules (Zentgraf *et al.*, 1975) supported by recent biochemical evidence: in long-term experiments it was shown that when radioactive precussor is administered to anemized pigeons the RNP particles extracted from the nuclei of maturing peripheral blood cells were found to be very weakly labeled even 24 hours after introduction of radioactivity. However, after 2–5 days (when the cells' nucleotide pool has been free of radioactivity for a long time) RNP particles contained enough label (Stvolinskaya and Gasaryan, 1980b). It is very likely that early erythroblasts form a category of highly stable RNA which remains in the nuclei as constituents of stable interchromatin RNP particles until maturation is completed.

Thus, analysis of hnRNA metabolism in differentiating erythroid cells reveals a remarkable feature, the existence of stable hnRNA derivatives in their nuclei. The available data make it possible to postulate at least two stable components: (1) stable (nonglobin) pre-mRNA homologous to cytoplasmic RNAs and (2) highly stable RNA which is maintained in a mature erythrocyte. The latter can hardly be regarded as pre-mRNA but rather as a structural component of the chromatin.

The function of the high-molecular stable nuclear RNA is as yet unknown. Some speculations as to the first of the two categories of stable RNA will be advanced in Section V,B.

Globin mRNA Precursors

Of the genes expressed in erythroid cells only the globin ones are being successfully studied. This is largely due to the possibility of using cDNA transcribed from globin mRNA for structural analysis and for hybridization. The available data are mostly based on a cDNA mixture transcribed from α and β globin mRNAs; cloned, i.e., individual cDNAs have only recently become accessible.

The basic idea that underlies the present-day concepts as to the molecular mechanisms of the expression of globin (and other) genes was advanced a long time ago as the pre-mRNA hypothesis. Over the last 10 years researchers have been striving to confirm it experimentally (see for revue Scherrer *et al.*, 1979). The globin system has served as the basic model for testing the hypothesis, and the history of these studies demonstrates the utmost complexity of the problem and warns one against jumping to conclusions as to the existence and the parameters of other pre-mRNAs.

Two experimental approaches are used at the moment to detect RNA molecules carrying globin-coding sequences in the nucleus or the cytoplasm. The first consists in hybridizing trace amounts of highly labeled cDNA with a large excess of cold RNA and estimating the concentration of globin sequences from the kinetics of hybridization. The main difficulty here is that the concentration in question is exceedingly low, virtually on the brink of the method's sensitivity, while the use of a very large RNA excess might involve contamination with mature globin mRNA through its aggregation with other RNA kinds. Therefore it was essential to such an experiment that the preparations should be proved free of aggregated admixtures of globin mRNA. The second approach consists in affinity chromatography of labeled RNA on cDNA-cellulose columns. So far this method has been used to look for pre-mRNA within pulse-labeled hnRNA of erythroid cells, i.e., among newly formed RNA. The two methods have got unequal capabilities, which must have been one of the reasons for the discrepant results (see Scherrer *et al.*, 1979).

Imaizumi *et al.* (1973) were the first to hybridize globin cDNA with various fractions of duck hnRNA obtained by centrifugation in sucrose gradient containing 99% DMSO and to report the presence of globin sequences not only within the low-molecular fractions but within those with a molecular weight of 5×10^6. Later a large series of studies was published based on the chromatography of pulse-labeled hnRNA (mainly from mammalian erythroid cells) on cDNA-cellulose. The size of the "pre-mRNA" detected in these studies did not exceed 16 S (14–16 S).

Now the existence of ~15 S (1500 nucleotides) precursors of globin mRNA in the nuclei of mammalian and avian erythroid cells is certain (see Scherrer *et al.*, 1979). It includes copies from two introns, the first consisting of about 100 base pairs and located at the 5′ region and the second 800 base pairs, at the 3′ end region. If we assume that the DNA fragments coding for 15 S pre-mRNA is the entire globin gene, it would mean that the globin gene has no other (out of mRNA) sequences but introns and exons. There is extensive evidence to the effect that this is not the case, at least in birds. After the above-cited study by Imaizumi and coworkers (1973) other reliable data have been reported indicating that the primary precursor of globin mRNA has a far larger size than 16 S. Williamson and Tobin (1977) investigated the matter in blood erythroid cells of 5.7- and 11-day-old chick embryos and adult anemized chicken. The authors found 28 S hnRNA (1.6×10^6 daltons) contained globin mRNA sequences, thereby confirming the data of Scherrer's group. Niessing (1978) fractionated hnRNA from immature blood forms of anemized ducks by chromatography on cross-linked sepharose in pure formamide at 46°C, i.e., under conditions precluding RNA aggregation. Two kinds of globin pre-mRNA were detected: 14–18 S and 23–28 S. Only the smaller precursor contained long poly(A). The author draws very cautious conclusions: he states that the 14–16 S component may be the real precursor of globin mRNA, while the existence of the 26–28 S precursor still has to be checked. It is probable that globin pre-mRNAs larger than 16 S are easily degraded during manipulations. This possibility is supported by the results of Agliano *et al.* (1979): only after the nuclei of duck erythroblasts were treated with aminomethyl trioxalene and UV irradiated, did it become possible to detect globin pre-mRNA with a high sedimentation coefficient (about 23 S). Precursors of globin mRNA sized ~28 S have been detected in nuclear RNA extracted from bone marrow and peripheral blood cells of the pigeon (Gasaryan *et al.*, 1980). The most conclusive data on the existence of pre-mRNA sized over 16 S was obtained in Scherrer's group. According to these data (Fig. 6), the primary transcript from the globin gene is ~15,000 nucleotides long, i.e., 10 times larger than the 15 S precursor (see Scherrer *et al.*, 1979; Reynaud *et al.*, 1980). Precursors of that size have not yet been reported by other authors. It should be borne in mind that pre-mRNA sized >15 S was practically always detected by the method of hybridization in solution, i.e., when a large RNA excess was involved. This is important if

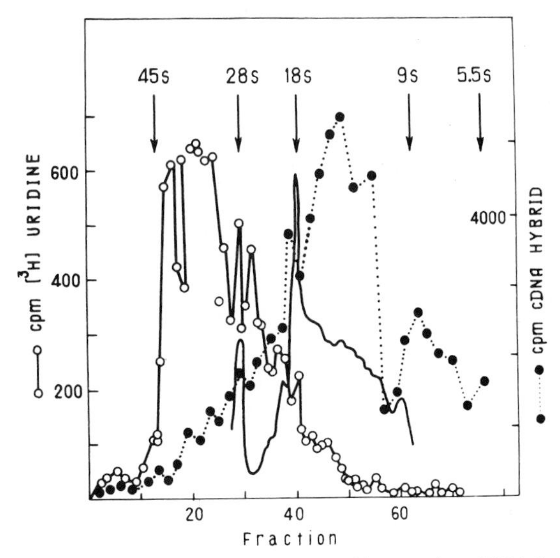

FIG. 6. Globin mRNA sequences in duck erythroblast nuclear RNA fractionated by gel electrophoresis in formamide at high temperature. Nuclear RNA from cells labeled 40 minutes with [^{3}H]uridine fractionated by electrophoresis in absolute formamide at 46°C (○); hybridization of the fractions to globin cDNA (●). (From Imaizumi, see Scherrer *et al.*, 1979.)

one recalls that the number of RNA molecules containing globin sequences is reportedly fewer the higher their molecular weight: 10,000 copies in the fraction of 10–15 S, 1–10 copies in fraction of 23–28 S, and ~1 copy in the 40–50 S fraction (Reynaud *et al.*, 1980).

Figure 7 shows a probable scheme of posttranscriptional stages in the formation of globin mRNA (Scherrer *et al.*, 1979). It is based on the supposition that the globin gene transcription unit is very long, about 12,000 base pairs. Within this "giant" gene ("transcripton") there is a section 1500–1600 base pairs long which includes three exons separated by two introns. The precursors have hairpins at the boundaries of these sections; during processing breaks occur in the region of the hairpins. The primary transcript is not polyadenylated and probably not yet methylated but contains "caps." It is unstable, and it gets transformed, as a result of processing, into a more stable polyadenylated and probably methylated precursor sized 1500–1600 nucleotides. The subsequent transformation of the precursor into mature mRNA (splicing) occurs in two stages: first the smaller intron's RNA copy is split out, then the RNA of the larger intron is split out and a mature globin mRNA is formed, as shown in rabbit (Grosveld *et al.*, 1981).

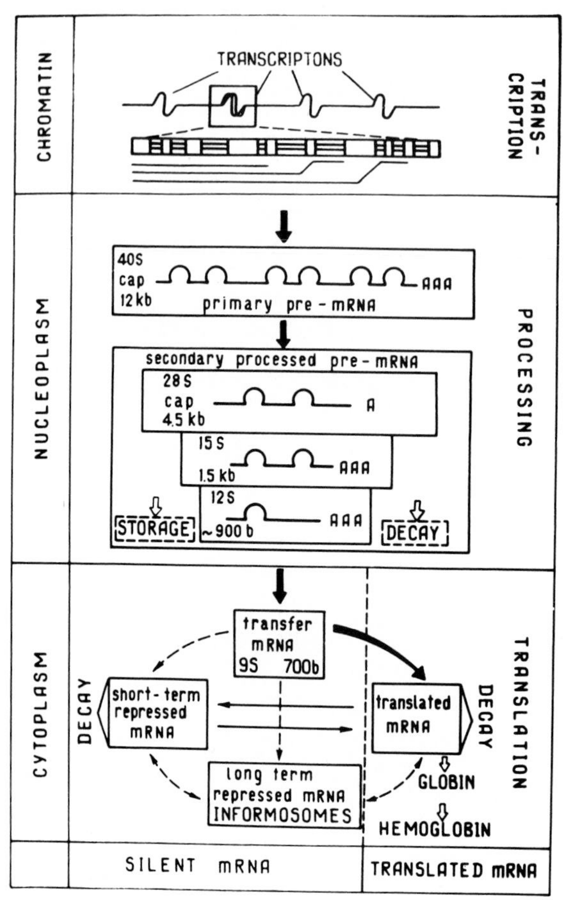

FIG. 7. A tentative flow diagram of globin messenger RNA formation and expression (explanation in the text). For details see Scherrer *et al.* (1979).

V. Possible Mechanisms Regulating Expression of the Genes in Differentiating Avian Erythroid Cells

In what way does an erythroid cell, in which tens of thousands of genes are expressed, accumulate an enormous amount of hemoglobin and gradually stop the synthesis of other proteins, having embarked upon terminal differentiation?

It has recently been shown (Sell *et al.*, 1979; Davidson and Britten, 1979) that the spectrum and amount of proteins synthesized in a cell are controlled by the mechanisms which condition the spectrum and amount of various mRNAs in the polysomes.

For the erythroid system the problem consists in assessing the relative role of transcriptional and posttranscriptional levels of regulation in conditioning the mRNA spectrum in the polysomes of the cells actively accumulating hemoglobin at intermediate stages of differentiation.

A. Transcriptional Level

At present it is impossible to assess with certainty whether there is a clear-cut qualitative differences between the sequences transcribed in various tissues (see Davidson and Britten, 1979; Sell *et al.*, 1979; Tobin, 1979). Even with regard to specialized genes there are contradictory data as to whether or not their transcription is confined only to the tissues in which they are normally expressed. Humphries *et al.* (1976) found the globin gene to be transcribed, albeit not very actively, in nonerythroid cells, whereas other authors arrived at the opposite result: no activity of globin genes was detected in oviduct (Bellard *et al.*, 1977, 1980) and the ovalbumin gene was found to be silent in the spleen and the liver (Ono and Getz, 1980) and in erythroid cells (Bellard *et al.*, 1977, 1980). Meanwhile, some difference between tissues was observed: more single-copy sequences are transcribed in cerebral cells than in the liver or kidneys (Chikaraishi *et al.*, 1978).

Erythroid system (and possibly other systems subject to terminal differentiation) is a patent case of gradual limitation of genome transcription. Unfortunately so far no one has reliably determined the complexity of the genome regions transcribed at the various stages of erythroid differentiation. Therefore the change in transcription ability in the cells has to be assessed on the basis of total RNA synthesis measurements (see Section IV,B) and the early evidence that the chromatin isolated from mature erythrocytes exhibits a lower template activity in transcription systems with bacterial RNA polymerase than the chromatin of immature erythroid cells (Gasaryan and Andreeva, 1972; Gasaryan *et al.*, 1973) or other active tissues (Seligy and Miyagi, 1969).

The mechanism underlying the limitation of transcription in erythroid cells, the extent to which this mechanism is locus-specific, and its relation to the formation of the polysomal mRNA spectrum in hemoglobin-accumulating cells are so far a complete blank. No site-specific factors capable of selectively inactivating the genes are found among structural components of the erythroid nuclei.

The suggestion that the H5 histone may be a factor which causes the additional condensation of the chromatin and inactivation of the genome in erythroid cells (Seligy and Neelin, 1970; Brasch *et al.*, 1971; Gasaryan and Andreeva, 1972; Gasaryan *et al.*, 1976a, 1978b) needs for confirmation data obtained using modern approaches. It seems, however, that H5 histone solely is hardly capable of selectively inactivating the genes. Besides, it is far from clear whether the massive condensation of the chromatin by the H5 histone is the primary cause of gene inactivation (Brasch *et al.*, 1971). Other structural changes (metilation, accumulation, or a loss of some specific nonhistone proteins) occurring in the chromatin may more selectively affect the conformation and the functional state of its various regions. Preparations of nonhistone proteins added to the chromatin increase the transcription of the globin genes by bacterial RNA polymerase (Chin *et al.*, 1975; Gadski *et al.*, 1978) and now a tentative search is in progress for proteins that can specifically affect the structure and function of globin genes (see, e.g., Ross *et al.*, 1979).

It has been reported that the globin (as well as ovalbumin) genes may differ in the degree of methylation (Van der Ploeg and Flavell, 1980), which might also be a structural basis for their different functional states.

One of the reasons for the decreasing level of transcription in a differentiating erythroid cell might be a drop in its RNA polymerase content. Indeed, the endogenous RNA polymerase activity is lower in isolated nuclei of mature forms than in immature ones (Appels and Williams, 1970; Gasaryan *et al.*, 1973; Scheintaub and Fiel, 1973). The activity of nucleolar RNA polymerase is still lower or practically nonexistent in reticulocytes, which correlates with the cessation of rRNA synthesis at this stage and the disappearance of the granuler component from the nucleolus (see above). Mature erythrocytes have no RNA polymerase A but keep an appreciable amount of chromatin-bound polymerases correlating with a certain level of hnRNA synthesis.

So although there is some imperfect correlation between the falling level of transcription *in vivo* and falling RNA polymerase activity in erythroid cells, it is not clear whether the change in polymerases is the cause of the transcription inactivation. It is quite possible that the activity and the amount of RNA polymerases fall as a secondary phenomenon because they are no longer needed and not as a transcription-limiting factor. Really, with regard to other eukaryotic cells RNA polymerase has not been suggested as a limiting factor (Roeder, 1976).

Anyway, genome transcription does get limited in erythroid cells, but whether this limitation ensures the timely exclusion of the products of thousands of genes from the polysomes at the intermediate stages of differentiation is far from clear.

B. Posttranscriptional Regulation

There is extensive, though mostly indirect, evidence to the effect that gene expression in higher organisms is largely controlled at the posttranscriptional level. Much of this evidence has been obtained in avian erythroid cells (Scherrer and Marcaud, 1968; Scherrer, 1973; Maundrel *et al.*, 1979; Tobin, 1979). Scherrer and Marcaud (1968) have advanced a concept of "cascade regulation," which proved to be a fruitful model of posttranscriptional regulation at the level of intranuclear selection and transfer of information from the genome to polysomes. Unfortunately there is no accurate and reliable information to date as to what kinds of mRNA and in what proportion come out of the nucleus into the cytoplasm in an erythroid cell. One problem that is now being experimentally tackled consists in using hybridization with cDNAs to evaluate the spectrum and proportion of various kinds of mRNA (*inter alia*, globin and nonglobin ones) in the nucleus and polysomes of immature cells (Lasky and Tobin, 1979; Knöchel *et al.*, 1979), i.e., at a stage when globin mRNAs are accumulated in the cytoplasm and most nonglobin ones disappear from it. The results of these, rare enough, studies are contradictory and do not yet allow any definite conclusions as to the level (transcriptional or posttranscriptional) at which the spectrum and proportion of mRNAs getting into the cytoplasm are formed.

Selection is also possible at the subsequent stages when the matured mRNA transferred to the polysomes; however, the matter remains to be elucidated in view of the obscure and apparently complex role of extrapolysomal mRNP. For instance, free mRNP particles contain both globin and nonglobin mRNA but in a different ratio than within the polysomes: globin mRNAs constitute 50–75% of polysomal mRNA and only 30% of the mRNA of free mRNP particles. However, the proportion is highly variable and the amount of globin mRNAs in free mRNP particles changes from 10 to 45% in the course of development (see Maundrel *et al.*, 1979).

The differential stability of mRNAs in the cytoplasm might be also important (see Kafatos, 1972): the lifetimes of globin and nonglobin mRNAs were shown to be different (Tobin, 1979).

Thus, it is not yet clear in what way the polysomal pattern of the mRNA is formed and maintained at every moment of erythroid development; even though there are very interesting observations and ideas concerning the problem sound evidence is still to be obtained. In this context one cannot bypass the interesting fact that the nuclei of avian erythroid cells contain considerable amounts of stable hnRNA with a property of pre-mRNA (see above). To explain this observation it can be supposed that the task of enriching the polysomes with globin (and with particular nonglobin) mRNAs (see, e.g., Thiele *et al.*, 1979) is fulfilled through the action of the factors discriminating posttranscriptionally the efficiency of the expression of nonglobin genes both in the nucleus and in the cytoplasm: in the cytoplasm the mRNAs to be eliminated are made less stable than the accumulating ones (see Kafatos, 1972; Tobin, 1979) whereas in the nucleus the precursors of these mRNAs are stabilized so as to prevent (or delay) their mRNA from maturing and getting into the cytoplasm. The outcome is the same: the polysomal concentration of nonglobin mRNAs decreases.

VI. Concluding Remarks

In spite of the great advantages of the erythroid model and similar ones for studying the regulation of gene activity in eukaryotes, specifically the mechanisms that change the composition of polysomal mRNA in the course of cell differentiation, we are still very far from solving these problems. One may expect genetic engineering to accelerate the progress of these studies: the cloning technique will be used to isolate the nonglobin genes expressed at early stages of erythroid differentiation, so as to follow their gradual inactivation by testing the nuclear and cytoplasmic RNAs of that and later stages.

Another possibility lies in further progress in the approaches to selective hydrolysis of active chromatin genes by nucleases. This technique might well serve to characterize not only the globin genes but many others in the chromatin of erythroid cells at different stages of differentiation. Besides, a combination of this technique with genetic engineering seems to promise insight into the conformational changes of intragenic regions and will hopefully provide a way to identify the protein components of these regions. These so far undeciphered conformational changes occurring in early embryogenesis may reflect the processes of the predetermination of stem cells which make the genes competent to the subsequent action of differentiation effectors. The method might make it possible to reveal structural changes in the globin and nonglobin genes in erythroid cells involved in the molecular mechanisms of their functioning, particularly the process of hemoglobin switching during embryogenesis.

One more final remark. As molecular biology grows more sensitive in tackling these problems, it is all the more important to link them up with the fine details of cell biological processes: the cell cycle, the change in morphology of the nucleus and the cytoplasm, and the specific developmental route characteristic of each type of erythropoiesis.

ACKNOWLEDGMENTS

The author is indebted to Dr. Tarantul V. Z., Dr. Stvolinskaya N. S., and Dr. Andreeva N. B. for stimulating discussions, to Dr. Stvolinskaya N. S. for providing originals of the microphotographs of Fig. 3, to Dr. Dubovaya V. I. for providing the data for Table III, and to Mrs. Chernorotova I. N. for technical assistance.

References

Adams, G. H. M., Vidali, G., and Neelin, J. M. (1970). *Can. J. Biochem.* **48,** 33–37.
Agliano, A. M., Nacci, A., Reymond, Ch., Appleby, D., and Spohr, G. (1979). *Eur. J. Biochem.* **95,** 203–213.
Allan, J., Staynov, D. Z., and Gould, H. (1980). *Proc. Natl. Acad. Sci. U.S.A.* **77,** 885–889.
Appels, R., and Williams, A. F. (1970). *Biochim. Biophys. Acta* **217,** 531–534.
Appels, R., Wells, J. R. E., and Williams, A. F. (1972). *J. Cell Sci.* **10,** 47–59.
Arthur, R. R., and Straus, N. A. (1978). *Can. J. Biochem.* **56,** 257–263.
Attardi, G., Parnas, H., Hwang, M.-I. H., and Attardi, B. (1966). *J. Mol. Biol.* **20,** 145–182.
Attardi, G., Parnas, H., and Attardi, B. (1970). *Exp. Cell Res.* **62,** 11–31.
Axel, R., Cedar, H., and Felsenfeld, G. (1973). *Proc. Natl. Acad. Sci. U.S.A.* **70,** 2029–2032.
Barrett, I., Marianka, D., Hamlyn, P. H., and Gould, H. J. (1974). *Proc. Natl. Acad. Sci. U.S.A.* **71,** 5057–5061.
Bellard, M., Gannon, F., and Chambon, F. (1977). *Cold Spring Harbor Symp. Quant. Biol.* **42,** 779–791.
Bellard, M., Kuo, M. T., Dretzen, G., and Chambon, P. (1980). *Nucleic Acids Res.* **8,** 2737–2750.
Berkowitz, E. M., and Doty, P. (1975). *Proc. Natl. Acad. Sci. U.S.A.* **72,** 3328–3332.
Bernhard, W. (1969). *J. Ultrastruct. Res.* **27,** 250–265.
Billett, M. A., and Hindley, J. (1972). *Eur. J. Biochem.* **28,** 451–462.
Bishop, J. O., Pemberton, R., and Baglioni, C. (1972). *Nature (London) New Biol.* **235,** 231–234.
Brasch, K., Seligy, V. L., and Setterfield, G. (1971). *Exp. Cell Res.* **65,** 61–72.
Briand, G., Kmiecik, D., Sautiere, P., Wouters, D., Borie-Loy, O., Biserte, G., Mazen, A., and Champagne, M. (1980). *FEBS Lett.* **112,** 147–151.
Brown, I. R., and Ingram, V. M. (1974). *J. Biol. Chem.* **249,** 3960–3972.
Bruns, G. A. P., and Ingram, V. M. (1973). *Philos. Trans. R. Soc. London Ser. B* **266,** N 877, 225–305.
Cameron, J. L., and Prescott, D. M. (1963). *Exp. Cell Res.* **30,** 609–612.
Chambon, P. (1977). *Cold Spring Harbor Symp. Quant. Biol.* **42,** 1209–1234.
Chapman, B. S., and Tobin, A. J. (1979). *Dev. Biol.* **69,** 375–387.
Chikaraishi, D. M., Deeb, S. S., and Sueoka, N. (1978). *Cell* **13,** 111–120.
Chin, J.-F., Tsai, Y.-H., Sakuma, K., and Hnilica, L. S. (1975). *J. Biol. Chem.* **250,** 9431–9433.
Chui, D., Brotherton, T., and Gauldie, J. (1979). *In* "Cellular and Molecular Regulation of Hemoglobin Switching," pp. 213–225. Grune Sttatton, New York.
Cortadas, J., Olofson, B., Meunier-Rotival, M., Macaya, G., and Bernardi, G. (1979). *Eur. J. Biochem.* **99,** 179–186.
Cummings, J. W., Liu, A. V., and Salser, W. A. (1978). *Nature (London)* **276,** 418–419.
Davidson, E. H., and Britten, R. J. (1979). *Science* **204,** 1052–1059.
Davidson, E. H., Galau, G. A., Angerer, R. C., and Britten, R. J. (1975). *Chromosoma* **51,** 253–259.
Dawid, I. B., and Wahli, W. (1979). *Dev. Biol.* **69,** 305–328.
Dick, C., and Johns, E. W. (1969). *Biochim. Biophys. Acta* **175,** 414–419.
Dodgson, J. B., Strommer, J., and Engel, J. D. (1979). *Cell* **17,** 879–887.
Eden, F. C., and Hendrick, J. P. (1978). *Biochemistry* **17,** 5838–5844.
Eden, F. C., Hendrick, J. P., and Gottlieb, S. S. (1978). *Biochemistry* **17,** 5113–5127.
Engel, J. D., and Dodgson, J. B. (1978). *J. Biol. Chem.* **253,** 8239–8246.
Epplen, J. T., Leipoldt, M., Engel, W., and Schmidtke, J. (1978). *Chromosoma* **69,** 307–321.
Epplen, J. T., Diedrich, U., Wagenmann, M., Schmidtke, J., and Engel, W. (1979). *Chromosoma* **75,** 199–214.
Felsenfeld, G. (1978). *Nature (London)* **271,** 115–122.
Fodor, E. J. B., and Doty, P. (1977). *Biochem. Biophys. Res. Commun.* **77,** 1478–1485.
Gadski, R. A., and Chi-Bom Chae (1978). *Biochemistry* **17,** 869–874.
Garel, A., Zalan, M., and Axel, R. (1977). *Proc. Natl. Acad. Sci. U.S.A.* **74,** 4867–4871.

Gasaryan, K. G. (1977). *Acta Biol. Med. Germ.* **35,** 295–303.
Gasaryan, K. G., and Andreeva, N. B. (1972). *FEBS Lett.* **27,** 263–266.
Gasaryan, K. G., and Kulminskaya, A. S. (1975). *Ontogenez (USSR)* **6,** 31–38.
Gasaryan, K. G., and Stvolinskaya, N. S. (1979). *Mol. Biol. Rep.* **5,** 233–236.
Gasaryan, K. G., Kirjanov, G. I., and Kulminskaya, A. S. (1970). *Biochimia (USSR)* **35,** 1238–1249.
Gasaryan, K. G., Kulminskaya, A. S., Ananjanz, T. G., and Kirjanov, G. I. (1971). *Ontogenez (USSR)* **2,** 263–275.
Gasaryan, K. G., Ananjanz, T. G., Fedina, A. B., and Andreeva N. B. (1973). *Mol. Biol. (USSR)* **7,** 73–83.
Gasaryan, K. G., Tarantul, V. Z., Baranov, Yu. N., and Lipasova, V. A. (1974). *Mol. Biol. (USSR)* **8,** 372–379.
Gasaryan, K. G., Andreeva, N. B., and Penkina, V. I. (1976a). *Differentiation* **5,** 21–28.
Gasaryan, K. G., Tarantul, V. Z., and Baranov, Yu. N. (1976b). *Differentiation* **6,** 41–46.
Gasaryan, K. G., Kuznetsova, E. D., Fetisova, I. V., and Tarantul, V. Z. (1977a). *FEBS Lett.* **77,** 251–254.
Gasaryan, K. G., Kuznetsova, E. D., Tarantul, V. Z., and Sivak, S. A. (1977b). *Chromosoma* **61,** 381–394.
Gasaryan, K. G., Tarantul, V. Z., and Lipasova, V. A. (1977c). *Mol. Biol. Rep.* **3,** 213–215.
Gasaryan, K. G., Kuznetsova, E. D., and Tarantul, V. Z. (1978a). *FEBS Lett.* **94,** 136–138.
Gasaryan, K. G., Vishnevskaya, T. Yu., and Andreeva, N. B. (1978b). *Differentiation* **10,** 123–127.
Gasaryan, K. G., Kuznetsova, E. D., Fetisova, I. V., and Tarantul, V. Z. (1979a). *Mol. Biol. (USSR)* **13,** 761–768.
Gasaryan, K. G., Tarantul, V. Z., Kuznetsova, E. D., Stvolinskaya, N. S., Nickolaev, A. I., Fetisova, I. V., Lipasova V. A., and Popov, L. S. (1979b). *Mol. Biol. (USSR)* **13,** 16–29.
Gasaryan, K. G., Dubovaya, V. I., Neznanov, N. S., Tarantul, V. Z., Skobeleva, N. A., and Frolova, L. Y. (1980). *Mol. Biol. (USSR)* **14,** 766–772.
Gasaryan, K. G., Goltsov, V. A., Kuznetsova, E. D., Tarantul, V. Z., and Popov, L. S. (1982). *Biochimia (USSR)* **47,** 71–80.
Ginder, G. D., Wood, W. J., and Felsenfeld, G. (1979). *J. Biol. Chem.* **254,** 8099–8102.
Grasso, J. A., and Woodard, J. W. (1966). *J. Cell Biol.* **31,** 279–294.
Grosveld, G. C., Koster, A., and Flavell, R. A. (1981). *Cell* **23,** 573–584.
Hughes, S. H., Stubblefield, E., Payvar, P., Engel, J. D., Dodgson, J. B., Spector, D., Cordell, B., Schimke, R. T., and Varmus, H. E. (1979). *Proc. Natl. Acad. Sci. U.S.A.* **76,** 1348–1353.
Humphries, S., Windass, J., and Williamson, R. (1976). *Cell* **7,** 267–277.
Imaizumi, M. T., Diggelmann, H., and Scherrer, K. (1973). *Proc. Natl. Acad. Sci. U.S.A.* **70,** 1122–1126.
Ingram, V. M. (1972). *Nature (London)* **235,** 338–339.
Jemenez, E. S., Gonzales, J. L., Dominguez, J. C., and Saloma, E. S. (1974). *Eur. J. Biochem.* **45,** 25–29.
Kabat, D. (1972). *Science* **175,** 134–140.
Kafatos, F. C. (1972). *Acta Endocrinol. Suppl.* **168,** 319–345.
Knöchel, W. (1975). *Biochim. Biophys. Acta* **395,** 501–508.
Knöchel, W., Render, T., and Grundmann, U. (1979). *Mol. Biol. Rep.* **5,** 161–164.
Kolodny, G. M., and Rosenthal, L. J. (1973). *Exp. Cell Res.* **83,** 429–433.
Kornberg, R. D. (1977). *Annu. Rev. Biochem.* **46,** 931–954.
Korvin-Pavlovskaya, E. G., Karalova, E. M., Kulminskaya, A. S., Magakyan, Yu. A., and Gasaryan, K. G. (1978). *Tsitologia (USSR)* **20,** 1016–1026.
Korvin-Pavlovskaya, E. G., Kulminskaya, A. S., Karalova, E. M., Magakjan, Yu. A., and Gasaryan, K. G. (1980). *Tsitologia (USSR)* **22,** 1413–1422.

Kulminskaya, A. S., Brodsky, V. Ya., and Gasaryan, K. G. (1978). *Ontogenez (USSR)* **9,** 601–608.
Kurochkin, S. N., Andreeva, N. B., Gasaryan, K. G., and Severin, E. S. (1977). *FEBS Lett.* **76,** 112–114.
Lasky, L., and Tobin, A. J. (1979). *Biochemistry* **18,** 1594–1598.
Lasky, L., Nozick, N. D., and Tobin, A. J. (1978). *Dev. Biol.* **67,** 23–39.
Levy, W. B., and Dixon, G. (1977). *Nucleic Acids Res.* **4,** 883–898.
Lewin, B. (1975). *Cell* **4,** 11–20.
Likowsky, Z., and Smetana, K. (1975). *Cell Tissue Res.* **162,** 271–277.
McGhee, J. D., and Ginder, G. D. (1979). *Nature (London)* **280,** 419–420.
Mahoney, K. A., Hyer, B. Y., and Chan, L. (1977). *Dev. Biol.* **56,** 412–416.
Man, N. T., and Cole, R. J. (1974). *Exp. Cell Res.* **83,** 328–334.
Maundrell, K., Maxwell, E. S., Civelli, O., Vincent, A., Goldenberg, S., Buri, J.-F., Imaizumi-Scherrer, M.-T., and Scherrer, K. (1979). *Mol. Biol. Rep.* **5,** 43–51.
Modak, S. P., Commelin, D., Grosset, L., Imaizumi, M. T., Monnal, M., and Scherrer, K. (1975). *Eur. J. Biochem.* **60,** 407–421.
Molgaard, H. V., Perucho, M., and Ruiz-Carrillo, A. (1980). *Nature (London)* **283,** 502–504.
Niessing, J. (1978). *Eur. J. Biochem.* **91,** 587–598.
Noll, M., and Kornberg, R. D. (1977). *J. Mol. Biol.* **109,** 393–404.
Ono, T., and Getz, M. J. (1980). *Dev. Biol.* **75,** 481–484.
Packman, S., Aviv, H., Ross, J., and Zeder, P. (1972). *Biochem. Biophys. Res. Commun.* **49,** 813–819.
Papies, B., Gross, Y., Coutelle, Ch., and Rosenthal, S. (1973). *Acta Biol. Med. Germ.* **31,** 543–560.
Puvion, E., and Bernhard, W. (1975). *J. Cell Biol.* **67,** 200–214.
Reynaud, C. A., Imaizumi-Scherrer, M. T., and Scherrer, K. (1980). *J. Mol. Biol.* **140,** 481–504.
Richards, R. I., Shine, J., Ullrich, A., Wells, J. R. E., and Goodman, H. M. (1979). *Nucleic Acids Res.* **7,** 1137–1146.
Ringerts, N. R., and Bolund, L. (1974). *In* "The Cell Nucleus" (H. Bush, ed.), Vol. III, pp. 417–446. Academic Press, New York.
Roeder, R. G. (1976). *In* "RNA-Polymerase" pp. 285–329. Cold Spring Harbor Lab., Cold Spring Harbor, New York.
Ross, D. A., Jackson, J. B., and Chae Chi-Bom (1979). *J. Cell Biol.* **83,** 146a.
Ruiz-Carrillo, A., Wangh, L. Y., and Allfrey, V. G. (1976). *Arch. Biochem. Biophys.* **174,** 273–290.
Scheintaub, H. M., and Fiel, R. J. (1973). *Exp. Cell Res.* **80,** 442–445.
Scherrer, K. (1973). *In* "Strategies for the Control of Gene Expression, 18th ONOZO Biological Conference" (A. Kohn and M. Revel, eds.), pp. 169–219.
Scherrer, K., and Marcaud, L. (1965). *Bull. Soc. Chim. Biol.* **47,** 1697–1705.
Scherrer, K., and Marcaud, L. (1968). *J. Cell. Physiol.* **72,** (Suppl.1), 181–212.
Scherrer, K., Marcaud, L., Zajdela, F., London, J. M., and Gross, F. (1966). *Proc. Natl. Acad. Sci. U.S.A.* **56,** 1571–1578.
Scherrer, K., Imaizumi-Scherrer, M. T., Reynaud, C., and Therwath. A. (1979). *Mol. Biol. Rep.* **5,** 5–28.
Schweiger, H. G. (1962). *Int. Rev. Cytol.* **13,** 135–201.
Seligy, V. L., and Miyagi, M. (1969). *Exp. Cell Res.* **58,** 27–34.
Seligy, V. L., and Neelin, J. M. (1970). *Biochim. Biophys. Acta* **213,** 380–390.
Sell, S., Thomas, K., Michalson, M., Salatrepat, J., and Bonner, J. (1979). *Biochim. Biophys. Acta* **564,** 173–178.
Shaw, B. R., Herman, T. M., Kovacic, R. T., Beaudreau, G. S., and Van Holde, K. E. (1976). *Proc. Natl. Acad. Sci. U.S.A.* **73,** 505–509.

Shimizu, K. (1976). *Dev. Biol.* **48,** 317–326.
Sinclair, G. P., and Brasch, K. (1975). *Rev. Can. Biol.* **34,** 287–303.
Small, J. V., and Davies, H. G. (1972). *Tissue Cell* **4,** 341–352.
Smetana, K., and Likovsky, Z. (1978). *Eur. J. Cell Biol.* **17,** 146–158.
Spohr, G., Imaizumi, M. T., and Scherrer, K. (1974). *Proc. Natl. Acad. Sci. U.S.A.* **71,** 5009–5013.
Spohr, G., Mirault, M.-E., Imaizumi, M. T., and Scherrer, K. (1976). *Eur. J. Biochem.* **62,** 313–322.
Stalder, J., Groudine, M., Engel, J. D., and Weintraub, H. (1980). *Cell* **19,** 973–980.
Steggles, A. W., Wilson, G. M., Kantor, J. A., Picciano, D. K., Falvey, A. K., and Anderson, W. F. (1974). *Proc. Natl. Acad. Sci. U.S.A.* **71,** 1219–1223.
Stvolinskaya, N. S., and Gasaryan, K. G. (1980a). *Mol. Biol. (USSR)* **14,** 1336–1342.
Stvolinskaya, N. S., and Gasaryan, K. G. (1980b). *Ontogenez (USSR)* **11,** 130–137.
Stvolinskaya, N. S., Princeva, O. Yu., and Gasaryan, K. G. (1981). *Tsitologia (USSR)* **23,** 254–257.
Sung, M. T. (1977). *Biochemistry* **16,** 286–290.
Sung, M. T., and Freedlender, E. F. (1978). *Biochemistry* **17,** 1884–1890.
Tarantul, V. Z., Baranov, Yu. N., Prima, V. I., Lipasova, V. A., and Gasaryan, K. G. (1973). *Mol. Biol. (USSR)* **7,** 849–858.
Tarantul, V. Z., Lipasova, V. A., Baranov, Yu. N., and Gasaryan, K. G. (1974). *Mol. Biol. (USSR)* **8,** 864–870.
Thiele, B. J., Belkner, J., Andree, H., Rapoport, T. A., and Rapoport, S. M. (1979). *Eur. J. Biochem.* **96,** 563–569.
Tobin, A. J. (1979). *Dev. Biol.* **68,** 47–58.
Tobin, A. J., Colot, H. V., Kao, J., Pine, K. S., Portnoff, S., Zagris, N. N., and Zarin, N. (1976). *In* "Eukaryotes at the Subcellular Level" (J. A. Last, ed.), pp. 211–255. Dekker, New York.
Tobin, A. J., Selvig, S. E., and Lasky, L. (1978). *Dev. Biol.* **67,** 11–22.
Tobin, R. S., and Seligy, V. L. (1975). *J. Biol. Chem.* **250,** 358–364.
Tooze, J., and Davies, H. G. (1967). *J. Cell Sci.* **2,** 617–638.
Van der Ploeg, L. H. T., and Flavell, R. A. (1980). *Cell* **19,** 947–958.
Villeponteaux, B., Lasky, L., and Harary, J. (1978). *Biochemistry* **17,** 5532–5536.
Weintraub, H. (1975). *In* "Results and Problems in Cell Differentiation" (J. Reinnart and H. Holtzer, eds.), Vol. VII. Springer-Verlag, Berlin and New York.
Weintraub, H. (1978). *Nucleic Acids Res.* **5,** 1179–1188.
Weintraub, H., and Groudine, M. (1976). *Science* **193,** 848–856.
Whitlock, J. P., Jr., and Simpson, R. T. (1976). *Biochemistry* **15,** 3307–3314.
Williams, A. F. (1972). *J. Cell Sci.* **11,** 771–776.
Williamson, P. L., and Tobin, A. J. (1977). *Biochim. Biophys. Acta* **475,** 366–382.
Yaguchi, M., Roy, C., and Seligy, V. L. (1979). *Biochem. Biophys. Res. Commun.* **90,** 1400–1406.
Yataganas, X., Gahrton, G., and Thorell, B. O. (1970). *Exp. Cell Res.* **62,** 254–261.
Zanjani, E., McGlave, P., Bhakthanathsalan, A., and Stammatoyannopoulos, G. (1979). *Nature (London)* **280,** 495–496.
Zasloff, M., and Camerini-Otero, R. D. (1980). *Proc. Natl. Acad. Sci. U.S.A.* **77,** 1907–1911.
Zentgraf, H., Scheer, U., and Franke, W. W. (1975). *Exp. Cell Res.* **96,** 81–95.

NOTE ADDED IN PROOF: In the recent work of Engel's group [Dolan, M., Sugarman, B. J., Dodgson, J. B., and Engel, J. D. (1981). *Cell* **24,** 669–677] an unusual arrangement of the β-type globin genes is found in chickens: $5' \ldots \rho - \beta^{H} - \beta - \varepsilon \ldots 3'$. In all mammals examined so far β-type genes are arranged in order corresponding to their sequential activation in ontogenesis: embryonic–fetal–adult.

INTERNATIONAL REVIEW OF CYTOLOGY, VOL. 74

Morphological and Cytological Aspects of Algal Calcification

MICHAEL A. BOROWITZKA

Roche Research Institute of Marine Pharmacology, Dee Why, N.S.W. Australia

I. Introduction

The deposition of calcium salts, generally called calcification, is a common phenomenon in many algal orders (see Table 1 in Borowitzka, 1977) and also in other plants (Pentecost, 1980). It occurs in aerial, soil, freshwater, and marine environments. The deposits are usually of calcium carbonate and may occur either extra-, inter-, or intracellularly. The calcium carbonate is deposited as either the aragonite or the calcite crystal isomorph, and mixtures of the two isomorphs normally do not occur (Borowitzka, 1977). In recent years there have been many advances in our understanding of the physiology, biochemistry, and especially the morphology of algal calcification, although few testable models of the calcification mechanism have been proposed (see reviews by Littler, 1976; Borowitzka, 1977; Pentecost, 1980). Borowitzka (1982) has reviewed the existing models of algal calcification.

In order to understand the physical and biochemical processes participating in calcification it is necessary to appreciate the anatomy and cytology of the calcareous algae and the development of their $CaCO_3$ deposits. Such studies provide the framework for the physiological and biochemical studies and allow models for the calcification mechanism(s) of the algae to be

ISBN 0-12-364474-7

developed. The diversity of algal $CaCO_3$ deposits, both in organization and development, provides wide scope for comparative studies, and such studies are valuable in elucidating the calcification mechanisms. It must also be recognized that the calcareous algae cannot be viewed in isolation. Comparative studies with related noncalcareous algae are also important.

It has become clear that the regular deposition of calcium carbonate, as opposed to transient deposits which can occur under certain abnormal conditions in many algae (see for example Dodson, 1974), has evolved independently in the various algal taxa. It is therefore not surprising that different algae appear to have different mechanisms of deposition. Borowitzka *et al.* (1974) divided the algae into three groups based on the crystallography, localization, and organization of their $CaCO_3$ deposits. This general grouping will be used in this article.

II. Extracellular and Intercellular Deposits

A. Extracellular Deposits

The giant freshwater coenocytic green alga *Chara* is a good example of the simplest kind of extracellular $CaCO_3$ deposits. In *Chara corallina* [=*C. australis* var *nobilis* (Proctor, 1980)] crystals of calcite are deposited in bands on the surface of the cell wall of the cylindrical internodal cell (Borowitzka *et al.*, 1974). These bands correspond to the localized alkaline (pH 9.5–10.5) regions (Lucas and Smith, 1973; Lucas, 1979). The transition from a noncalcareous region to a calcareous one is gradual whereas the observed pH shifts are quite abrupt. The deposit is of discrete pyramidal to rectangular calcite crystals of 6–10 μm width and 6–27 μm length (Fig. 1). They show no preferred orientation with respect to the cell wall and have a foliated structure made up of minute laths >1.2 μm long, 0.4–0.7 μm wide, and about 0.3 μm thick with variable crystallographic faces. The laths are joined side to side to form sheets which tip at a slight angle to the interior of the crystal, and as a result the surface pattern resembles a tiled roof (Borowitzka *et al.*, 1974). These foliated crystals probably develop as a three-dimensional dendrite system like the foliated layer of the shell of bivalve molluscs (Kennedy *et al.*, 1969). There is no organic material associated with the crystals and in thin section they can be seen to be formed wholly outside the cell wall.

Calcification in *Chara* and *Nitella* results from the precipitation of $CaCO_3$ from high Ca^{2+} waters due to the localized OH^- efflux at certain areas of the internodal cell which raises the local CO_3^{2-} ion concentration leading

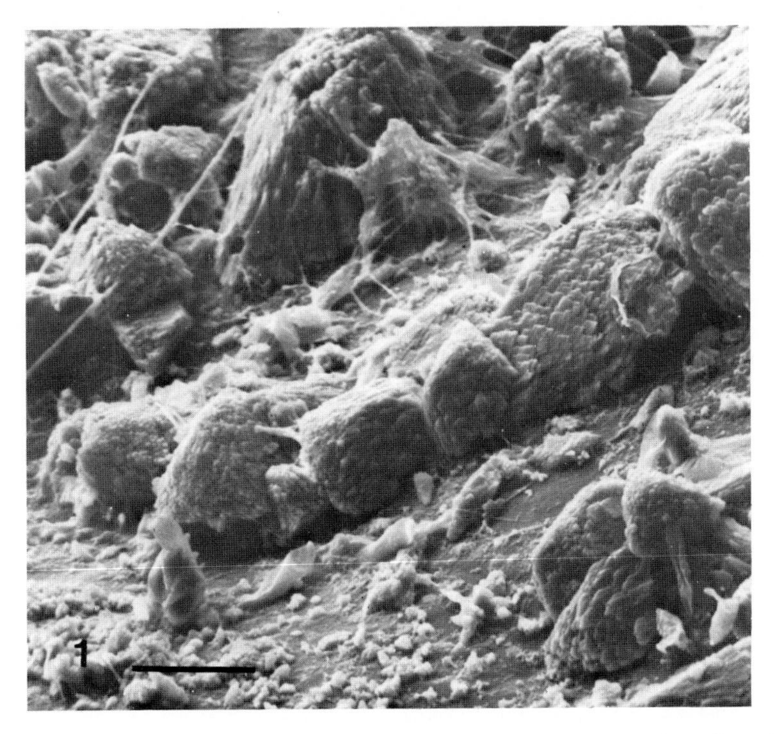

FIG. 1. *Chara Corallina*. Scanning electron micrograph of the calcite crystals on the surface of the internodal cell wall. The crystals show no organization with respect to the cell wall. Scale = 5 μm.

to $CaCO_3$ supersaturation (Spear *et al.*, 1969; Borowitzka, 1977; Lucas, 1979). The cytoplasmic ultrastructure of *Chara* has been studied by Chambers and Mercer (1964) and Pickett-Heaps (1967). There appears to be no cytological or morphological feature related to the acid/alkaline banding of the internodal cell surface. Barton (1965a) and Crawley (1965) observed a reticulate elaboration of the plasmalemma which Barton (1965b) called a charasome. It has been suggested that the charasome may be involved with the ionic fluxes responsible for the surface pH changes, however, the recent study of Franceschi and Lucas (1980) could not demonstrate any direct link between the occurrence of charasomes and the localized OH^- fluxes. Instead a role for the charasome in osmoregulation during rapid cell elongation has been proposed.

Localized OH^- efflux has also been demonstrated for the freshwater angiosperms *Potamogeton luceus* and *Elodea densa*. They take up bicarbonate ions at the lower leaf surface in the light. The carbon dioxide is assimilated and the excess hydroxyl ions are released from the upper leaf surface. In waters of sufficiently high calcium content, $CaCO_3$ is then precipitated on the leaf surface (Prins *et al.*, 1979, 1980). The leaf of *Elodea* is two cells thick, and that of *Potamogeton* three. The cells on the lower surface of the leaves of these plants appear to be transfer cells (Falk and Sitte, 1963; Pate and Gunning, 1972) as the plasmalemma surface area is greatly enlarged by extensive cell wall ingrowths. Transfer cells are usually associated with tissues where large-scale solute transport occurs. Prins *et al.* (1979) suggest that the plasmalemma in the wall ingrowth contains proton pumps, which are partially responsible for the uptake of HCO_3^- at this surface. Helder *et al.* (1980) also report that the upper cell wall of *Potamogeton lucens* is thicker and stains more densely than the lower leaf surface cell wall and thus provides a greater resistance to HCO_3^- diffusion. It remains to be seen whether these structural modifications are related to the polarized HCO_3^-/OH^- fluxes which result in the precipitation of $CaCO_3$ in these plants.

Calcification in the oogonial cells of characean algae may be both intracellular and extracellular (Horn af Rantzien, 1959; Daily, 1975). Dyck (1969) studied the development of these deposits in *Chara* by light and electron microscopy and showed that the $CaCO_3$ deposited in the cell walls of the oogonium had some degree of orientation similar to that of the wall fibrils. These deposits have been identified as calcite in *Chara globularis* (Horn af Rantzien, 1959). In *Nitellopsis obtusa* oogonia, deposition of the calcite starts along the adaxial walls of the spiral cells and lateral walls, finally filling the cell. Calcification ceases when the cell dies (Daily, 1973). No physiological or detailed ultrastructural studies of characean oogonial calcification have been carried out.

Surface encrustations of calcite also occur in the freshwater representatives of the genera *Chaetomorpha*, *Gongrosira*, and *Oedogonium* (Flajs, 1977a; Golubic and Fischer in Degens, 1976). These deposits, which consist of variously sized and randomly oriented calcite crystals, seem to be the direct result of localized pH changes caused by photosynthetic carbon dioxide uptake in waters highly saturated with calcium and carbonate.

In the marine environment the simplest type of extracellular calcification occurs in the brown alga *Padina* (Borowitzka *et al.*, 1974). The deposits of aragonite occur as fine needles in concentric bands on the surface of the alga (Miyata *et al.*, 1977). The crystals increase in density with thallus age and sometimes appear as "pillow-shaped" aggregates (Flajs, 1977a). Each "pillow" corresponds to an underlying cell. This "pillow" morphology may however be a result of the rather crude specimen preparation method used. In the older parts of the *Padina* thallus the aragonite crystals appear more granular or micritic (Borowitzka *et al.*, 1974; Miyata *et al.*, 1977). This material is probably formed by the dissolution and reprecipitation of the needle-like aragonite.

All the algae mentioned so far have one thing in common; calcium carbonate is precipitated on the surface of their cell walls due to surface pH changes. The surface alkalinization of the microenvironment near the cell wall is produced either by localized OH^- efflux or by shifts in the inorganic carbon equilibrium due to photosynthetic CO_2 uptake (see Appendix in Borowitzka and Larkum, 1976a; and Borowitzka, 1982). This phenomenon may be achieved by different ionic fluxes, but has the same general effect: $CaCO_3$ precipitation.

B. Intercellular Deposits

Intercellular deposits are a special case of extracellular deposits; the $CaCO_3$ is formed between cells but outside the cell wall. The best studied and largest group of calcareous algae with intercellular deposits of $CaCO_3$ are the marine algae of the orders Codiales (e.g., *Halimeda, Udotea*) and Dasycladales (e.g. *Neomeris*). The aragonite $CaCO_3$ of these algae is in the form of fine needles (Fig. 2) (Marszalek, 1971; Borowitzka *et al.*, 1974;

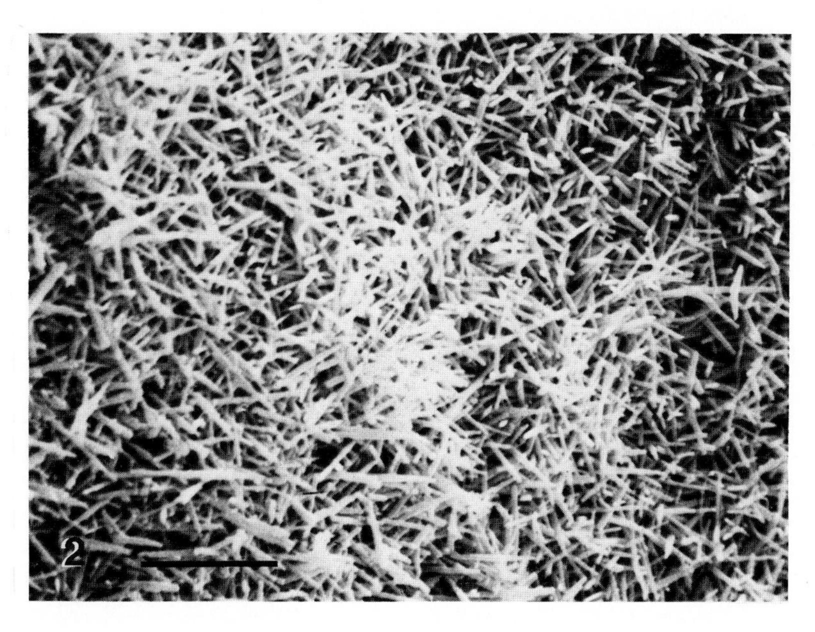

FIG. 2. *Neomeris annulata.* Scanning electron micrograph of aragonite needles filling the intercellular space. Scale = 5 μm.

Flajs, 1977a). In *Halimeda* the aragonite needles are formed within the intercellular spaces of the thallus. These spaces are isolated from the external medium by a layer of appressed utricles (Wilbur *et al.*, 1969; Borowitzka and Larkum, 1977). This means that, although the aragonite is wholly extracellular, it is still separated from the external seawater by a layer of cells. The aragonite needles have parallel sides for most of their length; they tend to taper slightly near the ends but rarely come to a point (Borowitzka *et al.*, 1974). They are 0.08–0.6 μm wide, up to 10 μm long, and less than 0.01 μm thick, depending upon species and probably also conditions of growth (Marszalek, 1971; Borowitzka *et al.*, 1974).

Calcification in *Halimeda cylindracea* begins after the filaments of the newly forming segment have begun to adhere to form the external utricular layer. Adhesion occurs by the fusion of the outer osmiophilic wall layers (cuticles) (Borowitzka and Larkum, 1977). The crystals are first deposited among the fibrils of the pilose outer layer of the walls facing into the intercellular space (Wilbur *et al.*, 1969; Borowitzka and Larkum, 1977). This outer layer stains with ruthenium red and some, but not all, of the fibrils are digested by the proteolytic enzymes pepsin and trypsin (Borowitzka and Larkum, 1977) suggesting that they are largely polysaccharide with some proteinaceous components. The small crystal nuclei rapidly enlarge from <10 nm diameter to needles up to 10 μm long, and soon fill the intercellular space. Although the needles have no specific orientation, the needle density is always highest near the cell wall (Borowitzka *et al.*, 1974; Flajs, 1977a). Secondary deposits of $CaCO_3$ on the surface of existing aragonite crystals have also been observed in older segments of *Halimeda tuna* (Borowitzka and Larkum, 1977).

There is some uncertainty whether the aragonite needles are associated with organic material. Borowitzka and Larkum (1977) found no evidence for an "organic sheath" around the aragonite needles of EDTA decalcified sections of *Halimeda* thalli. Nakahara and Bevelander (1978), however, found what appears to be organic material coating *Halimeda incrassata* aragonite needles. It is well known that aragonite and calcite both adsorb organic molecules, especially certain amino acids (Suess, 1970) and the material observed by Nakahara and Bevelander (1978) may be of this kind. It is however equally possible that this organic material is associated with the aragonite only during the early stage of deposition, disintegrating soon thereafter, as occurs in corals (Johnston, 1980). Since the presence or absence of organic material in association with the $CaCO_3$ is of significance to the possible mechanism of calcification, this question needs to be resolved.

Calcification in *Halimeda* is dependent on the anatomy of the thallus, especially the adhesion of the utricles which serve to isolate the intercellular (interutricular) space from the external medium, and the fact that CO_2 can be taken up from this intercellular space during photosynthesis (Borowitzka and Larkum, 1976a). It may be fortuitous that the chloroplasts reach structural, and presumably functional, maturity at about the time of utricle adhesion and the onset of calcification (Borowitzka and Larkum, 1977). On the other hand, both photosynthetic CO_2 uptake and the mature anatomy are required for $CaCO_3$ precipitation. Adhesion of the utricles in *H. cylindracea* is by fusion of the outer cuticle, however, in some species such as *H. macrophysa* the peripheral utricles are free and it would be interesting to study the development of the thallus and the aragonite deposits in these species (Hillis-Colinvaux, 1980). In all species studied the chloroplasts are concentrated in the outermost regions of the peripheral utricles, with only a few plastids along the inner walls of the utricles (Palandri, 1972; Colombo *et al.*, 1976a; Borowitzka and Larkum, 1977) and the arrangement of the plastids varies somewhat with sea depth (Colombo *et al.*, 1976b; Colombo and Orsenigo, 1977). The starch-containing amyloplasts (Borowitzka and Larkum, 1974) are located deeper in the thallus below the chloroplasts (Borowitzka and Larkum, 1977; Colombo and Orsenigo, 1977). The cytoplasmic ultrastructure of *Halimeda* species does not appear to be specialized for the calcareous nature of the alga, rather the features are related to the coenocytic nature of the alga. The ultrastructure of *Halimeda* is quite similar to other related coenocytic algae such as *Caulerpa* (Dawes and Rhamstine, 1967; Dawes and Barilotti, 1969) and *Codium* (Hawes, 1979).

The fine structure of the aragonite deposits in the red algae of the order Nemaliales is basically the same as in the Codiales (Borowitzka *et al.*, 1974; Flajs, 1977a). In *Liagora* the aragonite crystals are also precipitated within an intercellular space in semiisolation from the external medium, and the deposits are wholly extracellular. No organization of the aragonite needles is apparent. In *Galaxaura* calcification is extracellular, with apparently some deposition of crystals also in the cell wall (Flajs, 1977a; Okazaki and Furuya, 1977). The crystals appear to be aragonite and/or calcite depending upon season (Okazaki and Furuya, 1977). Unlike *Neomeris* the aragonite needles tend to form in small bundles. Calcification is greatest at the exterior of the thallus and decreases deeper in the thallus. In scanning electron micrographs of the *Galaxaura* medulla small clumps of $CaCO_3$ can be observed in the extensive intercellular spaces between the filaments (Figs. 3 and 4) and small crystals are seen in transmission electron micrographs (Fig. 5). They do not appear to be associated with the cell wall and lie free in the intercellular spaces. Further detailed structural studies are required.

The aragonite needles of *Neomeris annulata* (Borowitzka *et al.*, 1974), *Cymopolia barbata* (Böhm *et al.*, 1978), and *Udotea flabellum* (Marszalek,

1971) are also completely extracellular. The situation in the disc-like attachment stage of *Derbesia clavaeformis* is less clear. The aragonite deposits are extracellular, but occur between the cells and within an outer sheath (MacRaild and Womersley, 1974). Many questions about the development and nature of the $CaCO_3$ deposits of these algae remain to be answered.

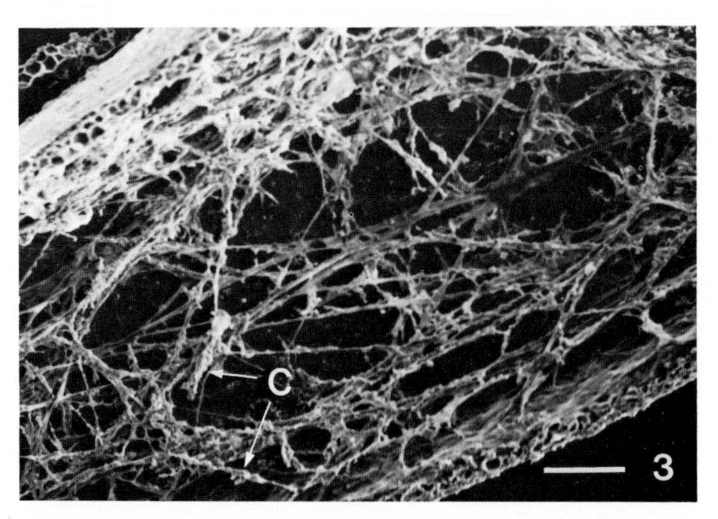

FIG. 3. *Galaxaura* sp. Scanning electron micrograph of transverse section of thallus. Small bundles of $CaCO_3$ (C) crystals can be seen adhering to the filaments in the medullary region of the thallus. Scale = 50 μm.

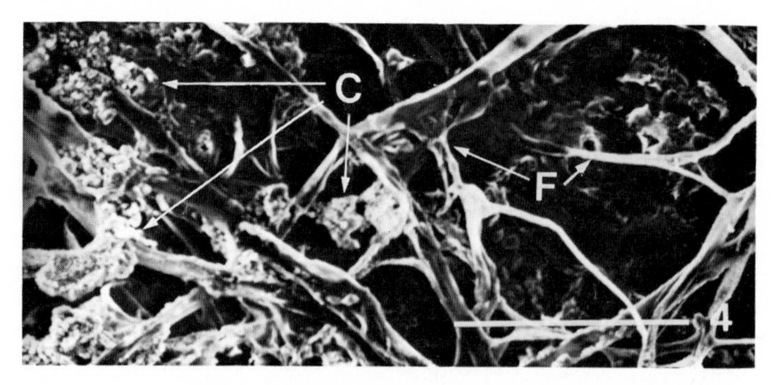

FIG. 4. *Galaxaura* sp. Higher magnification detail of intercellular $CaCO_3$ (C). F, Filament. Scale = 50 μm.

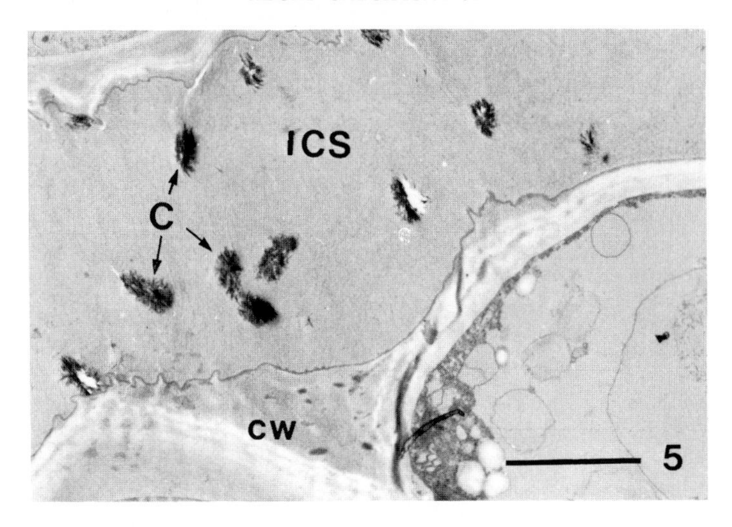

FIG. 5. *Galaxaura* sp. Transmission electron micrograph of the intercellular space (ICS) in the medullary region showing small $CaCO_3$ crystals (C) lying outside the cell walls (CW). Scale = 5 μm.

C. SHEATH CALCIFICATION

The aragonite deposits of *Rhipocephalus phoenix*, *Penicillus pyriformis*, *Udotea cyanthiformis*, and *U. conglutinata* are formed outside the cell wall, but in a sheath of unknown chemical composition (Böhm *et al.*, 1978). This "sheath calcification," which is well developed in the capitulum of *Penicillus* and *Rhipocephalus* has a number of characteristic features: (1) the aragonite needles are arranged in bundles in layers parallel to the filament axis within a porous organic multilayered sheath which forms an integral part of the outer filament wall; (2) these filaments show vacuolar inclusions of calcium oxalate; and (3) deposits of $CaCO_3$ in between the capitular filaments and the outside of the sheath are rare if not totally absent.

The role of the organic "sheath" in calcification and its chemical composition is still unknown. It is possible that the sheath is chemically inert with respect to calcification and may just act as a barrier to diffusion of inorganic carbon, in a way similar to the appressed utricles of *Halimeda* (Borowitzka and Larkum, 1976a,b,c). Reduced diffusion of inorganic carbon combined with photosynthetic CO_2 uptake can lead to localized changes in pH resulting in aragonite precipitation (Borowitzka and Larkum, 1976b). A more direct involvement of this organic material such as in epitaxial nucleation is also possible. It is interesting to note that bundles of aragonite needles also occur in the calcareous tubes of the worm *Salmacina incrustans* where the aragonite is deposited in a mucopolysaccharide-protein gel (Haas, in Flajs, 1977a). The possible metabolic interrelationship between the vacuolar calcium oxalate and "sheath calcification" needs to be examined.

The genus *Udotea* is of greatest interest. Thallus anatomy and calcification pattern vary from species to species. *Udotea petiolata* is not calcified and has no apparent vacuolar calcium oxalate. The blade however has a continuous cortical layer similar to *Halimeda* which isolates the intercellular space from the external medium (Hamel, 1930; Colombo, 1978; Colombo and DeCarli, 1980). *Udotea conglutinata* and *U. cyanthiformis* exhibit "sheath calcification" and contain calcium oxalate. However, the capitulum of *U. flabellum*, which has a fully developed cortical surface similar to *Halimeda*, has acicular aragonite deposits between the thick irregularly branched filaments and no "sheath calcification" (Marszalek, 1971; Böhm *et al.*, 1978). Vacuolar calcium oxalate is absent in this alga. The study of Böhm and co-workers (1978) shows that cortical calcification in *U. flabellum* occurs between the organic layer of the terminal filament wall, resembling the "sheath calcification" of other *Udotea* species. Two types of calcareous deposits of possibly different origin have been recognized in this region, an inner layer of crystals with a fairly smooth appearance, and a layer of larger, coarser crystals of a secondary deposit which forms between the inner filament wall and the outer sheath (see Fig. 31 in Böhm *et al.*, 1978). *U. gigantea* may represent a species with calcareous deposits even closer to *Halimeda* (Flajs, 1977a). The genus *Udotea* therefore presents a continuum from the typical intercellular calcification of *Halimeda* and related algae, to a well-developed "sheath calcification" system as in *Udotea cyanthiformis*. Comparative studies of calcification in these species as well as transmission electron microscopic studies combined with histochemical studies of the "sheath" and the cell wall could be of great interest. The genus *Bornetella* also seems to show similar variation in its $CaCO_3$ deposits as *Udotea* (Solms-Laubach, 1893). The cytoplasmic ultrastructure of *Udotea* is very similar to that of *Halimeda* (compare Borowitzka and Larkum, 1977; and Colombo, 1978).

Similar variation in thallus form and calcification as between *Halimeda* and *Udotea* and related species can also be observed in the fossil green algal genera *Calcifolium*, *Ivanovia*, *Anchicodium*, and *Eugonophyllum* (Wray, 1977). *Calcifolium* appears to be a monostromatic "sheath" calcifying alga (Maslov, 1956) whereas the others have medullary and cortical regions of varying complexity (Konishi and Wray, 1961). Examination of fossil green algae (Wray, 1977) suggests that the *Halimeda* calcification type represents a later, evolutionarily more advanced system.

The desmid *Oocardium stratum* forms crystals of calcite oriented around the gelatinous stalks (Wallner, 1933; Mathews *et al.*, 1965; Flajs, 1977a). This might be a specialized modification of "sheath calcification."

Despite being one of the most intensively studied algae, little is known of calcification in *Acetabularia. Acetabularia* forms calcareous deposits on the walls of the cap and the stalk (Solms-Laubach, 1887; Leitgeb, 1888; Berger *et al.*, 1974). Solms-Laubach (1887) showed that the $CaCO_3$ is deposited within a mucilage layer (sheath?) on the outside of the cell wall. Flajs (1977a) showed that these 5-μm-long aragonite needles are arranged in small bundles suggestive of the arrangement of the products of "sheath calcification" in *Udotea* species. Figure 6 is a transmission electron micrograph of the cell wall (and mucilage layer?) of an *Acetabularia calyculus* stem showing the extensive deposits of $CaCO_3$ in the wall. The deposits appear to be layered and there is an outer $CaCO_3$-free wall layer. Further extensive studies are required to describe fully the development and organization of these deposits. Flajs (1977a) has also detected the presence of calcium oxalate. The reproductive cysts of various species of *Acetabularia chalmasia* and *Acicularia*, known as calcispheres are also calcified (Marszalek, 1975; Bailey *et al.*, 1976).

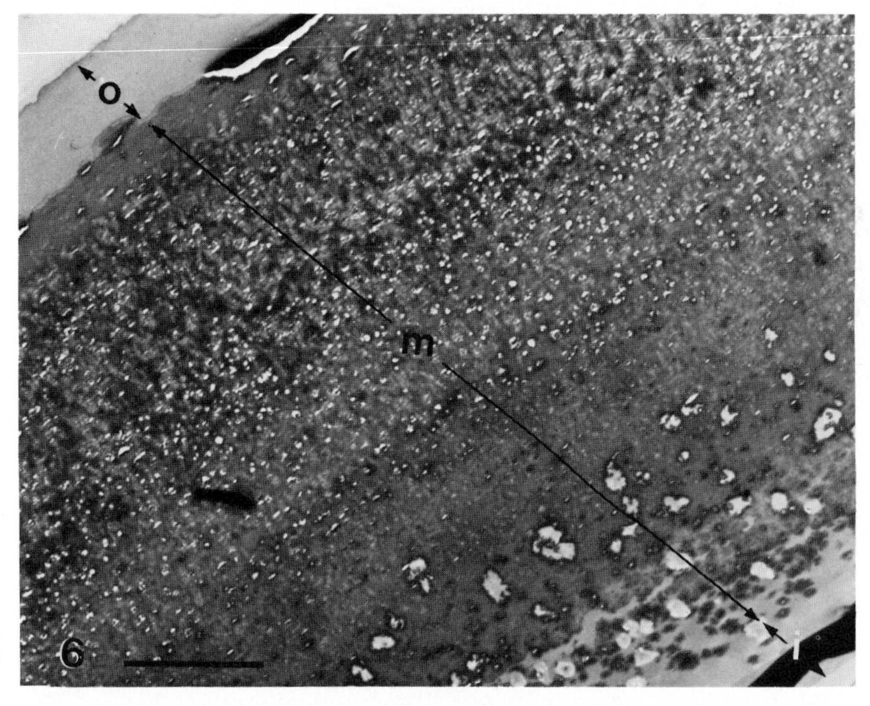

FIG. 6. *Acetabularia calyculus.* Transmission electron micrograph of a section across the stalk cell wall of the heavily calcified plant. The middle region of the cell wall (m) contains extensive $CaCO_3$. There is a $CaCO_3$-free outer layer (o) and also a $CaCO_3$-free inner layer (i). The organization and density of the $CaCO_3$ varies from the exterior to the interior. The black band on the interior of the wall is an artifact; a fold in the section. Scale = 5 μm.

These hollow spheres, which contain the developing gametes, are 140–185 μm in a diameter with a 10–25 μm thick wall composed of granular to acicular aragonite. They have a 40 μm circular opening through which the mature gametes are released. The difference in crystal organization, between vegetative cells and reproductive cysts, may be due to the difference in cell wall chemical composition. The cell wall of the diploid stem and cap is composed of rhamnose, xylose, and glucuronic acid with galactose and its 4-*O*-methyl derivative as the major sugars (Zetsche, 1967; Smestad and Percival, 1972). In the haploid cysts, however, cellulose is the major constituent (Nisizawa *et al.*, 1974). Similar differences in cell wall composition between the diploid and haploid stages of siphonous green algae have been observed by Neumann (1969) in *Derbesia/Halicystis* and by Rietema (1971) in *Bryopsis hypnoides*. These algae do not calcify (see MacRaild and Womersley, 1974, for a possible exception), so that it does not appear that the cell wall per se is responsible for deposition of the $CaCO_3$. It is more likely that the cell wall excerts a controlling influence on the organization of the $CaCO_3$ crystals. Eiseman (1970) has also shown that environment affects calcification of the cysts.

D. Blue-Green Algae

The blue-green algae (Cyanophyta) present, at least in part, what appears to be a special case of "sheath" calcification. Calcification occurs on or in the mucilagenous sheaths of these algae (Black, 1933; Monty, 1967; Shinn *et al.*, 1969; Riding, 1977a) and in the marine environment is usually the crystal isomorph aragonite. In freshwater blue-green algae, calcite has been reported to encrust the filaments, probably forming both within the sheath and outside it (cf. Irion and Müller, 1968; Golubic, 1973; Borowitzka *et al.*, 1974; Flajs, 1977a; Pentecost, 1978). Riding (1977a) recognizes two distinct types of calcified blue-green algal filaments: micrite-impregnated sheath and microspar-encrusted sheaths. These result from either two different mechanisms or two different rates of calcification. The calcified sheaths of *Plectonema* found at Aldabra are filled with irregular bundles of needle-like crystals, (Riding, 1977a, Figs. 2 and 3), whereas the *Rivularia rufescens* collected in a freshwater stream in Austria is surrounded by granular to massive rhombohedral calcite crystals (Flajs, 1977a). These crystals are always outside the sheath (Pentecost, 1978). The walls and sheaths of blue-green algae contain acidic polysaccharides and pectin (Leak, 1967) made up of a wide range of sugar moieties (Metha and Vaidya, 1978) and have been shown to bind calcium ions (Somers and Brown, 1978); however, there have been no transmission electron microscopic studies of the relationship between the $CaCO_3$ crystals and the sheath organic material.

In the marine environment blue-green algal mats not only deposit $CaCO_3$ per se, but also trap sedimentary $CaCO_3$ and at times form large structures known as stromatolites (Walter, 1976; Golubic and Focke, 1978). Blue-green algae also bore into limestone and are an important factor in the biodestruction of limestone deposits (Golubic, 1973). They show a wide diversity of structure, but in common with their calcareous relatives they have a acidic mucopolysaccharide sheath, and it has been suggested that this material may be involved in the mechanism of carbonate dissolution (Campion-Alsumard, 1979). These observations suggest that it is not the sheath which is responsible for the deposition (or dissolution) of the $CaCO_3$ but some particular aspects of the physiology of the alga. The green endolithic alga *Oestrobium* sp. not only grows within limestone sediments, but also calcifies upon death, both within the filament and on the surface of the filament (Kobluk and Risk, 1977). Calcium carbonate precipitation in this alga is not dependent on the metabolism of the alga, rather metabolism inhibits calcification. The deposition of the $CaCO_3$, however, seems to be influenced or controlled by organic materials in the algal thalli.

III. Calcified Cell Walls

The Corallinaceae (Rhodophyta, Cryptonemiales) are cosmopolitan in distribution (Adey and McIntyre, 1973; Johansen, 1974) and have an extensive fossil record (Wray, 1977). Their $CaCO_3$-impregnated cell walls make them important sediment and reef-forming organisms. The cell walls of vegetative cells, with the exception of the walls of the genicula (those cells forming the articulations in coralline algae such as *Corallina*), hair cells (e.g., in *Fosliella*), and the walls of some cells surrounding the conceptacles, contain extensive deposits of the calcite crystal isomorph of $CaCO_3$ mixed with some magnesium carbonate (Fig. 7). The amount of $CaCO_3$ in the cell wall varies with species, cell type, and age of the alga. All calcified cell walls also contain small amounts of organic material making up less than 0.35% (w/w) (Bignot, 1978).

In germinating tetraspores calcification begins in the middle lamella (Furuya, 1960; Vesk and Borowitzka, unpublished results). The first formed crystals tend to lie parallel to the cells along the middle lamella, but show little other organization. In fully decalcified sections, the middle lamella generally appears as a zone of more darkly staining fibrillar material. The middle lamella has been suggested to consist of pectic material whereas the rest of the cell wall consists of cellulose and some protein. There are, however, few histochemical data available, the most extensive studies being those of Yendo (1904) and Matty and Johansen (1981). The increased electron density may also be purely a result of a greater concentration of organic material along the junction of the cell walls of adjacent cells. The stainability of the middle lamella varies greatly between species and within a thallus

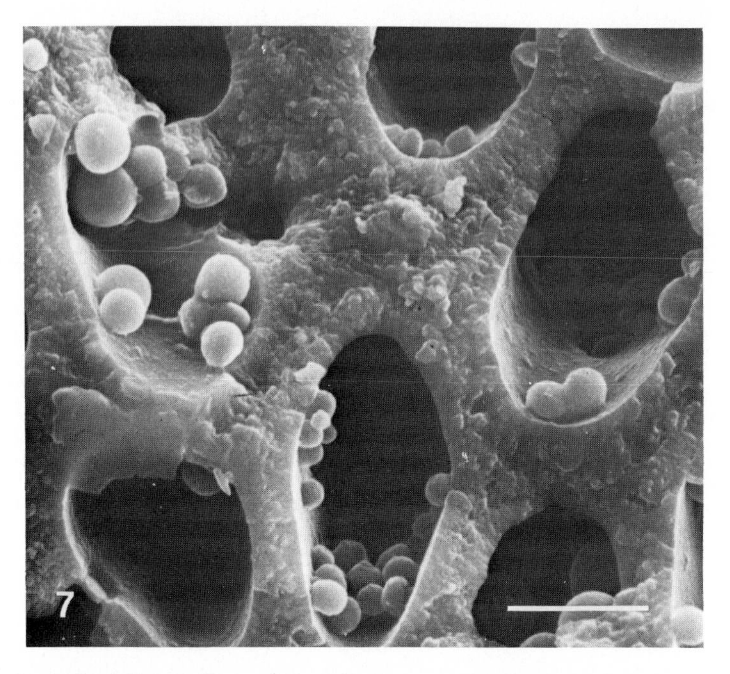

FIG. 7. *Lithophyllum molluccense*. Scanning electron micrograph of a fractured part of the thallus showing the $CaCO_3$-impregnated cell walls. Note the orientation of the calcite crystals surrounding the cell cavity. The round structures are remnants of cell organelles. Scale = 5 μm.

(cf. Giraud and Cabioch, 1977, 1979; Cabioch and Giraud 1978a,b; Borowitzka and Vesk, 1978, 1979).

With continued growth, the space between the middle lamella and the plasmalemma becomes filled with $CaCO_3$. X-Ray crystallography has shown that calcite crystals deposited near the plasmalemma are generally oriented with their *c* axis (i.e., the long axis) at right angles to the plasmalemma (Baas-Becking and Galliher, 1931). This orientation can also be seen in scanning electron micrographs (Fig. 7) (Borowitzka *et al.*, 1974; Flajs, 1977a,b), in replicas of fractured thalli (Giraud and Cabioch, 1979), in non- or partially decalcified thin sections under the transmission electron microscope (Figs. 8 and 9) (Borowitzka and Vesk, 1978, 1979; Giraud and Cabioch, 1979), and in polarized light (Pobeguin, 1954). This orientation is much more definite in heavily calcified genera such as *Lithophyllum* and *Lithothamnium* (Fig. 7) and may be rather difficult to discern in lightly calcified specimens (e.g., Garbary and Veltkamp, 1980). In some species such as *Goniolithon strietum* these oriented crystals appear to lie in concentric layers around the cell lumens (Flajs, 1977a).

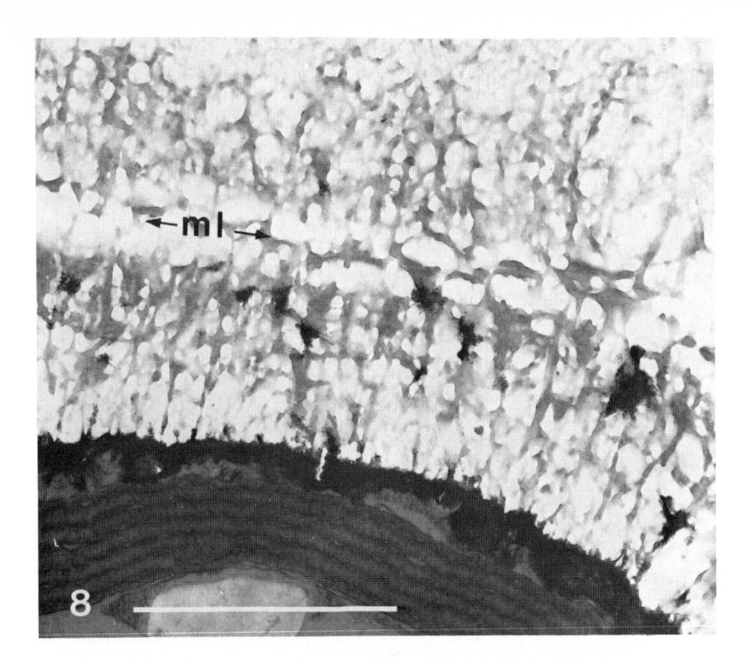

FIG. 8. *Jania capillacea*. Transmission electron micrograph of an undecalcified thallus. The calcite has dissolved out of the section during poststaining leaving holes (clear spaces) in the section where the crystals were. Note the small crystals radiating away from the cell and the larger crystals along the middle lamella (ml). Scale = 5 μm.

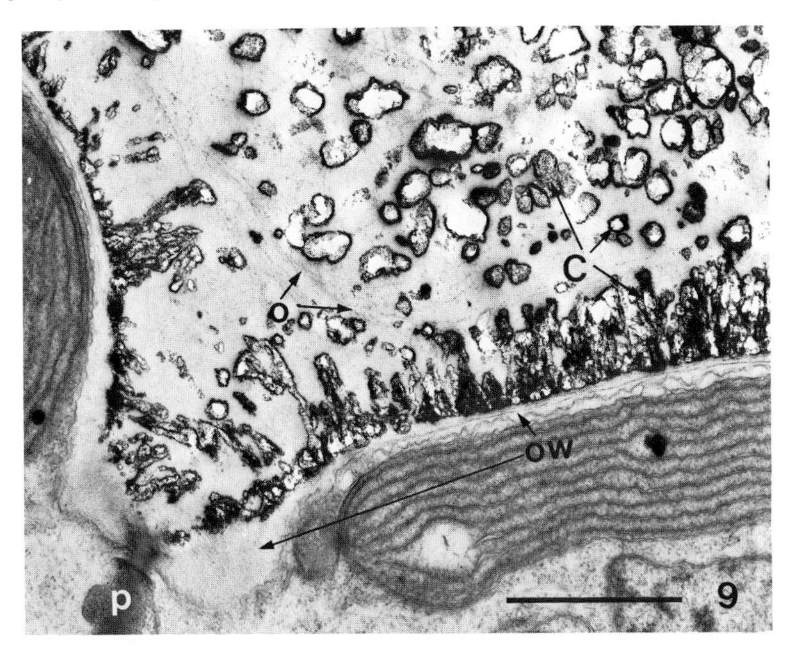

FIG. 9. *Lithothrix aspergillum*. Transmission electron micrograph of a partially decalcified thallus showing the oriented crystals near the cell, and the organic wall material (O) between the $CaCO_3$ (C). Note also the $CaCO_3$-free wall layer (OW) just outside the plasmalemma and the more extensive $CaCO_3$-free wall regions near the pit connection (p). Scale = 1 μm.

The calcite crystals are always in contact with the organic material of the cell wall (Fig. 9) (e.g., polysaccharides, proteins etc.) and on decalcification the organic cell wall fibrils retain their organization with electron transparent spaces where the calcite crystals were located (Fig. 10). From limited studies done on the early stages of calcification in the germinating tetraspores of *Haliptilon* (=*Corallina*) *cuvieri* it appears that the organic cell wall fibrils play some part in orienting the calcite crystals, however, much of the orientation of the fibrils seen in decalcified sections is due to the crystals displacing the fibrils during growth. The detailed histochemical studies of Yendo (1904) and Matty and Johansen (1981) and the electron microscopic studies of Bailey and Bisalpultra (1970) and Borowitzka and Vesk (1978) on *Corallina* spp. suggest that calcification in these algae is initiated along the pectin-containing middle lamella, and that later oriented crystal growth occurs among the cellulose fibrils of the inner cell walls. It is however not known whether this cellulose is directly involved in the formation and organization of the calcite crystals, or whether the "active" components are other polysaccharides and proteins which occur in low concentrations in the cell wall.

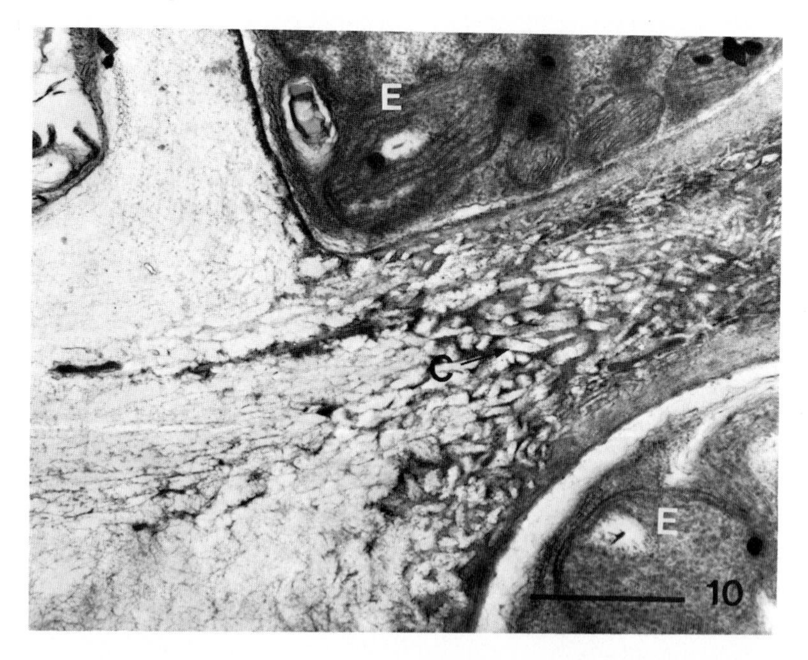

FIG. 10. *Haliptilon cuvieri*. Fully decalcified section through the cell wall between two epidermal cells (E). Calcification is least near the thallus exterior (right-hand side of micrograph), and increases further into the thallus (toward the left-hand side of micrograph). The shapes of individual crystals can be seen by the clear spaces (C) left in the organic wall material which surrounded the crystals. Scale = 1 μm.

Amphiroa foliacea cell walls show a complex binding pattern of Ca^{2+} (Borowitzka, 1979) which suggests the presence of at least two Ca^{2+} binding components which are presumed to be the $-COO^-$ sites of cellulose and the $-O-SO_2-O^-$ sites of the pectic substances. In the brown alga *Laminaria saccharina* the sulfated polysaccharides are located in the middle lamella (Callow and Evans, 1976). It is inviting to speculate that the wall fibrils act as a matrix for epitaxial nucleation and crystal growth (Degens, 1976), however, much more data on the early stages of calcification and on the chemical composition of the organic components of the coralline cell wall are required before any firm conclusions can be arrived at.

The degree of calcification and the organization of the calcite crystals are not constant throughout the thallus. X-Ray microprobe analysis (Borowitzka and Nockolds, unpublished results) has shown that the average $CaCO_3$ density in the cell walls of *Amphiroa foliacea* increases with age of the intergeniculum (segment). Moberly (1968) and Kolesar (1978) have also shown that the level of magnesium varies with the skeletal growth rate. The outer cell wall of coralline algal thalli is largely organic being made up mainly of pectin substances (Matty and Johansen, 1981). Giraud and Cabioch (1979) using careful fixation, embedding, and sectioning have shown that this wall also contains small, scattered, needle-like calcite crystals. These crystals occur throughout the external cell walls of the epidermal cells, but they do not appear to penetrate into the cell wall ingrowths which are characteristic of coralline epidermal cells (Giraud and Cabioch, 1979). The density of the $CaCO_3$ deposits varies greatly between species and within a thallus.

Close examination of the published micrographs of Bailey and Bisalpultra (1970; their Fig. 5) and Borowitzka and Vesk (1978, 1979) show the lacunae left in the organic material of the external cell wall by the $CaCO_3$ crystals which were removed during decalcification and specimen preparation. There is a clear gradient in $CaCO_3$ crystal density from the outside of the thallus to the interior. This can be seen in the lateral cell walls of epidermal cells (Fig. 10). Near the thallus exterior individual crystals of 8 nm thickness and up to 120 nm length can be seen in *Haliptilon curvieri* (Fig. 10) (Borowitzka and Vesk, 1978).

The cell walls are never completely impregnated with $CaCO_3$ and Pit connections are surrounded by a ring of noncalcareous wall (Fig. 9). In some algae such as *Corallina vancouverensis* and *Lithothrix aspergillum* there is a thin layer of organic material just exterior to the plasmalemma (Bailey and Bisalpultra, 1970; Borowitzka and Vesk, 1979) and calcification does not extend to the plasmalemma (Fig. 9).

The genicular cell walls of articulated coralline algae show a complex layering of organic material. These walls are free of $CaCO_3$ except for the ends of those genicular cells which join with the calcified intergeniculum. These cells have completely calcified cell walls at the geniculum–intergeniculum junction and can be completely noncalcified elsewhere. In *Haliptilon* the single layer of long genicular cells is calcified at both ends where it passes into the intergeniculum. In *Calliarthron* where the geniculum is multilayered only the ends of those cells abutting onto the intergeniculum are calcified. In *Lithothrix* where the geniculum is overgrown by intergenicular cells some of the peripheral genicular cells may be wholly or partially calcified. The genicular cells represent a fascinating problem for the study of calcification mechanisms.

Genicular cell walls show histochemical differences to those of the intergenicular cell walls (Yendo, 1904; Matty and Johansen, 1981) and they have two quite distinct regions: a dense concentrically fibrillar wall immediately next to the cell and a less dense mucilagenous region between the two adjacent cell walls (Bailey and Bisalpultra, 1970; Borowitzka and Vesk, 1978). The dense fibrillar wall layer represents the cell wall proper. The fibrils are arranged predominantly concentrically around the long axis of the genicular cells. The density of the fibrils varies giving the impression of growth rings, and increases with age (Borowitzka and Vesk, 1979). Between this wall and the intracellular mucilage there usually is a thin layer of more osmiophilic material (Borowitzka and Vesk, 1978, 1979). In *Lithothrix aspergillum*, $CaCO_3$ crystals are sometimes observed in the mucilage filled intercellular spaces. These $CaCO_3$ deposits consist of discrete large crystal units which appear to have developed from smaller crystal nuclei by syntaxial growth, resulting in the fan-like arrangement seen in thin sections (Borowitzka and Vesk, 1979). In the SEM they appear very similar in structure to the calcite crystals found on the outer surface of *Chara* cell walls (Borowitzka *et al.*, 1974).

These structural studies of coralline calcification suggest that a number of different stages and processes of $CaCO_3$ crystal growth occur in the coralline algal thallus. These include the initial deposition of small crystals (crystal nuclei) along the middle lamella in growing tissues and germinating spores and the subsequent formation of oriented crystals near the plasmalemma. This is followed by infilling of new $CaCO_3$, probably by epitaxial nucleation, between the existing crystals to produce the massive deposits of genera such as *Lithothamnium* or *Porolithon*. The *inter*cellular crystals of *Lithothrix aspergillum* genicula represent a "secondary" stage of calcification associated with living tissues (Borowitzka and Vesk, 1979), whereas the *intra*cellular aragonite and calcite observed in some thalli represent another "secondary" stage of calification associated with dead tissues (Alexandersson, 1974; Bignot, 1978). These deposits are probably formed by dissolution and recrystallization of the cell wall $CaCO_3$ following the death of the cells.

The primary areas of calcification in *Corallina officinalis* appear to be the walls of the terminal meristem and the epithallial cells. Andrake and Johansen (1980) found that the vital dye Alizarin Red S was deposited predominantly in these regions. Alizarin Red S has been used extensively as a histological stain for calcium and it is usually bound at the sites of calcium deposition, irrespective of whether calcium carbonate (Lamberts, 1978) or calcium phosphate (Cameron, 1930) is being deposited. Alizarin staining has been used with variable success to measure algal calcification or growth rates (Stearn *et al.*, 1977; Andrake and Johansen, 1980; Wefer, 1980). Alizarin Red can also be used in conjunction with Alizarin Blue to measure relative growth rates over time (Barnes, 1972); however, since little is known of how these dyes work, nothing can be inferred from the staining data as to the mechanism of calcification. The dyes however are quite a valuable tool in studies of *in situ* growth rates, and once more is known of the chemistry of the staining reaction they may provide important data on the calcification mechanism.

The cellular fine structure of coralline algal cells is generally similar to that of other red algae. One of the most striking features of coralline algal ultrastructure, other than the $CaCO_3$ deposits, is the structure of the epidermal cells. The epidermal cells show a distinct apico-basal polarity and do not contain starch. Their most conspicuous features are the extensive ingrowths of the cell wall from the exterior of the thallus into the cells (Giraud and Cabioch, 1976, 1977, 1979; Cabioch and Giraud, 1978a,b; Borowitzka and Vesk, 1978, 1979). The degree of elaboration and organization of these ingrowths varies between species. Giraud and Cabioch (1976, 1977) interpreted these ingrowths as Golgi vesicles secreting cell wall material to the exterior of the cell. Borowitzka and Vesk (1978, 1979), however, have not been able to observe any active Golgi bodies, which are usually associated with secreting cells (e.g., Ramus, 1972; Benner and Schnepf, 1975), in epidermal cells. Furthermore, studies with colloidal lanthanum as a tracer for the extracellular synplast, as well as serial sectioning have shown that these wall ingrowths are in direct communication with the outer cell wall and that there are no apparent Golgi "vesicles" in the cells (Borowitzka and Vesk, unpublished results). The vesicles of Giraud and Cabioch (1976, 1977) probably represent section through the wall ingrowths.

The epidermal cells with their extensive cell wall ingrowths resemble the transfer cells of higher plants which function in facilitating short distance solute transport by increasing the total area of plasmalemma–cell wall contact (cf. Gunning and Pate, 1969; Pate and Gunning, 1972). Epidermal transfer cells have also been reported in other aquatic plants (Falk and Sitte, 1963; Gunning and Pate, 1969; Jagels, 1973) but have not been found in any other algae. In analogy with higher plant transfer cells, the coralline epidermal cells may function in (1) the transport of nutrients from the seawater to the thallus, and/or (2) transport of cell wall or mucilage polysaccharides to the to the outside of the thallus. It is also interesting to note that the lower leaf cells of *Elodea* and *Potamogetum* which take up HCO_3^- also have wall ingrowths (Falk and Sitte, 1963; Prins *et al.*, 1979).

The epidermal cells have numerous well-developed chloroplasts which are often found in close association with the wall ingrowths, but contain no starch. It is possible that the photosynthate produced by the chloroplasts is either immediately metabolized to supply energy for the active transport of solutes into or out of the cells, or that the photosynthate is transported to cells deeper within the thallus for storage as starch. The heavy impregnation of the cell walls with $CaCO_3$ must inhibit the diffusion of nutrients to the cells lower in the thallus as well as reducing the amount of light reaching the chloroplasts of these cells. Specialized transport cells therefore may be an essential adaptation of coralline algae to their calcareous habit. This hypothesis is supported by the observation that where epidermal cells are damaged by grazers, new epidermal cells are formed by the subtending meristematic cells (Borowitzka and Vesk, 1978). Coralline algae do not appear to produce any substances which prevent overgrowth by other algae and invertebrates, and regular grazing which reduces this overgrowth appears to be essential to the survival of the alga.

Transport of nutrients and photosynthate to the cells deeper in the thallus must be via the pit connections. The ultrastructure of coralline pit connections is similar to that of other red algae (compare micrographs in Bouck, 1962; Bisalpultra *et al.*, 1967; Duckett *et al.*, 1974; Pueschel, 1977; Feldmann *et al.*, 1977; Peyriere, 1977) and to date evidence for an intercellular communication function for pit connections remains largely circumstantial (Pueschel, 1980a,b). LaVelle (1979) has shown that photosynthate is translocated from subapical intergenicular to the apical geniculum in *Calliarthron tuberculosum*, and although the evidence is not conclusive, the most likely path for this translocation is via the pit connections. Hartmann and Eschrich (1969) and Turner and Evans (1978) have also demonstrated translocation of organic compounds in other red algae without identifying the path of translocation.

Another apparent cytological adaptation of coralline algae to their calcareous habit can be seen in the distribution of chloroplasts and starch. Most of the plastids are localized in the outer cell layers whereas the starch is generally localized deeper within the thallus (Borowitzka and Vesk, 1978, 1979; Matty and Johansen, 1981). Light penetration through $CaCO_3$ is quite limited (Pentecost, 1978) and starch grains in the outer cells would only serve to further reduce the amount of light incident on the chloroplast thylakoids.

Algae with heavily calcified cell walls (exoskeletons?) also face limitations to cell expansion and growth. The subapical meristematic cells are generally only lightly calcified and may divide and enlarge by usual means. However, the formation of genicula and of conceptacles presents a different problem. In *Lithothrix aspergillum* the upper ends of the genicular cells have partially calcified cell walls whereas the lower ends penetrate only briefly into the intergenicular $CaCO_3$. These genicular cells increase in length by elongation at the lower uncalcified end only, presumably by localized addition of new cell wall material as occurs in other red algae (Waaland and Waaland, 1975). The lack of $CaCO_3$ in the lower parts of the geniculum may be a result of this elongation. Cell wall formation and extension in higher plants is usually associated with acidification of the area immediately next to the region of wall formation (Marré *et al.*, 1975), and such acidification could be responsible for the lack of calcification. It seems reasonable to propose that some similar process of localized acidification of the extracellular milieu occurs during the formation of conceptacles resulting in decalcification and followed by cellular disintegration and new cell growth. Conceptacle formation has only been studied with the light microscope (e.g., Johansen, 1969; Lebednik, 1978) and little is known of the cytology of this complex process.

There have been almost no studies of the morphology of the $CaCO_3$ deposits of the red algal genus *Peyssonellia* which deposits aragonite. The SEM studies of Flajs (1977a) show that the aragonite is apparently deposited in the cell wall near the middle lamella. Decalcified sections of *Peyssonellia* (Borowitzka, unpublished results) show the presence of loose fibrillar material in the intercellular spaces between the cell walls, and it is within this material that the $CaCO_3$ is deposited. This material appears very similar to the intercellular material in *Lithothrix* genicula (Borowitzka and Vesk, 1979).

The recent studies of the cytology of coralline algae and their associated $CaCO_3$ deposits have provided much information on the development of the calcareous deposits. However, further histochemical studies combined with physiological studies are required to determine whether the organic cell wall material is involved only in crystal nucleation and in determining the crystal isomorph deposited or whether it also plays an active role in the later stages of calcification.

IV. Intracellular Deposits

The unicellular algae of the suborder Coccolithophorineae (Chrysophyta, Prymnesiophyceae) form variously shaped and calcified plates known collectively as coccoliths (see Okada and McIntyre, 1977; Bold and Wynne, 1978, for details of taxonomy). This article will concentrate only on the structural aspects and the reader is referred to the reviews of Paasche (1968), Klaveness and Paasche (1979), and Borowitzka (1982) for physiological and biochemical aspects. The coccoliths are formed intracellularly but eventually become lodged around the outside of the cell (Fig. 11). Dixon in 1900 first suggested the intracellular origin of coccoliths from studies of sectioned and stained cells, and his observations were supported by many subsequent workers (e.g., Lohmann, 1902; Paasche, 1962). These observations have since been confirmed by electron microscopy and many coccolithophores have been studied in varying detail. These include *Calyptrosphaeria sphaeroidea* (Klaveness, 1973), *Coccolithina leptopora* (Blackwelder *et al.*, 1979), *Coccolithus pelagicus* (Manton and Leedale, 1963, 1969), *Cricosphaeria roscoffensis* (Gayral and Fresnel, 1976), *C. gayraliae* (Beuffe, 1978), *Emiliania huxleyi* (Klaveness 1972a,b; Klaveness and Paasche, 1971), *Hymenomonas carterae* (Manton and Leedale, 1969; Pienaar, 1969a,b, 1971; Leadbeater, 1970; Outka and Williams, 1971), *Hymenomonas globosa* (Gayral and Fresnel, 1976), *Hymenomonas lacuna* (Pienaar, 1976), *Ochrosphaeria verrucosa* (West, 1969), and *Pleurochrysis scherffelii* (Leadbeater, 1971).

The life cycles of most coccolithophorids are unknown. However, those

FIG. 11. *Cyclococcolithus leptoporus.* Scanning electron micrograph showing the numerous coccoliths surrounding the cell. The overlapping individual calcite crystals making up the coccoliths can be seen (micrograph courtesy of Dr. G. Hallegraeff). Scale = 5 μm.

studied to date show a wide range of modes of vegetative and sexual reproduction (Paasche, 1968; Lefort, 1975; Klaveness and Paasche, 1979). Production of calcareous coccoliths occurs only in some stages of the life cycle. For example *Pleurochrysis* has a predominant nonmotile stage with wholly organic scales (Brown, 1969). There are also two different motile flagellated stages one of which forms both organic surface scales and calcareous coccoliths (Parke, 1961; Leadbeater, 1971). The planktonic *Emiliania huxleyi* has uncalcified organic scales in the motile stage whereas the nonmotile stage carries coccoliths (Klaveness and Paasche, 1971; Klaveness, 1972b). The benthic phase of many genera also produce calcareous elements in their mucilagenous sheath (Parke, 1971) and these deposits may be wholly extracellular unlike the coccoliths.

Coccoliths are built up of single crystals of calcium carbonate, usually the calcite crystal isomorph (Paasche, 1968; Blackwelder *et al.*, 1976; see Borowitzka, 1977, for data on $CaCO_3$ crystal isomorphs). These individual crystals can easily be recognized in scanning or electron micrographs of most species, but are most prominent in members of the Calyptrosphaeraceae where the coccoliths (holococcoliths) are built up entirely of microcrystals with regular crystallographic features (Gaarder, 1962; Okada and McIntyre, 1977). The overlapping plates of the coccoliths of *Cyclococcolithus leptoporus* each consisting of a single calcite crystal can clearly be seen (Fig. 11). Each crystal unit is enclosed in an organic envelope (Outka and Williams, 1971; Klaveness, 1973, 1976; Beuffe, 1978) and some organic material (polysaccharide) appears to be occluded in the crystal matrix (Klaveness, 1976; DeJong *et al.*, 1979).

The cell exerts precise control over crystal growth. The holococcoliths of *Crystallolithus hyalinus* [the flagellate stage of *Coccolithus pelagicus* (Parke and Adams, 1960)] are built up of approximately 80 rhombohedral calcite crystal units arranged to form the oval saucer-shaped coccolith (Gaarder and Markali, 1956). The coccoliths (placoliths) of *E. huxleyi* have 20–40 structural units consisting of an upper and a lower element together with a sector of the central connecting tube (Braarud *et al.*, 1953). Each unit is a single calcite crystal (Watabe, 1967).

All coccoliths studied so far, with the notable exception of *E. huxleyi* (Klaveness, 1972a), are formed on a flat organic scale or baseplate. Outka and Williams (1971) studying *Hymenomonas carterae* were the first to describe in detail the sequential assembly of coccoliths. The coccolith baseplate and the unmineralized body scales are both formed in Golgi vesicles at the same time. In *H. lacuna* the coccolith baseplates are formed in the periphery of the Golgi system while the unmineralized body scales are formed within the normal array of the Golgi system (Pienaar, 1976). The formation of different organic scales within the same Golgi system is not unusual. For example Moestrup and Thomsen (1974) showed that the green flagellate

Pyramimonas orientalis produced at least five different scales simultaneously in the Golgi. Development of the coccolith proceeds with the appearance of "coccolithosomes." The coccolithosomes are minute, highly stainable bodies which are formed in ears or pockets of the coccolith vesicle, from which they apparently move into the vesicle and form into a morphologically determined shape, called the "organic matrix" by Manton and Leedale (1969) or "rim sheath" by Outka and Williams (1971), along the organic baseplate rim. It has been impossible to determine if coccolithosomes are synthesized at the ends of the same cisterna that contains the baseplate, or in separate vesicles. If the latter, then the vesicles must eventually fuse with the base vesicles (Outka and Williams, 1971). In *Pleurochrysis*, amorphous polysaccharide originating from budding vesicles of the rough endoplasmic reticulum migrates to, and fuses with, the periphery of the distal Golgi vesicles thereby covering the scale surface with amorphous, acidic polysaccharide (Brown and Romanovicz, 1976). The role of the vescile membrane in influencing the shape and size of the growing matrix is uncertain. Once formation of the organic matrix or rim sheath is complete, the spaces delineated by the organic matrix are filled in with $CaCO_3$. On completion, the coccolith is excreted from the cell and lodges in the mucilagenous layer around the cell.

Coccolithogenesis proceeds in a similar manner to the above in many of the species studied so far, although some small differences have been observed. In *Coccolithus pelagicus* for instance, each coccolith is formed in a Golgi cisterna adjacent to the vesicle stack; the coccolith vesicle is T-shaped and the stem of the T retains its position in the stack. Only one coccolith is formed at a time (Manton and Leedale, 1969). The silica frustules of diatoms are also formed within vesicles in the cytoplasm (see Schmid *et al.*, 1981, for review) and this kind of intracellular formation of skeletal hard parts showing great organization has great similarity to coccolithogenesis.

Emiliania huxleyi differs from the above examples in that its coccoliths do not have an organic baseplate. The pioneering study of Wilbur and Watabe (1963) showed that the coccolith was formed in separate "organic matrix region." Coccolith formation takes place in a vesicle separate from the Golgi stack but possibly connected to the endoplasmic reticulum (Klaveness, 1972a). The coccolith vesicle takes on the shape of the coccolith and becomes filled with the organic matrix material. Using very careful specimen preparation to minimize dissolution of the $CaCO_3$, Klaveness (1976) was able to show that matrix formation and mineral deposition are simultaneous events in *E. huxleyi*. There is no baseplate in this species, and each calcite crystal is surrounded at all times by a complementary matrix and grows from a small rhombohedral crystal into its final complicated shape. These small rhombohedral calcite crystals soon join into a ring approximately oriented for growth along the axes. Contrary to the coccolithophorid species where coccolith formation is preceded by formation of a rigid organic scale, lateral growth of the coccolith may occur in *E. huxleyi* (Klaveness, 1976). The completed coccolith is enclosed within the matrix envelope. Each element (there are 30–40 per coccolith) is enclosed in a separate matrix compartment. This envelope persists after the coccolith is expelled from the cell (Braarud *et al.*, 1953) and may even be recognized in fossil coccoliths (Hamano and Honjo, 1969).

Klaveness and Paasche (1979) hold the view that matrix formation (excluding the baseplate) and mineralization are simultaneous events and that shape is determined by the coccolith vesicle membrane functioning as a local mold during morphogenesis (cf. also Faure-Fremiet *et al.*, 1968; Klaveness, 1976). Although there is not always a direct and close correspondence between the shape of the coccolith to be formed and the coccolith vesicle membrane, leading to some doubt as to the moulding function of the membrane (cf. Outka and Williams, 1971), there is some indirect support for this hypothesis (Outka and Williams, 1971). The organic matrix contained in the coccolith vesicle is made up of an acidic sulfated polysaccharide (Fichtinger-Schepman *et al.*, 1979). Such molecules are known to swell in the presence of glutaraldehyde, the fixative used in the electron microscopic studies, and may therefore distort the shape of the coccolith vesicle during specimen preparation for microscopy. For example, osmium-fixed material of the siliceous body scales of *Paraphysomonas* appear to be formed in cytoplasmic vesicles of exactly the same shape as the scale (Hibberd, 1979). Any lack of correspondence between the shape of the nascent coccolith and the coccolith vesicle membrane may therefore be artifactual. This point could possibly be clarified by a study using freeze-substitution rather than normal fixation.

The completed coccoliths are released from the anterior end of the cell (Ariovich and Pienaar, 1977), however, little is known of the process of intracellular transport and extrusion of the completed coccolith. Plasma rotation has been demonstrated in *Pleurochrysis* (Brown, 1974) and this appears to be the means by which coccoliths and scales are deposited evenly over the surface of the cell in these forms. The peripheral sac is considered to be the "ball-bearing" upon which the protoplast rotates (Brown, 1974; Brown *et al.*, 1973), and this model could apply to a number of other Prymnesiophytes. However, the situation is not quite as clear-cut since the single peripheral sac of *Pleurochrysis* may in other species be represented by more than one cavity (Klaveness, 1973) and appears to be absent in yet other species (Klaveness and Paasche, 1979). Once extruded from the cell, the coccolith lodges in the external mucilage. If cell division occurs the existing external coccoliths redistribute themselves over both daughter cells (Stacey and Pienaar, 1980).

In some species such as *Calyptrosphaeria* the cell surface is covered by a "skin" which lies external to the coccolith layer (Klaveness and Paasche, 1979). The holococcoliths of the *Cristallolithus* phase of *Coccolithus pelagicus* also lie below such a "skin." The function of this "skin," which appears to be pectic in nature, is unknown.

One difficulty with proposing a generalized structural model for coccolith formation is the fact that so few of the 250+ species of coccolithophores (Loeblich and Tappan, 1966) have been studied. This is due to the fact that cultures, particularly of the oceanic planktonic species, have as yet not been achieved. If one considers the extreme diversity of coccolith morphology (cf. Kamptner, 1940; Okada and McIntyre, 1977, and Figs. 11 and 12) studies of coccolithogenesis in genera such as *Papposphaeria* and *Pappomonas* (Tangen, 1972; Manton *et al.*, 1976), *Rhabdospheria* (Leadbeater and Morton, 1973) or *Wigwamma* (Manton *et al.*, 1977; Thomsen, 1980) could be of the greatest value to our understanding of coccolithogenesis. In *Papposphaeria lepida* for example, the 4.5–7.0 μm diameter cell forms coccoliths of 2.0–3.9 μm length (Tangen, 1972), i.e., the coccolith must occupy a very large part of the cell volume. The work of Pienaar and Norris (1979) on *Chryso-*

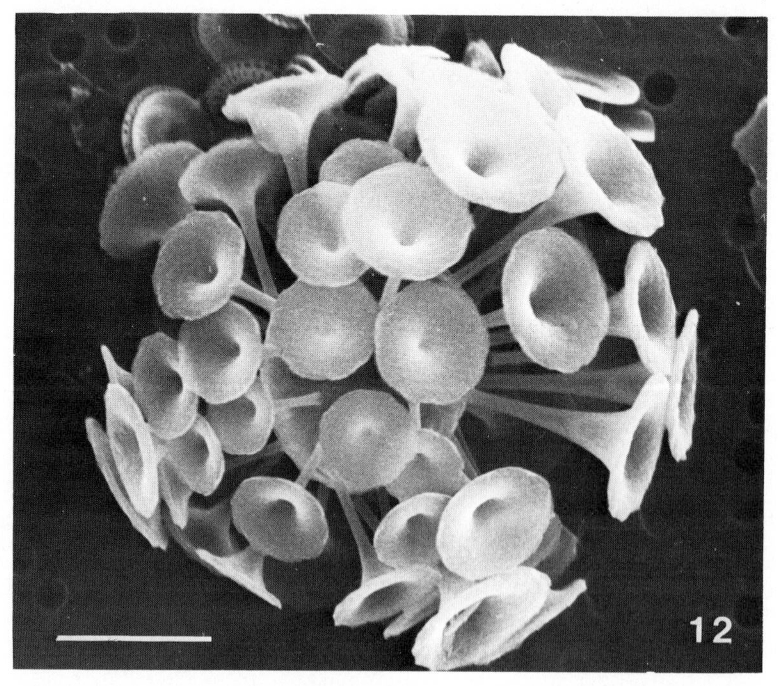

FIG. 12. *Discosphaeria tubifera.* Scanning electron micrograph showing the large trumpet-like processes on the coccoliths (micrograph courtesy of Dr. G. Hallegraeff). Scale = 5 μm.

chromulina spinifera which produces an organic spine scale which is often more than three times as long as the cell shows that this scale is produced within a (Golgi?) vesicle within the cell. The scale is apparently folded, and it probably unfolds once assembly is almost complete. It may be that at this stage the emerging distal region of the spine is still covered with a layer of cytoplasm. Coccolithophorids which have two layers of different coccoliths are often encountered in genera such as *Syracosphaeria* (Gaarder and Heimdal, 1977). Detailed ultrastructural studies of coccolithogenesis in such species should provide important data on the mechanism of coccolith formation.

Clocchiatti (1971) has also found some coccolithophores which have coccoliths characteristic of both *Gephyrocapsa oceanica* and *Emiliania huxleyi*, an observation which strongly suggests that these two species are cospecific, and that the production of a particular kind of coccolith may be under environmental control. However none of the existing cultures of these algae is known to deposit such a mixture of coccoliths.

Within the Chrysophyta in the class Chrysophyceae there exist some genera such as *Chrysococcus* and *Pseudokephyrion* which have a calcified lorica (Bourrelly, 1963). Nothing is known as to the mode of formation of these structures.

V. Non-$CaCO_3$ Deposits

Calcium carbonate is not the only mineral deposit formed by algae. Aside from the large and delicately sculpted deposits of silica formed by diatoms and some other algae (Schmid *et al.*, 1981), algae have been shown to deposit calcium oxalate, barium sulfate, and more rarely other mineral salts.

Calcium oxalate is deposited within the vacuole of a number of Dasycladalean and Caulerpalean algae of the genera *Acetabularia, Bornetella, Cymopolia, Penicillus, Rhipocephalus,* and *Udotea* (Leitgeb, 1888; Solms-Laubach, 1893; Friedmann *et al.*, 1972; Turner and Friedmann, 1974; Böhm *et al.*, 1978), and the Cladophoralean *Apjohnia laetevirens* (Dawes, 1969). In the former the vacuolar calcium oxalate seems to always be associated with "sheath" calcification of the cell walls (Böhm *et al.*, 1978). Calcium oxalate has also been reported in dinoflagellates (Taylor, 1968). Calcium oxalate crystals occur frequently in various higher plants (Arnott and Pautard, 1970; Fischer and Hinz, 1974). They form within chambers (Kristallvakuole), often within vacuoles. There is some disagreement as to the origin and nature of these vacuoles (cf. Wattendorf, 1978). In *Acacia* the crystal sheath consists of cellulose–pectic wall material and in other higher plants the crystals are surrounded by membranes or membrane-like material

(Arnott, 1973), whereas in *Penicillus capitatus*, the only alga where electron microscopic data are available, the crystal chamber is said to be made up of microtubules (Turner and Friedmann, 1974). The recent studies of Menzel and Grant (1981) on *Caulerpa* suggest that these "microtubules" are actually vacuolar proteinaceous material. This is interesting in light of the observation of Buttrose and Lott (1978) who found calcium oxalate within protein bodies of *Eucalyptus*. The origin and possible function of these crystals and their possible functional relationship to "sheath calcification" is in need of detailed study.

The rhizoids of the green alga *Chara* contain statolith vacuoles filled with crystallites of $BaSO_4$ (Schröter *et al.*, 1975). Similar crystals have also been observed in *Spirogyra* (Kreger and Boere, 1969). In *Chara* the $BaSO_4$-containing vesicles appear to be formed from the endoplasmic reticulum. Vesicles of smooth endoplasmic reticulum bud off from the rough endoplasmic reticulum. During formation of the statolith the inner endoplasmic reticulum membrane is lost and the radially arranged $BaSO_4$ crystals are surrounded by a single membrane (Schröter *et al.*, 1973, 1975).

VI. Crystal Formation

An examination of the development and ultrastructure of algal calcium carbonate deposits shows two major types: those which are associated with organic material throughout their development and those which are not. Further differentiation can be made on whether these deposits are intracellular or extracellular. Crystal formation basically requires two steps: crystal nucleation and crystal growth (crystallization). Kinetic studies of calcium carbonate precipitation show that nucleation in seawater is a slow process possibly requiring up to 10^5 years in normal seawater (Pytkowicz, 1965). Conversion of the seawater bicarbonate to carbonate could significantly speed up the rate of nucleation, i.e., only 500 hours if all the bicarbonate were converted to carbonate. Nucleation is therefore a major rate-limiting step in calcification. Photosynthetic removal of CO_2 from a semi or wholly isolated compartment will raise the concentration of carbonate (Borowitzka and Larkum, 1976a; Borowitzka, 1977) and many calcareous algae such as *Halimeda* and *Galaxaura* have the appropriate anatomical features which allow such a process to take place. Others such as *Chara* secrete OH^- ions thus increasing the local carbonate ion concentration. The presence of an organic matrix or nucleation site would also serve to accelerate $CaCO_3$ crystal nucleation. Such material may exist in the pilose layer of the *Halimeda* cell wall where the initial aragonite crystal nuclei form, on the base plate of coccolithophorids and the cell walls of coralline algae where calcite crystals are deposited. For example, Okazaki and co-workers (1979) have shown that acid-insoluble residues of the calcite depositing alga *Serraticardia maxima* caused spontaneous precipitation of calcite and vaterite in a solution of $Ca(HCO_3)_2$. The acid-insoluble residue of the noncalcareous alga *Chondrus verrucosa* caused the precipitation of calcite, whereas the residues of the aragonite depositing algae *Galaxaura fastigiata* and *G. falcata* induced the precipitation of aragonite only. The presence of calcium binding material has been demonstrated in many calcareous algae (e.g., Böhm, 1973; Borowitzka and Larkum, 1976a; Borowitzka, 1979; Misonou *et al.*, 1980). A significant proportion of the soluble protein of the organic matrix of mollusc shells is composed of a repeating sequence of aspartic acid separated by either glycine or serine. This negatively charged aspartic acid may function as a template for epitaxial nucleation of calcium carbonate crystals (Weiner and Hood, 1975). If these amino acids are arranged in a β-sheet configuration the distance from one aspartic acid residue to the next is 0.695 nm. The Ca–Ca distance of the unit cells of aragonite and calcite ranges from about 0.3 to 0.65. The outer cell wall layers of many algae also contain up to 70–80% protein (Hanic and Craigie, 1969) whose amino acid composition is unknown. The methylated, acidic polysaccharide associated with the coccoliths of *Emiliania huxleyi* seems to have a similar function (Fichtinger-Schepman *et al.*, 1979). It is also likely that a proteinaceous or possibly an acid mucopolysaccharide component in algal cell walls or mucilages (i.e., in the coralline algae, the blue-green algae and in the green "sheath" calcifiers), acts as such a matrix. It is therefore important to establish the chemical nature of both the major and minor cell wall and mucilage components of calcareous algal cell walls. Comparative studies with related noncalcareous algae may also be of use.

Once crystal nuclei are supplied then continued crystal growth may proceed without an organic mediator. Studies on the kinetics of calcium carbonate crystallization in dilute freshwater systems show that at moderate to low precipitation rates, calcite grows by a surface-controlled mechanism such as spiral dislocation growth (cf. Nancollas and Reddy, 1971; Wiechers *et al.*, 1975). This means that the precipitation rate is linearly proportional to the product of calcium and carbonate ion concentrations and is not diffusion limited. In saline solutions this process can still occur although magnesium ions depress calcite precipitation rates as do some other ions. These ions also influence the crystal isomorph which is precipitated (Kitano, 1964; Bischoff and Fyfe, 1968). Spiral dislocations have been observed in electron micrographs of growing aragonite crystals in some molluscs (Towe and Hamilton, 1968; Simkiss and Wada, 1980) but not as yet in algae.

Organic molecules may not only induce calcium carbonate crystal formation by providing nucleation sites and act in determining the crystal isomorph precipitated, but they may also inhibit crystal formation. For example, the high concentrations of phenolic substances released by brown algae (Sieburth and Jensen, 1969) are known to inhibit calcite precipitation (Reynolds, 1978). They may therefore be responsible for the singular lack of calcareous Phaeophyceae. Phosphates are also known to inhibit calcification (Simkiss, 1964).

The possible mechanisms of algal calcification and the associated metabolic processes are discussed in detail by Borowitzka (1982).

VII. General Discussion

The deposition of calcium carbonate by algae, although a widespread phenomenon, must have evolved separately in the various divisions and possibly orders at different times. Much of the fossil history of algae is of course that of the calcareous algae since these are best preserved. Calcareous organisms in general begin to predominate the fossil record in the lower Cambrian, and Lowenstam and Margulis (1980) suggest that this is due to the origin of refined mechanisms of intracellular calcium ion concentration coupled with natural selection by predators. Aside from the calcareous blue-green algae (Riding, 1977b) the earliest calcareous algae such as the dasyclads (Herak *et al.*, 1977) and the Solenoporaceae (Poignant, 1977) show quite complex anatomy, and their calcareous nature may relate more to the fact of having evolved an anatomy which encourages $CaCO_3$ precipitation due to CO_2 removal from a semiisolated compartment. The remains of surface calcifiers such as *Padina* would of course be much harder to recognize. The Corallinaceae appear to have arisen somewhat later in the Devonian (Wray, 1977). Flajs (1977a) attempted to see whether there was a phylogenetic derivation of the structural types of algal $CaCO_3$ deposits, however any similarity between algal groups probably represents convergent evolution rather than phylogenetic derivation.

The calcareous habit may have evolved due to a number of different selective pressures. In the coccolithophores the formation of coccoliths may be a means of neutralizing excess OH^- ions formed during HCO_3^- uptake for photosynthesis (Paasche, 1964). The formation of the more complex coccoliths may also be a means of better controlling the sinking rate of these algae. The formation of the extracellular calcite deposits in *Chara* also seems to be a by-product of HCO_3^- uptake in photosynthesis (Lucas, 1979) and this may be a general feature of algal calcification (Borowitzka, 1982). Calcareous skeletons may also reduce grazing pressure, although it must be noted that all calcareous algae are grazed by fish or invertebrates, and that some such as the crustose coralline algae appear to have evolved quite complex interrelationships with some of these grazers (Borowitzka, 1981). The skeletons of the coralline algae also allow them to grow in high wave energy habitats, such as the windward crests of reefs. Without their calcitic skeletons reef-crest forming algae such as *Porolithon onkodes* could not grow in such a high wave energy environment.

Whatever the evolutionary pressures which led to the appearance of the various calcareous algae, their importance to the modern environment cannot be denied, especially in reef habitats. The aragonitic algae such as *Halimeda* are major carbonate sediment formers and may produce about 2 gm $CaCO_3$ m^{-2} per month (Neumann and Land, 1975) in the Bahamas. The crustose coralline algae are essential to maintain the windward reef crest and function as sediment consolidators and cementers (Stearn *et al.*, 1977). The calcareous algae are of course also important in many other marine and freshwater ecosystems and it is important therefore to understand the mechanism of calcification and the physiology of these algae if we are to ever understand and possibly manage and preserve these ecosystems.

As stated previously algal calcification may be a by-product of bicarbonate uptake in photosynthesis, at least in some algae such as the Characeae, corallines, and coccolithophores and possibly even *Halimeda*. The differences between the algae represent evolutionary divergence. The ability to take up bicarbonate provides algae with extra carbon for photosynthesis (Lehman, 1978) and is especially important in alkaline waters. The question therefore becomes why other bicarbonate-utilizing algae, or algae with similar anatomy to *Halimeda* do not calcify. As pointed out by Borowitzka (1982) algae in general secrete a wide range of organic molecules and it is possible that these molecules inhibit $CaCO_3$ nucleation and therefore calcification. It is therefore important to study not only calcareous algae but also their noncalcareous relatives.

References

Adey, W. H., and McIntyre, I. G. (1973). *Bull. Geol. Soc. Am.* **84,** 883–904.

Alexandersson, T. (1974). *J. Sediment. Petrol.* **44,** 7–26.

Andrake, W., and Johansen, H. W. (1980). *J. Phycol.* **16,** 620–622.

Ariovich, D., and Pienaar, R. N. (1979). *Br Phycol. J.* **14,** 17–24.

Arnott, H. J. (1973). *In* "Biological Mineralization" (I. Xipkin, ed.), pp. 609–627. Wiley, New York.

Arnott, H. J., and Pautard, F. G. E. (1970). *In* "Biological Calcification: Cellular and Molecular Aspects" (H. Schraer, ed.), pp. 375–446. Appleton, New York.

Baas-Becking, L. G. M., and Galliher, E. W. (1931). *J. Phys. Chem.* **35,** 467–479.

Bailey, A., and Bisalpultra, T. (1970). *Phycologia* **9,** 83–101.

Bailey, G. P., Rezak, R., and Cox, E. R. (1976). *Phycologia* **15,** 7–18.

Barnes, D. J. (1972). *Proc. R. Soc. London Ser. B* **182,** 331–350.

Barton, R. (1965a). *Nature (London)* **205,** 201.

Barton, R. (1965b). *Planta* **66,** 95–105.
Benner, U., and Schnepf, E. (1975). *Protoplasma* **85,** 337–349.
Berger, S., Sandakhchiev, L. S., and Schweiger, H. G. (1974). *J. Microsc.* **19,** 89–104.
Beuffe, H. (1978). *Protistologica* **16,** 451–458.
Bignot, G. (1978). *Rev. Micropaleontol.* **20,** 177–186.
Bisalpultra, T., Rusanowski, P. C., and Walker, W. S. (1967). *J. Ultrastruct. Res.* **20,** 277–289.
Bischoff, J. L., and Fyfe, W. S. (1968). *Am. J. Sci.* **266,** 65–79.
Black, M. (1933). *Philos. Trans. R. Soc. London B* **222,** 165–192.
Blackwelder, P. L., Brand, L. E., and Guillard, R. L. (1979). *Scanning Electron Microsc.* **1979/II,** 417–420.
Blackwelder, P. L., Weiss, R. E., and Wilbur, K. M. (1976). *Mar. Biol.* **34,** 11–16.
Böhm, E. L. (1973). *Int. Rev. Ges. Hydrobiol.* **58,** 117–126.
Böhm, L., Fütterer, D., and Kaminski, E. (1978). *J. Phycol.* **14,** 486–493.
Bold, H. C., and Wynne, M. J. (1978). "Introduction to the Algae." Prentice-Hall, New York.
Borowitzka, M. A. (1977). *Oceanogr. Mar. Biol., Annu. Rev.* **15,** 189–223.
Borowitzka, M. A. (1979). *Mar. Biol.* **50,** 339–347.
Borowitzka, M. A. (1981). *Endeavour* **5,** 99–106.
Borowitzka, M. A. (1981). *Adv. Phycol.* **1** (in press).
Borowitzka, M. A., and Larkum, A. W. D. (1974). *Protoplasma* **81,** 131–144.
Borowitzka, M. A., and Larkum, A. W. D. (1976a). *J. Exp. Bot.* **27,** 864–878.
Borowitzka, M. A., and Larkum, A. W. D. (1976b). *J. Exp. Bot.* **27,** 879–893.
Borowitzka, M. A., and Larkum, A. W. D. (1976c). *J. Exp. Bot.* **27,** 894–907.
Borowitzka, M. A., and Larkum, A. W. D (1977). *J. Phycol.* **13,** 6–16.
Borowitzka, M. A., and Vesk, M. (1978). *Mar. Biol.* **46,** 295–304.
Borowitzka, M. A., and Vesk, M. (1979). *J. Phycol.* **15,** 146–153.
Borowitzka, M. A., Larkum, A. W. D., and Nockolds, C. E. (1974). *Phycologia* **13,** 195–203.
Bouck, G. B. (1962). *J. Cell Biol.* **12,** 553–569.
Bourrelly, P. (1963). *Ann. N.Y. Acad. Sci.* **108,** 421–429.
Braarud, T., Gaarder, K. R., Markali, J., and Nordli, E. (1953). *Nytt Mag. Bot.* **1,** 129–134.
Brown, R. M. (1969). *J. Cell Biol.* **41,** 109–123.
Brown, R. M. (1974). Wiss. Film C1071. Inst. Wiss. Film, Göttingen.
Brown, R. M., and Romanovicz, D. K. (1976). *Appl. Polym. Symp.* **28,** 537–585.
Brown, R. M., Herth, W., Franke, W. W., and Romanovicz, D. K. (1973). *In* "Biogenesis of Plant Cell Wall Polysaccharides" (F. Loewus, ed.), pp. 207–257. Academic Press, New York.
Buttrose, M. S., and Lott, J. N. A. (1978). *Can. J. Bot.* **56,** 2083–2091.
Cabioch, J., and Giraud, G. (1978a). *Phycologia* **17,** 369–381.
Cabioch, J., and Giraud, G. (1978b). *C. R. Acad. Sci. Paris D* **286,** 1783–1785.
Callow, M. E., and Evans, L. V. (1976). *Planta* **131,** 155–157.
Cameron, G. R. (1930). *J. Pathol. Bacteriol.* **33,** 929–955.
Campion-Alsumard, T. (1979). *Oceanol. Acta* **2,** 143–156.
Chambers, T. C., and Mercer, F. V. (1964). *Aust. J. Biol. Sci.* **17,** 372–387.
Clocchiatti, M. (1971). *C. R. Acad. Sci. Paris D* **273,** 318–321.
Colombo, P. M. (1978). *Phycologia* **17,** 227–235.
Colombo, P. M., and DeCarli, M. E. (1980). *Cytobios* **27,** 147–155.
Colombo, P. M., and Orsenigo, M. (1977). *Phycologia* **16,** 9–17.
Colombo, P. M., Rascio, N., Solazzi, A., and Tolomio, C. (1976a). *Mem. Biol. Mar. Oceanogr.* **6,** 183–195.
Colombo, P. M., Orsenigo, M., Solazzi, A., and Tolomio, C. (1976b). *Mem. Biol. Mar. Oceanogr.* **6,** 197–208.
Crawley, J. C. W. (1965). *Nature (London)* **205,** 200–201.

Daily, F. K. (1973). *Bull. Torr. Bot. Club.* **100,** 75–78.
Daily, F. K. (1975). *Phycologia* **14,** 331–332.
Dawes, C. J. (1969). *Phycologia* **8,** 77–84.
Dawes, C. J., and Barilotti, D. C. (1969). *Am. J. Bot.* **56,** 8–15.
Dawes, C. J., and Rhamstine, E. C. (1967). *J. Phycol.* **3,** 117–126.
Degens, E. T. (1976). *Top. Curr. Chem.* **64,** 1–112.
DeJong, E., van Rens, L., Westbroek, P., and Bosch, L. (1979). *Eur. J. Biochem.* **99,** 559–567.
Dixon, H. H. (1900). *Proc. R. Soc. London* **66,** 305–315.
Dodson, J. (1974). *Hydrobiologia* **44,** 247–255.
Duckett, J. G., Buchanan, J. S., Peel, M. C., and Martin, M. T. (1974). *New Phytol.* **73,** 497–507.
Dyck, L. A. (1969). Ph.D. thesis, Washington University, St. Louis, Missouri.
Eiseman, N. J. (1970). *Phycologia* **9,** 45–47.
Falk, H., and Sitte, P. (1963). *Protoplasma* **57,** 290–303.
Faure-Fremiet, E., Andre, J., and Ganier, M. (1968). *J. Microsc.* **7,** 693–704.
Feldmann, J., Feldmann, G., and Guglielmi, G. (1977). *Rev. Algol. N. S.* **12,** 11–30.
Fichtinger-Schepman, A. M. J., Kamerling, J. P., Vliegenthart, J. F. G., DeJong, E. W., Bosch, L., and Westbroek, P. (1979). *Carbohydr. Res.* **69,** 181–189.
Fischer, H., and Hinz, U. (1974). *Protoplasma* **81,** 349–362.
Flajs, G. (1977a). *Paleontographica* **160,** 69–128.
Flajs, G. (1977b). *In* "Fossil Algae" (E. Flügel, ed.), pp. 225–231. Springer-Verlag, Berlin and New York.
Franceschi, V. R., and Lucas, W. J. (1980). *Protoplasma* **104,** 253–271.
Friedmann, E. I., Roth, W. C., Turner, J. B., and McEwen, R. S. (1972). *Science* **177,** 891–893.
Furuya, K. (1960). *Bot. Mag. Tokyo* **73,** 355–359.
Gaarder, K. R. (1962). *Nytt Mag. Bot.* **10,** 35–51.
Gaarder, K. R., and Heimdal, B. R. (1977). *"Meteor" Forschungs-ergeb. Ser. D* **24,** 54–71.
Gaarder, K. R., and Markali, J. (1956). *Nytt Mag. Bot.* **5,** 1–5.
Garbary, D., and Veltkamp, C. J. (1980). *Phycologia* **19,** 49–53.
Gayral, P., and Fresnel, J. (1976). *Phycologia* **15,** 339–355.
Giraud, G., and Cabioch, J. (1976). *Phycologia* **15,** 405–414.
Giraud, G., and Cabioch, J. (1977). *Rev. Algol. N. S.* **12,** 45–60.
Giraud, G., and Cabioch, J. (1979). *Biol. Cell.* **36,** 81–86.
Golubic, S. (1973). *In* "The Biology of Blue-green Algae" (V. G. Carr and B. A. Whitton, eds.), pp. 434–472. Blackwell, Oxford.
Golubic, S., and Focke, J. W. (1978). *J. Sediment. Petrol.* **48,** 751–764.
Gunning, B. E. S., and Pate, J. S. (1969). *Protoplasma* **68,** 107–133.
Hamano, M., and Honjo, S. (1969). *J. Geol. Soc. Jpn.* **75,** 607–614.
Hamel, G. (1930). *Rev. Algol.* **5,** 404–417.
Hanic, L. A., and Craigie, J. S. (1969). *J. Phycol.* **5,** 89–102.
Hartman, T., and Eschrich, W. (1969). *Planta* **85,** 303–312.
Hawes, C. R. (1979). *New Phytol.* **83,** 445–450.
Helder, R. J., Boerma, J., and Zanstra, P. E. (1980). *Proc. K. Ned. Akad. Wet. Ser. C* **83,** 151–166.
Herak, M., Kochansky-Devide, V., and Gusic, I. (1977). *In* "Fossil Algae" (E. Flügel, ed.), pp. 143–153. Springer-Verlag, Berlin and New York.
Hibberd, D. J. (1979). *Arch. Protistenk.* **121,** 146–154.
Hillis-Colinvaux, L. (1980). *Adv. Mar. Biol.* **17,** 1–327.
Horn af Rantzien, H. (1959). *K. Sven. Vetensk. Akad. Ark. Bot. Ser. 2* **4,** 165–332.
Irion, G., and Müller, G. (1968). *In* "Recent Developments in Carbonate Sedimentology in Central Europe" (G. Müller and G. M. Friedman, eds.), pp. 151–171. Springer-Verlag, Berlin and New York.

Jagels, R. (1973). *Am. J. Bot.* **60,** 1003–1009.
Johansen, H. W. (1969). *Univ. Calif. Publ. Bot.* **49,** 1–98.
Johansen, H. W. (1974). *Oceanogr. Mar. Biol., Annu. Rev.* **12,** 77–124.
Johnston, I. S. (1980). *Int. Rev. Cytol.* **67,** 171–214.
Kamptner, E. (1940). *Naturh. Mus. Wien. Ann. Anz.* **51,** 54–149.
Kennedy, W. J., Taylor, J. D., and Hall, A. (1969). *Biol. Rev.* **44,** 499–530.
Kitano, Y. (1964). *In* "Recent Researches in the Fields of Hydrosphere, Atmosphere and Nuclear Geochemistry" (Y. Miyake and T. Koyama, eds.), pp. 305–319. Marutzen, Tokyo.
Klaveness, D. (1972a). *Protistologia* **8,** 335–346.
Klaveness, D. (1972b). *Br. Phycol. J.* **7,** 309–318.
Klaveness, D. (1973). *Nytt Mag. Bot.* **20,** 151–162.
Klaveness, D. (1976). *Protistologia* **12,** 217–224.
Klaveness, D., and Paasche, E. (1971). *Arch. Mikrobiol.* **75,** 382–385.
Klaveness, D., and Paasche, E. (1979). *In* "Biochemistry and Physiology of Protozoa" (M. Levandowsky and S. H. Hutner, eds.), 2nd Ed., Vol. I., pp. 191–213. Academic Press, New York.
Kobluk, D. R., and Risk, M. J. (1977). *J. Sediment. Petrol.* **47,** 517–528.
Kolesar, P. T. (1978). *J. Sediment. Petrol.* **48,** 815–820.
Konishi, K., and Wray, J. L. (1961). *J. Paleontol.* **35,** 659–666.
Kreger, D. R., and Boere, H. (1969). *Acta Bot. Neerl.* **18,** 143–151.
Lamberts, A. E. (1978) *In* "Coral Reefs: Research Methods" (D. R. Stoddart and R. E. Johannes, eds.), pp. 523–527. UNESCO, Paris.
LaVelle, J. M. (1979). *Mar. Biol.* **55,** 37–44.
Leadbeater, B. S. C. (1970) *Br. Phycol. J.* **5,** 57–69.
Leadbeater, B. S. C. (1971). *Ann. Bot.* **35,** 429–439.
Leadbeater, B. S. C., and Morton, C. (1973). *Nova Hedwigia* **24,** 207–233.
Leak, L. V. (1967). *J. Ultrastruct. Res.* **21,** 61–74.
Lebednik, P. A. (1978). *Phycologia* **17,** 388–395.
Lefort, F. (1975). *Cah. Biol. Mar.* **16,** 213–229.
Lehman, J. T. (1978). *J. Phycol.* **14,** 33–42.
Leitgeb, H. (1888). *Sitzungs-ber. Akad. Wiss. Wien. Math. Naturwiss. Kl., Abt. 1* **96,** 13–37.
Littler, M. M. (1976). *Micronesica* **12,** 27–41.
Loeblich, A. R., and Tappan, H. (1966). *Phycologia* **6,** 81–216.
Lohmann, H. (1902). *Arch. Protistenk.* **1,** 89–165.
Lowenstam, H. A., and Margulis, L. (1980). *BioSystems* **12,** 27–41.
Lucas, W. C. (1979). *Plant Physiol.* **63,** 248–254.
Lucas, W. C., and Smith, F. A. (1973). *J. Exp. Bot.* **24,** 1–14.
MacRaild, G. N., and Womersley, H. B. S. (1974). *Phycologia* **13,** 89–93.
Manton, I., and Leedale, G. F. (1963). *Arch. Mikrobiol.* **47,** 115–136.
Manton, I., and Leedale, G. F. (1969). *J. Mar. Biol. Assos. U.K.* **49,** 1–16.
Manton, I., Sutherland, J., and McCully, M. (1976). *Br. Phycol. J.* **11,** 225–238.
Manton, I., Sutherland, J., and Oates, K. (1977). *Proc. R. Soc. London Ser. B* **197,** 145–168.
Marré, E., Lado, P., Radi-Caldogno, F., Colombo, R., Cocucci, M., and De Michelis, M. I. (1975). *Physiol. Veg.* **13,** 797–811.
Marszalek, D. S. (1971). *Scanning Electron Microsc. 1971* Pt. I, pp. 273–280.
Marszalek, D. S. (1975). *J. Sediment. Petrol.* **45,** 266–271.
Maslov, V. P. (1956). *Acad. Sci. USSR Inst. Geol. Sci. Trud.* **160,** 1–301.
Mathews, H. L., Prescott, G. W., and Obenshain, S. S. (1965). *Proc. Soil Sci. Soc. Am.* **29,** 729–732.
Matty, P. J., and Johansen, H. W. (1981). *Phycologia* **20,** 46–55.

Menzel, D., and Grant, B. (1981). *Protoplasma* **107,** 47–61.
Metha, V. B., and Vaidya, B. A. (1978). *J. Exp. Bot.* **29,** 1423–1430.
Milliman, J. D. (1977). *In* "Fossil Algae" (E. Flügel, ed.), pp. 232–247. Springer-Verlag, Berlin and New York.
Misonou, T., Okazaki, M., Furuya, K., and Nisizawa, K. (1980). *Jpn. J. Phycol.* **28,** 105–112.
Miyata, M., Okazaki, M., and Furuya, K. (1977). *Bull. Jpn. Soc. Phycol.* **25,** 1–6.
Moberly, R. (1968). *Sedimentology* **11,** 61–82.
Moestrup, O., and Thomsen, H. A. (1974). *Protoplasma* **81,** 247–269.
Monty, C. L. V. (1967). *Ann. Soc. Geol. Belg.* **90,** 55–100.
Nakahara, H., and Bevelander, G. (1978). *Jpn. J. Phycol.* **26,** 9–12.
Nancollas, G. H., and Reddy, M. M. (1971). *J. Colloid Interface Sci.* **37,** 824–830.
Neumann, K. (1969). *Helgol. Wiss. Meeresunters* **19,** 355–375.
Neumann, A. C., and Land, L. S. (1975). *J. Sediment. Petrol.* **45,** 763–786.
Nisizawa, K., Kuroda, K., Tomita, Y., and Shimahara, H. (1974). *Bot. Mar.* **17,** 16–19.
Okada, H., and McIntyre, A. (1977). *Micropaleontol.* **23,** 1–55.
Okazaki, M., and Furuya, K. (1977). *3rd Int. Symp. Mech. Biomineralizat. the Invertebr. Plant., Kashikojima, Japan* (Abstract).
Okazaki, M., Misonou, T., and Furuya, K. (1979). *Bull. Tokyo Gakugei Univ. Ser. 4* **31,** 215–221.
Outka, D. E., and Williams, D. C. (1971). *J. Protozool.* **18,** 285–297.
Paasche, E. (1962). *Nature (London)* **193,** 1094.
Paasche, E. (1964). *Physiol. Plant. (Suppl.)* **3,** 5–82.
Paasche, E. (1968). *Annu. Rev. Microbiol.* **22,** 71–85.
Palandri, M. (1972). *Caryologia* **25,** 211–235.
Parke, M. (1961). *Br. Phycol. Bull.* **2,** 47–55.
Parke, M. (1971). *Proc. Plankton. Conf., 2nd, Rome* pp. 929–937.
Parke, M., and Adams, I. (1960). *J. Mar. Biol. Assos. U.K.* **39,** 263–274.
Pate, J. S., and Gunning, B. E. S. (1972). *Annu. Rev. Plant. Physiol.* **23,** 173–196.
Pentecost, A. (1978). *Proc. R. Soc. London Ser. B* **200,** 43–61.
Pentecost, A. (1980). *Int. Rev. Cytol.* **62,** 1–27.
Peyriere, M. (1977). *Rev. Algol. N. S.* **12,** 31–43.
Pickett-Heaps, J. D. (1967). *Aust. J. Biol. Sci.* **20,** 539–551.
Pienaar, R. N. (1969a). *J. Cell Sci.* **4,** 561–567.
Pienaar, R. N. (1969b). *J. Phycol.* **5,** 321–331.
Pienaar, R. N. (1971). *Protoplasma* **73,** 217–224.
Pienaar, R. N. (1976). *J. Mar. Biol. Assos. U.K.* **56,** 1–11.
Pienaar, R. N., and Norris, R. E. (1979). *Phycologia* **18,** 99–108.
Pobeguin, T. (1954). *Ann. Sci. Nat. Bot.* **15,** 29–109.
Poignant, A.-F. (1977). *In* "Fossil Algae" (E. Flügel, ed.), pp. 177–189. Springer-Verlag, Berlin and New York.
Prins, H. B. A., Snel, J. F. H., Helder, R. J., and Zanstra, P. E. (1979). *Hydrobiol. Bull.* **13,** 106–111.
Prins, H. B. A., Snel, J. F. H., Helder, R. J., and Zanstra, P. E. (1980). *Plant Physiol.* **66,** 818–822.
Proctor, V. W. (1980). *J. Phycol.* **16,** 218–233.
Pueschel, C. M. (1977). *Protoplasma* **91,** 15–30.
Pueschel, C. M. (1980a). *Phycologia* **19,** 210–217.
Pueschel, C. M. (1980b). *Science* **209,** 422–423.
Pytkovicz, R. M. (1965). *J. Geol.* **73,** 196–199.
Ramus, J. (1969). *J. Cell Biol.* **41,** 340–345.
Ramus, J. (1972). *J. Phycol.* **8,** 97–111.
Reynolds, R. C. (1978). *Limnol. Oceanogr.* **23,** 585–597.

Riding, R. (1977a). *Palaeontology* **20,** 33–46.
Riding, R. (1977b). *In* "Fossil Algae" (E. Flügel, ed.), pp. 202–211. Springer-Verlag, Berlin and New York.
Rietema, H. (1971). *Acta Bot. Neerl.* **20,** 291–298.
Schmid, A. M., Borowitzka, M. A., and Volcani, B. E. (1981). *In* "Cytomorphogenesis in Plants" (O. Kiermayer, ed.), pp. 63–97. Springer-Verlag, Berlin, and New York (in press).
Schröter, K., Rodriguez-Garcia, M. I., and Sievers, A. (1973). *Protoplasma* **76,** 435–442.
Schröter, K., Läuchli, A., and Sievers, A. (1975). *Planta* **122,** 213–225.
Shinn, E. A., Lloyd, R. M., and Ginsburg, R. N. (1969). *J. Sediment. Petrol.* **39,** 1202–1228.
Sieburth, J. M., and Jensen, A. (1969). *J. Exp. Mar. Biol. Ecol.* **3,** 275–289.
Simkiss, K. (1964). *Biol. Rev.* **39,** 487–505.
Simkiss, K., and Wada, K. (1980). *Endeavour N. S.* **4,** 32–37.
Smestad, B., and Percival, E. (1972). *Carbohydr. Res.* **25,** 299–304.
Solms-Laubach, H.v. (1887). "Einleitung in die Paläophytologie." Felix, Leipzig.
Solms-Laubach, H.v. (1893). *Ann. Jard. Bot. Buitenzorg* **11,** 61–97.
Somers, G. F., and Brown, M. (1978). *Estuaries* **1,** 17–28.
Spear, D. G., Barr, J. K., and Barr, C. E. (1969). *J. Gen. Physiol.* **54,** 397–414.
Stacey, V. J., and Pienaar, R. N. (1980). *Br. Phycol. J.* **15,** 365–376.
Stearn, C. W., Scoffin, T. P., and Martindale, W. (1977). *Bull. Mar. Sci.* **27,** 479–510.
Suess, E. (1970). *Geochim. Cosmochim. Acta* **34,** 157–168.
Tangen, K. (1972). *Norw. J. Bot.* **19,** 171–178.
Taylor, D. L. (1968). *J. Mar. Biol. Assos. U.K.* **48,** 349–366.
Thomsen, H. A. (1977). *Bot. Nat.* **130,** 147–153.
Thomsen, H. A. (1980). *Br. Phycol. J.* **15,** 335–342.
Towe, K. M., and Hamilton, G. H. (1968). *Calc. Tissues Res.* **1,** 306–318.
Turner, C. H. C., and Evans, L. V. (1978). *Br. Phycol. J.* **13,** 51–55.
Turner, J. B., and Friedmann, E. I. (1974). *J. Phycol.* **10,** 125–134.
Waaland, S. D., and Waaland, J. R. (1975). *Planta* **126,** 127–138.
Wallner, J. (1933). *Planta* **20,** 287–293.
Walter, M. R. (1976). *In* "Stromatolites" (M. Walter, ed.), pp. 87–112. Elsevier, Amsterdam.
Watabe, N. (1967). *Calc. Tissues Res.* **1,** 114–121.
Wattendorf, J. (1978). *Protoplasma* **95,** 193–206.
Wefer, G. (1980). *Nature (London)* **285,** 323–324.
West, J. A. (1969). *Phycologia* **8,** 187–192.
Weiner, S., and Hood, L. (1975). *Science* **190,** 987–989.
Wiechers, H. N., Sturrock, P., and Marais, G.v.-R. (1975). *Water Res.* **9,** 835–845.
Wilbur, K. M., and Watabe, N. (1963). *Ann. N.Y. Acad. Sci.* **109,** 82–112.
Wilbur, K. M., Hillis-Colinvaux, L., and Watabe, N. (1969) *Phycologia* **8,** 27–35.
Wray, J. L. (1977). "Calcareous Algae." Elsevier, Amsterdam.
Yendo, K. (1904). *J. Coll. Sci. Imp. Univ. Tokyo* **19,** 15–45.
Zetsche, K. (1967). *Planta* **76,** 326–334.

INTERNATIONAL REVIEW OF CYTOLOGY, VOL. 74

Naturally Occurring Neuron Death and Its Regulation by Developing Neural Pathways

TIMOTHY J. CUNNINGHAM

Department of Anatomy, The Medical College of Pennsylvania, Philadelphia, Pennsylvania

I. Introduction

Understanding any ontogenetic process requires both sufficient descriptive information and an appreciation of the role the phenomenon plays in assuring the final form and function of the mature individual. For processes such as cell division and cell migration, the second of these requirements is easily fulfilled because their value is obvious. For other phenomena, such as natural cell death, the contributions made are sometimes less apparent. Indeed, there are examples from several organ systems, including the nervous system, where cell death plays a recognized role in organogenesis. For example, the death of interdigital tissue during the development of the forelimb provides for the appropriate construction of the hand (Saunders, 1966); the death of certain cells in the optic cup allows blood vessels into the eye and axons to grow to the brain (Silver and Hughes, 1973). However, most of the cell death which occurs during the development of the nervous system has no such obvious outcome. In many areas of the nervous system, neurons

ISBN 0-12-364474-7

complete their final mitosis and migrate to the appropriate location only to have one-half or more of the cells degenerate. Prior to the period of cell death, all neurons begin to differentiate normally and each appears to have as much chance to survive as the next. The cell becomes vulnerable when it begins to contact other cells in order to lay down functional pathways.

A recognition of the neuron's dependence on these other cells appears to be the key to understanding why the neuron dies. Yet, it is still not entirely clear how neuron death ultimately contributes to the normal functioning of the mature nervous system. Does neuron death eliminate cells which form insufficient numbers of synapses with connecting structures? Does neuron death remove mismigrated or grossly misprojecting cells, i.e., developmental errors which would be incompatible with normal functioning? Or perhaps neuron death is responsible for the ordered connectivity found in the nervous system, serving to eliminate neurons which violate the rules of neural specificity. While each of these functions of neuron death has been suggested (and there is some evidence for each) no one explanation seems entirely adequate.

In this article, I will first describe degenerating neurons and show how the morphological expression of cell death is closely related to a neuron's state of maturity. The timing of natural cell death also will be considered in order to define where natural neuron death fits into other aspects of neural development and neuron differentiation. Experimental studies will show the dependence a neuron has on the cell populations with which it ultimately connects. Finally, the possible functions of cell death will be considered in an attempt to provide a more comprehensive explanation for this phenomenon.

II. Morphology of Degenerating Cells

There have been several descriptions of degenerating neurons in the developing nervous system of normal mammalian and nonmammalian vertebrates (e.g., Cunningham *et al.*, 1980, 1981; Giordano *et al.*, 1980; Chu-Wang and Oppenheim, 1978a; Sohal and Weidman, 1978; Panesse, 1976; Pilar and Landmesser, 1976; O'Connor and Wyttenback, 1974; Cantino and Daneo, 1972). These studies are usually initiated with the hope of providing fine-structural information which might reflect the underlying mechanisms of the degeneration process. As will be seen later, such morphological correlates may be possible.

Based on these investigations, it is now clear that there are two distinct types of naturally degenerating neurons on morphological grounds. In the first, the initial and most dramatic changes are found in the nucleus where the chromatin is atypically condensed (Fig. 1). The cytoplasm is immature

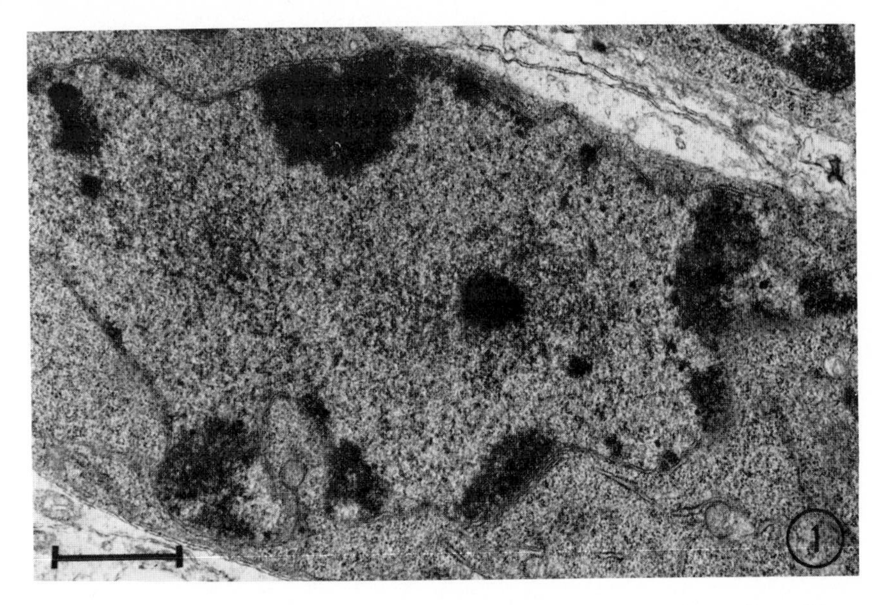

FIG. 1. Example of "Nuclear" type degeneration. Degenerating neuron in ganglion cell layer of normal 7-day-old rat. Chromatin is atypically condensed and cytoplasm is filled with free monosomes but few cisternae. Bar represents 1 μm.

with little or no rough endoplasmic reticulum (RER) and numerous free monosomes. In later stages of degeneration, the nuclear chromatin condenses further and the nuclear membrane breaks down (Fig. 2). In the second type of neuron death the initial changes do not appear in the nucleus but in the cytoplasm. There is an abundance of intracytoplasmic cisternae, including Golgi and RER, but these cisternae are dilated compared to adjacent cells which appear normal (Fig. 3). The cell is uniformly electron dense and appears shrunken. Eventually, the nuclear chromatin will condense in this cell also, and there will be further disintegration of the cytoplasm.

Pilar and Landmesser (1976) aptly term the two morphological types of neuron death "nuclear" and "cytoplasmic," respectively, because these terms describe the most dramatic changes in the neurons. Since their study other investigators have shown that one or the other or both of these types of neuron death may appear during normal development, depending on the species and area of the nervous system examined. It was the observations of Pilar and Landmesser, however, which first indicated that the type of morphology a cell would exhibit during degeneration was related to the state of maturity of that cell at the time of its death. They find that naturally degenerating chick ciliary ganglion cells show dilated cytoplasmic cisternae

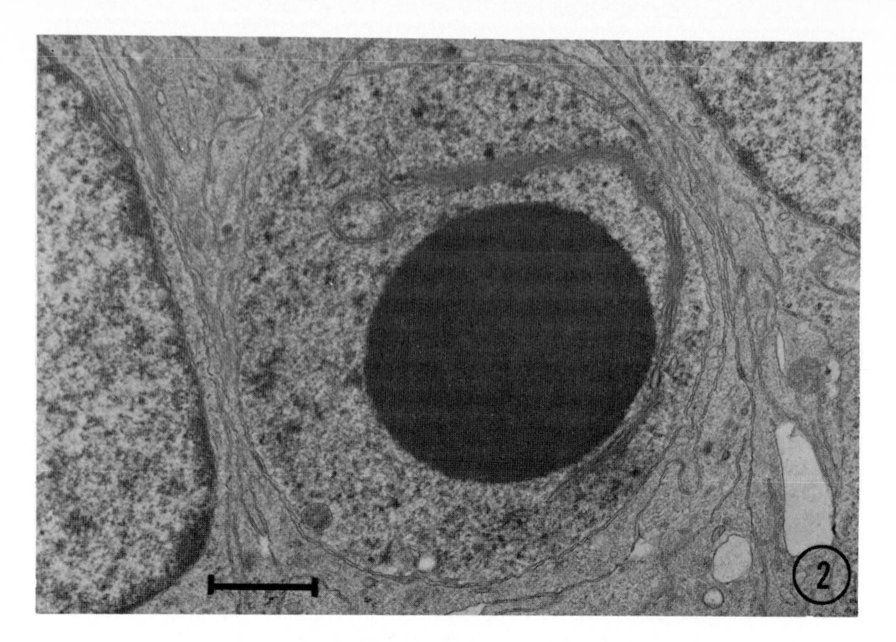

FIG. 2. Late stage of retinal ganglion cell degeneration. Nuclear membrane has broken down and chromatin appears to coalesce. Bar represents 1 μm.

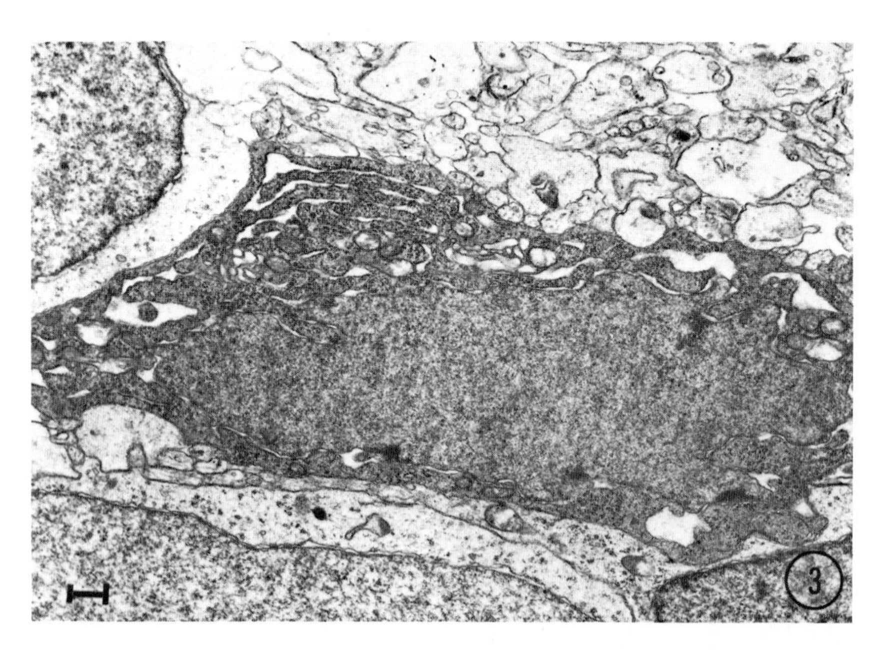

FIG. 3. Example of "cytoplasmic" type degeneration. Typical degenerating neuron in superficial layers of the superior colliculus of normal 6-day-old rat. Cells show overall increase in electron density and cytoplasmic cisternae are dilated. Bar represents 1 μm.

and increased electron density, i.e., the cytoplasmic type degeneration. When all the target structures for the neurons are removed prior to innervation, they find the expected enhancement of neuron death in the ganglion (see below), but surprisingly, the cells show condensed nuclear chromatin and immature cytoplasm, i.e., the nuclear type degeneration. Since it is known that an abundance of RER indicates a more mature neuron (Tennyson and Mytilineous, 1979; Wechsler and Meller, 1967), and since the development of this organized protein synthetic machinery correlates with the onset of synthesis of transmitter-associated enzymes in the ganglion (Chiappinelli *et al.*, 1976), the results lead to an important conclusion. The presence of a target is essential not only for neuron survival, but also for structural and correlated biochemical maturation. When the neurons are deprived of their target, this mature protein synthetic machinery never develops, and as a result, the ganglion neurons die in a relatively immature state. The naturally degenerating neurons, on the other hand, have been allowed to make sufficient contact with the target and therefore have developed the structures associated with synthesis of protein for export. Thus, when these cells die they are more mature (Pilar and Landmesser, 1976).

Unfortunately, all areas of the nervous system do not give such straightforward results. In chick spinal cord, motor neurons show the two types of degeneration, but both appear normally and both appear after target deprivation (Chu-Wang and Oppenheim, 1978a). In the duck trochlear nucleus only the cytoplasmic type of cell death is found normally *and* after target deprivation (Sohal and Weidman, 1978). In the normal neonatal rat superior colliculus both types appear but the vast majority show the cytoplasmic type of degeneration (Giordano *et al.*, 1980), while in the retina of normal rats of the same ages there is only the nuclear type of degeneration (Cunningham *et al.*, 1980, 1981). Therefore, normally degenerating neurons do not always show the cytoplasmic or more mature form of degeneration and target deprived cells do not always show the nuclear or less mature form of degeneration. These findings lead to the conclusion that some neurons can die normally in a less differentiated state while other neurons can be deprived of their target structures and still mature.

It is not unreasonable to suppose that neurons could die naturally at different developmental stages and later it will be suggested that the underlying causes of the death of less mature cells are different from that of more mature cells. The neurons which mature in the absence of their target structures present a more difficult problem. Either neurons are able to achieve maturity without outside help, or structures other than the target can trigger the maturation of an organized system of RER.

If the protein synthetic machinery of a neuron does not develop entirely independent of target structures, then it may be that the afferent input is important for some neurons. We recently tested this proposition by removing the eyes from newborn rats and examining the superficial layers of their superior colliculi at times when large numbers of degenerating cells appear in normal rats (Giordano and Cunningham, 1981; and in preparation). Although a small percentage of neurons may be target deprived because of cell loss in other visual centers, the major effect of eye removal is deafferentation. In normal colliculi, 98% of the degenerating neurons show the cytoplasmic type of cell death, but following enucleation about this same percentage show the nuclear type degeneration (Fig. 4). In this case, it appears that afferents are able to exert the same controls over neuronal maturation as target structures. Furthermore, afferents are likely to be responsible for the maturation of those neurons which achieve maturity in the absence of targets (see above).

The ideas of Pilar and Landmesser (1976) appear to be fundamentally correct. A neuron will show one or the other type of degeneration depending on its maturity at the time it is selected to die. In some cases, contact with target structures may be important to the cell in achieving this maturity, and in other cases, afferent contact controls this phase of maturation. Nuclear

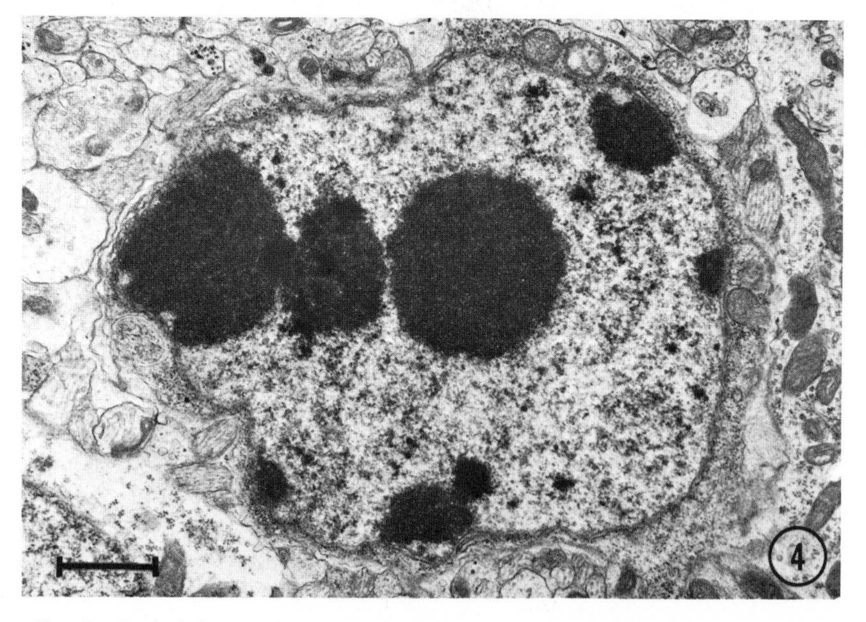

FIG. 4. Typical degenerating neuron in superior colliculus of 6-day-old rat after bilateral enucleation at birth. Neurons now show "nuclear" type degeneration (see text). Bar represents 1 μm.

type degeneration occurs when insufficient contact is made and as a result, the cytoplasm is unable to mature. Cytoplasmic type degeneration is found when the cell makes sufficient contact with its target or its input to trigger the formation of an organized system of RER for the production of enzymes related to synaptic transmission.

III. Timing of Neuron Death: Studies of Normal Development

Ever since the study of Hamburger and Levi-Montalcini (1949) in which they documented the phenomenon of naturally occurring neuron death in chick spinal ganglia, neuroembryologists have attempted to correlate the timing of cell death with the timing of other aspects of neural development and neuron differentiation. In this way they could determine which developmental processes are likely to be most directly correlated with neuron death. It is now clear that a large proportion of neurons which are destined to die naturally have a remarkable capacity to differentiate normally. As will be outlined below, most neurons which die have already migrated to their appropriate positions, developed dendritic and axonal processes, received some synaptic connections from afferent sources, and have sent axons into their target structures.

A. Cell Migration

The majority of neurons which degenerate naturally during development do so after they have migrated to their appropriate positions. Nevertheless, examination of regions where neurons are actively dividing but have not yet migrated suggests some cells may be lost prior to migration. In neuroblastic zones of the infant rat brain, degenerating cells are frequently encountered (Cunningham *et al.*, 1980, and unpublished observations). It is assumed that these cells are recently postmitotic. Chick dorsal root ganglion cells also may degenerate immediately after division (Hamburger *et al.*, 1981; Panesse, 1976). Since it is unlikely that such cells have contacted afferents or target structures and since these contacts appear to be critical to neuron survival (see below), the factors involved in the death of young daughter cells are probably different from those involved in the neuron death which occurs somewhat later in development. As suggested by Hamburger *et al.* (1981) the reasons for this cell degeneration probably have more to do with the successful completion of mitosis (even though cell division may appear to be completed).

The death of neurons which are in the process of migrating or of those which have migrated to inappropriate positions has also been suggested to occur (Hughes, 1973; Clarke and Cowan, 1976). Clarke and Cowan (1976) labeled neurons of the isthmo-optic nucleus (ION) by injection of retrogradely transported horseradish peroxidase (HRP) into the targets of these neurons. After injections early in development, they find a few cells which are some distance from the normal area occupied by the nucleus. Most such abnormally positioned cells disappear later in development which suggests that the cells migrate inappropriately and therefore are eliminated. However, the number of such malpositioned cells is very small (2–3% of the ION neurons) and the disappearance of these cells constitutes only a small percentage of the total number of cells which migrate appropriately and later die (60% of the ION neurons, Clarke *et al.*, 1976). Mismigration and subsequent degeneration may also account for our observations in the retina of neonatal rats. At times when a number of degenerating cells appear in the ganglion cell layer, a few degenerating cells, identical in appearance to degenerating ganglion cells, are found in the inner plexiform layer (Cunningham, Mohler, and Green, unpublished observations). This layer is predominantly acellular in the mature eye. Therefore, the dying cells may represent ganglion cells which did not reach the innermost parts of the retina during migration.

While it is not clear whether the factors involved in the death of mismigrated cells are different from those responsible for the death of appropriately migrated cells, the results of Clarke and Cowan (1976) are notable because they suggest that the atypical position of the neuron does not interfere with other important aspects of differentiation, i.e., the cell still forms an axon and sends it to the correct target. Later I will argue that despite this target contact, the abnormal position of the neuron still represents a disadvantage during development because mismigration should interfere with the cell's ability to achieve an appropriate complement of afferents. In other words, it is reasonable to suppose that these inputs are directed to the area of the brain where the cell should be.

B. Development of Dendrites and Outgrowth of Axons

After migration but before the period of cell death, individual neurons in a particular population provide no obvious anatomical clues as to which will degenerate. It is sometimes difficult to determine what aspects of differentiation occur prior to the natural death of an individual cell because different neurons in the population may be in different stages of maturation. One approach to this problem has been to measure the progress of development in a neuron population in which neuron death has been greatly enhanced. In most cases, this is accomplished by removal of target structures for neurons in a particular area prior to the time these targets are innervated.

The assumption behind this approach is that accentuated neuron loss due to target removal is actually an exaggerated version of normal neuron loss and that the factors responsible for the death of individual cells are the same in both situations (indeed, the timing of cell death is similar in both cases). The advantage to this type of experiment, at least for some areas of the nervous system, is that target removal results in complete or nearly complete loss of the innervating neuronal population whereas only a proportion of cells are lost in normal animals. Therefore, when sampling individual target-deprived cells as to their degree of differentiation, one can be reasonably certain that the neuron will in fact die.

Using this approach, Oppenheim *et al.* (1978) removed the limb bud in the chick embryo before the limb was innervated by spinal motor neurons. Prior to the period of cell death, they examined several developmental parameters including the elaboration of dendritic processes by the spinal motor neurons. They find that some neurons destined to be eliminated could in fact develop normally appearing dendritic processes. Similar findings have been reported for the chick ciliary ganglion and the chick neural retina after target removal early in development (Landmesser and Pilar, 1974a; Hughes and LaVelle, 1975). However, the fact that neurons can develop dendritic processes in the absence of target structures is not altogether surprising since afferents to neurons have been more strongly implicated in regulation of dendritic outgrowth than have targets (see review by Cunningham and Murphy, 1979). As will be shown below, target removal has no detectable effect on the growth of at least some afferents to the deprived cells, and so these growing axons are still able to exert their effects on dendritic development.

The outgrowth of axons by neurons destined to die normally or after induced cell death is perhaps more interesting since for some neurons the interactions of axon terminals with their target structures may be decisive in ensuring cell survival (see below). It is now known that neurons which undergo natural cell death cannot only sprout axons but send these axons to appropriate target structures. While this was suspected to be the case almost from the beginning of studies of naturally occurring neuron death, e.g., see Levi-Montalcini (1950), the first convincing demonstration of axonal innervation prior to cell death came with the application of retrograde tracing techniques. For example, Clarke and Cowan (1976) were able to label all neurons in the isthmo-optic nucleus of the normal chick by injection of HRP into the eye prior to the period of cell death. Since over half the cells die naturally, all or nearly all of the neurons which later degenerate have sent axons to their target zone. Similar findings using similar techniques have been reported for spinal motor neurons of the chick (Chu-Wang and Oppenheim, 1978b). A failure of axonal elongation and growth to target structures is therefore not responsible for the death of neurons since both surviving and dying cells accomplish this outgrowth. Unfortunately, the HRP labeling studies indicate only that the axons have reached the target zone so there is still no direct evidence that neurons which degenerate actually form synaptic connections with their target structures.

C. Receipt of Synaptic Connections

Electron microscopic examination of several areas of the nervous system of several species shows that some cells which are destined to die receive morphologically normal synaptic connections prior to their death (Giordano *et al.*, 1980; Oppenheim *et al.*, 1978; Knyihar *et al.*, 1978; Pilar and Landmesser, 1976; Landmesser and Pilar, 1974a). In the ciliary ganglion of the embryonic chick (Landmesser and Pilar, 1974a,b) and in the superior colliculus of the neonatal rat (Giordano *et al.*, 1980) typical synapses have been observed on the soma of neurons in early stages of natural degeneration. The synapses formed on ciliary ganglion neurons which later will die because of target deprivation are functional by electrophysiologic criteria (Landmesser and Pilar, 1974a,b). This suggests that the synaptic connections formed on naturally degenerating cells also are functional or were so at an earlier time. It is not known how many synapses form on cells which will degenerate compared to those which will not, but it is clear that however many connections do form, these alone are not sufficient to reverse the fate of a particular neuron.

We studied the temporal relationship between afferent synaptogenesis and neuron death in the superficial layers of the rat superior colliculus. Synaptogenesis in the superior colliculus of rats is relatively protracted with the very first connections from the eye forming on the sixteenth day of gestation and a burst of synaptogenesis on days 13 or 14 postnatal which is around the time of normal eye opening (Lund and Bunt, 1976; Lund and Lund, 1972). The timing of cell death in the superior colliculus does not mirror that of synaptogenesis. There was little cell death after postnatal day 8, with the vast majority of degenerating profiles appearing during the first week following birth. In fact, the data suggested that the peak of cell death was between 5 and 8 days postnatal, about a week before the peak of synaptogenesis.

In the superior colliculus therefore, some degenerating cells have already received synaptic connections and the majority of collicular neurons die well before afferent synaptogenesis is completed. These results might be interpreted to reflect little or no relationship between a neuron's afferents and its ability to survive. However, experimental evidence suggests that afferents do influence neuron survival during the normal development of this region (see below). Since the timing of synaptogenesis and neuron death are somewhat disparate, the basis of this control may not be the *final* number of synapses an individual neuron receives. It is just as likely that some of the early forming synapses exert this influence, or alternatively, the growing optic afferents effect neuron survival but do so prior to synaptogenesis. The peak of cell death in the colliculus does coincide with a substantial increase in the size of the optic fiber layer which suggests that there may be a wave of afferent influx around this period (Cunningham, unpublished observations). Furthermore, if this control is mediated by the transfer of life-sustaining trophic factors (as has been suggested for target control, see below), then it is conceivable that afferents could transfer these factors prior to forming synaptic connections.

IV. Control of Neuron Death: Experimental Studies

From the above discussion it should be clear that a large part of the development of a particular neuron or a particular area of the nervous system is completed before the time of natural cell death. Special attention was given to axon growth into a target region and the invasion of afferents because the timing of these events appears to correlate closely with the timing of the naturally occurring neuron death. Indeed, a great deal of experimental evidence indicates that the factors involved in neuron survival have to do with other connecting cell populations and the signals these cell populations send through synaptic connections and/or growing axons. However, it is widely believed that the critical signals come solely from a neuron's *target* structures, while the importance of *afferents* to the survival of a particular neuron has received considerably less attention. In some cases afferent control is dismissed altogether (e.g., see Jacobson, 1978) despite a number of classical studies which, although inconclusive, point to afferent influences on normal cell death for certain areas of the nervous system (see Jacobson, 1970, for review). In this section I will consider the role of target structures on neuron survival, review some recent experiments which demonstrate a similar role for afferent input, and briefly discuss the evidence which suggests that the dependence a neuron has on these connecting cell populations is mediated by neurotrophic factors.

A. Importance of Target Structures

As already noted, removal of targets for certain populations of neurons accentuates the normal attrition of neurons during development (see Jacobson, 1978; Cowan, 1973; and Prestige, 1970, for extensive reviews). In many cases, the loss of neurons in the innervating population is virtually complete. The conclusion from this kind of experiment is that during *normal* development neurons are competing for target space and those which fail to achieve adequate contact with the target die. Removal of target structures puts all neurons in the position of not achieving sufficient contact and therefore the cell death is greatly accentuated. While these results are consistent with the idea of a primary role for target structures in cell survival they are not altogether conclusive. For example, in normal animals neurons could die quite independently of target structures, the latter serving only a maintenance function to those neurons which survive the period of naturally occurring neuron death. More convincing evidence for direct involvement of targets in neuron survival would be obtained if additional targets are made available to a particular neuron population during development and these are capable of supporting additional cells. Assuming all neurons (including those which would normally die) have equal access to the extra target structures, then the probability that an individual neuron will contact a target is increased and the probability that the neuron will die is reduced. As a result, cells which would normally degenerate should be saved.

A clear demonstration of such an effect was accomplished first by Hollyday and Hamburger (1976) who transplanted supernumerary limbs to the chick embryo. While the results were somewhat variable they found increases of 11–27% in the number of spinal motor neurons on the side of the transplant. These investigators were able to show conclusively that the excess results from "hypothanasia" of neurons (a reduction in normally occurring neuron death), rather than from some other mechanism such as recruitment of additional neurons from neuroblastic zones. In the regions studied 40–50% of the neurons normally die so these increases in cell number constitute the rescue of 30–50% of the naturally degenerating population. A saving of from 54–74% of the naturally degenerating neuronal population has now been reported for other neural systems in birds after transplantation of additional target structures (Narayanan and Narayanan, 1978; Boydston and Sohal, 1979).

In amphibia, limb transplants are also able to rescue spinal motor neurons which would normally die (Hollyday and Mendell, 1976). However, the number of naturally degenerating neurons which are saved is smaller than that reported for chicks. In fact, supernumerary limbs in amphibia, on the average, save about half as many cells as do extra targets in birds (see Oppenheim, 1981, for review). One explanation for the low percentage of rescued cells in these experiments might be that surgically added limbs in amphibia are less available for innervation than similar limbs in birds. The extra limb would then be able to prevent the death only of those motor neurons which manage to gain access to the transplant. When this problem is seemingly eliminated by examination of frogs with congenital supernumerary limbs, the rescue rate is comparable to that obtained in the limb transplant experiments (Lamb, 1979a; Pollack, 1969). We are therefore left with somewhat of a paradox, especially in amphibia; the effect of limb removal is almost complete neuron death but limb transplants are able to save only a fraction of the normally degenerating cells. It is clear therefore that target structures are essential for neuron survival, but a failure to compete for target contact may not be the only explanation for neuron death (see below).

It has been possible to show the importance of competition for target contact by experimental manipulations other than transplantation. Pilar *et al.* (1980) sectioned two of the three nerve trunks which emanate from neurons in the chick ciliary ganglion and innervate structures in the eye. Although the lesions were made prior to the period of cell death, they eliminate those neurons which contribute axons to the sectioned nerves because of a direct retrograde reaction and degeneration. The death of neurons which contribute to the surviving branch is quantitatively reduced (39% rescued) compared to the death of these same neurons in normal chicks. They find evidence also that the remaining neurons expand their innervation territory within the target perhaps by sprouting additional collateral branches. The neurons thereby have more potential sites of target contact and may have more access to trophic factors involved in supporting the life of the cell.

An increase in innervation territory and subsequent rescue of neurons may explain also the intriguing results of the experiments in which pharmacological blockade of neuromuscular function results in a significant reduction in cell death in the innervating motor neuron population. This effect was discovered by Pittman and Oppenheim (1978) and has later been repeated and confirmed (Pittman and Oppenheim, 1979; Creazzo and Sohal, 1978; Laing and Prestige, 1978). Application of α-bungarotoxin, α-cobratoxin, curare, or Botulinum toxin to the embryo results in variable but significant decreases in motor neuron death. These results have been interpreted as a reflection of the need for functional neuromuscular activity to initiate the naturally occurring cell death which is found in spinal motor neurons. It is known however that the blocking agents which are used in these experiments have other effects besides paralysis of the muscle. For example, there is likely to be a prevention of the normal decrease in acetylcholine receptors which occurs during development of the muscle (Gordon *et al.*, 1975). Indeed, Pittman and Oppenheim (1979) find that after curare treatment, the motor axons sprout extra collaterals increasing their innervation territory within the muscle which is consistent with the increased number of receptors.

Therefore, the results of pharmacological blockade experiments and the studies of Pilar *et al.* (1980) suggest that natural cell death can be prevented by allowing neurons to increase their innervation territories and thereby increase their probability of target contact. This implies that the *availability* of target structures is a critical factor at least for these neuron populations. A further implication is that competition for target space also plays a role. Yet other experiments suggest that successful competition for target contact is not the sole determinant of neuron survival. For example, Lamb (1979b) attempted to rescue neurons in the caudal part of the spinal cord of *Xenopus* tadpoles by removing competing rostral spinal segments. The caudal neurons send axons to knee flexors early in development but then die leaving only the innervation of more rostral neurons. Removal of the latter cells should give the caudal neurons a competitive advantage and prevent their death. However, they still degenerate after removal of rostral neurons leaving the knee flexors uninnervated. These and more recent results (Lamb, 1980, see below) have caused Lamb to question the validity of the competition hypothesis.

B. Importance of Afferents

As stated above, recent developmental literature tends to ignore the role of afferents in ensuring neuron survival despite the evidence that afferents are essential to other aspects of normal neural development. I have already shown that deafferentation may have similar effects on neuron differentiation as those described after target removal. Furthermore, there are several reports of neuron loss in developing neural systems of both mammals and nonmammals after removal of afferent input (see Jacobson, 1970, 1978; Cowan, 1970, for reviews). Unfortunately, in most of these cases a relationship between neuron loss after deafferentation and naturally occurring cell death has not been established and so very few investigators have concluded that the mechanisms which underly cell death in the two situations are similar. This is especially the case for the central nervous system of mammals, because there are a number of infant deafferentation studies but little is known about naturally occurring death in the systems examined. Nevertheless, experiments with the infant rodent superior colliculus provide some evidence for a temporal relationship between normal and deafferentation-induced cell death.

DeLong and Sidman (1962) find a 40% neuron loss in the mouse superior colliculus after eye removal at birth.[1] Their results indicate further that this

[1] The superior colliculus of the rodent is unlike those small, morphologically homogeneous neuron groups which are commonly used to study neuron death in nonmammals. For this reason neuron numbers cannot be reliably estimated in conventionally stained material. For example, there are several problems with cell identifications (e.g., neurons vs glia) especially after lesions. To avoid these difficulties DeLong and Sidman (1962) specifically labeled neurons by injections of tritiated thymidine at gestational ages when the vast majority of cells which take up the label are neurons. Since the label is virtually permanent unless the cell degenerates, neuron attrition following eye removal can be studied simply by counts of labeled cells.

deafferentation-induced neuron death occurs between postnatal day 4 and day 7. In more recent studies of the superior colliculi of normal rats and hamsters, large numbers of degenerating cells are found around these ages (Giordano *et al.*, 1980; Berg and Finlay, 1979; Arees and Ångström, 1977; see above), and it is reasonably certain that natural neuron loss also occurs in the colliculus of the normal mouse.[2]

Even if the temporal relationship between deafferentation and natural neuron death was more firmly established, by the same arguments applied to neuron loss after target removal, such a relationship does not demonstrate that afferents are critical to neuron survival during normal development. More convincing evidence can be provided by demonstrating the prevention of cell death by transplantation of additional afferent sources. However, classical experiments in which such attempts were made (see Jacobson, 1970, for review) are confounded by interpretational difficulties. In some cases, naturally occurring neuron loss is not demonstrated in the region under study. In other cases, transplantation of additional afferent sources does produce hypertrophy and hyperplasia in the recipient zone but the exact elements of nervous tissue which contribute to these increases are unclear (e.g., neurons, glia, neuropil).

We became interested in testing the proposition that increased optic afferents could rescue cells in the infant rat superior colliculus (Cunningham *et al.*, 1979). Since transplantation of additional peripheral organs (such as an extra eye) is not feasible in neonatal mammals, it was necessary to take advantage of the abnormal retinal projections which are demonstrable after lesions to the infant rodent visual system (e.g., Schneider, 1973; Cunningham, 1972). These lesions obviously deprive some of the subcortical visual centers of their input and/or target structures, yet at the same time other visual centers may receive an abnormally increased input. We find evidence for such an increased input to the nucleus of the optic tract (NOT) and to the superficial layers of superior colliculus following early destruction of the dorsal lateral geniculate nucleus by posterior cortical lesions. It appears that in the absence of the geniculate, a major target for optic axons, the axons increase their projection to the NOT and superior colliculus thereby providing these nuclei with additional input (Fig. 5). Using the same experimental approach as DeLong and Sidman (1962; see footnote 2) we find increases in neuron

[2] This should not be taken to mean that all transneuronal degenerative changes following either deafferentation or target removal represent enhanced naturally occurring neuron death. In the first place, transneuronal degenerative changes have been described following lesions in adult animals, at times expected to be well after the developmental period of natural neuron loss (Cowan, 1970). Second, although normal neuron death seems to occur in most areas of the nervous system there are some areas which show little or no natural cell death. Nevertheless, neuron death after deafferentation or target removal still occurs (see Parks, 1979; Armstrong and Clark, 1979; Harkmark, 1954).

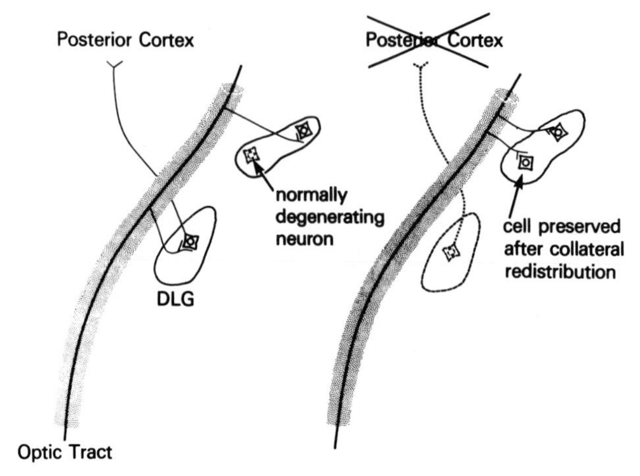

FIG. 5. Schematic representation of neuron rescue in the visual system of rats. In normal rats, a collateralized optic pathway develops to the dorsal lateral geniculate nucleus and other visual centers (represented by unlabeled outline). Some neurons in these areas degenerate naturally. Posterior cortex lesions at birth destroy the geniculate in 96 hours. Collaterals appear to redistribute to some of these regions and rescue cells which would normally die (from Cunningham *et al.*, 1979).

numbers averaging 39% in NOT and 15% in the colliculus. The total magnitude of naturally occurring neuron loss is not yet known for these regions so the percentage of cells rescued by the increased input is also not known.

C. Conclusions from Experimental Studies

Clearly there is more experimental work which establishes the importance of target structures to neuron survival than similar work devoted to the importance of afferent input. The reasons for this imbalance may be largely technical. It is usually much easier to deprive a developing neuron completely of target structures than to deprive it completely of afferent input. For example, limb removal constitutes total peripheral target deprivation for spinal motor neurons. These cells receive most of their afferent input from interneurons in the spinal cord and specific destruction of the latter is not feasible at present. Even when deafferentation is possible as in the rodent superior colliculus, it constitutes only partial removal of afferents because the colliculus receives input from a variety of other sources. Thus, it is easy to see how target structures could be viewed as a more powerful determinant of neuron survival during normal development.

Some of the experimental evidence does indicate that enhancement of target contact prevents neuron death. Still, the results from transplantation studies are somewhat disappointing. Limb transplants usually rescue only a percentage of the naturally degenerating population, and in some cases, this percentage is quite small. Other experimental evidence, mostly from amphibia, seems to conflict with the general idea of target competition. Because of these results, and in light of the recent evidence which suggests that in some cases afferent input may exert the same controls as targets, the idea that targets have the major or only role in determining which cells survive the developmental period may require revision (see below).

D. Possibility of Life-Sustaining Trophic Factors

The fact that neurons which survive the period of naturally occurring neuron death make contact with other cell populations is by itself considered to be insufficient in sustaining the life of the cell. For several decades, neuroembryologists have speculated upon the existence of life-sustaining factors located within connecting cells and the intercellular transfer of these factors. The assumption is that if a particular neuron has access to sufficient amounts of such a substance then it will live. These speculations have been encouraged significantly by the discovery of nerve growth factor (NGF) (Levi-Montalcini and Hamburger, 1951). NGF not only enhances the outgrowth of neuronal processes from sympathetic and sensory ganglion cells, but also increases neuron numbers within the ganglia (see Stoeckel and Thoenen, 1975; Harper and Thoenen, 1980, for reviews). Furthermore, NGF is made in the peripheral targets of these neurons and transported back to the cell body (the presumed site of action). There is now good evidence from studies on chick sensory ganglia (Hamburger *et al.*, 1981) and on rodent sympathetic ganglia (Hendry and Campbell, 1976; Kessler and Black, 1980) that the increased ganglion cell numbers found after NGF treatment are due to a prevention of cell death.

Strong evidence for substances which are able to prevent the death of other cell types is not available. Nevertheless, experiments on neurons developing *in vitro* provide additional encouragement. The outgrowth of cell processes from cultured spinal motor neurons of tadpoles (Pollack, 1980) or of rats (Dribin and Barrett, 1980) is enhanced by factors derived from muscle. Furthermore, increased survival of cultured chick ciliary ganglion neurons is obtained following application of factors derived from the target structures of these neurons (Nishi and Berg, 1979).

So far attempts to find such substances for a particular neuron type have concentrated on its target structures because most of the available evidence stresses the importance of targets to neuron survival. With increasing evidence that afferents also are involved in naturally occurring neuron death, the possibility exists that similar substances are present in those cells which provide input to the neuron. Thus, the anterograde transfer of trophic material may have to be considered also despite the complexity this adds to the search.

V. Conclusions: The Function of Naturally Occurring Neuron Death

A common explanation for neuron death during development can be found in the expression "there is safety in numbers." The implication is that neurons are overproduced initially because there are life threatening situations ahead in which some cells are bound to be lost. Part of understanding the overall function of natural neuron death depends on determining which aspects of the developing environment make the cells vulnerable and how some cells manage to escape. For this reason, I have concentrated so far on those factors which appear to contribute to a particular neuron's ability to survive the period of natural neuron death. A number of experiments suggest that target structures are critical to neuron survival and other experiments suggest a similar role for afferents. Unfortunately, an understanding of why a neuron will survive or die in specific systems gives little insight into the adaptive value of naturally occurring neuron death. How does the natural death of neurons contribute to the successful completion of development and therefore ensure the appropriate functioning of the mature nervous system?

A. Simple Competition

Much of the experimental work reviewed above is consistent with the idea that neurons are overproduced because later they will compete for contact with other cells and some will fail. Such competition ensures that surviving neurons will obtain sufficient numbers of afferent or target contacts (or sufficient amounts of trophic substances which will eventually ensure these contacts). Neurons which fail are eliminated because their continued existence would most likely interfere with normal functioning in the more mature nervous system.

While the idea of simple competition for input or target space is indeed attractive, there are experimental situations where it has proven inadequate. It must be concluded that the *sole* function of neuron death is not simply the elimination of neurons that are unable to form or achieve enough connections.

B. Correction of Projection Errors and Synaptic Specificity

If we are forced to abandon the idea that neuron death serves only to ensure that the cells will have sufficient numbers of synaptic connections, then the next obvious suggestion is that in some cases it is not the actual number of connections which determines neuron survival but the appropriateness of these connections. One function of neuron death then is the elimination of neurons which form or receive inappropriate connections, i.e., neurons which violate the rules of neuronal specificity. There is evidence that neurons with gross errors in their connectivity die during development, but as for the elimination of neurons which have mismigrated (see above) the death of abnormally projecting cells is overshadowed by the cell death which is found among neurons that project correctly. Early in development, some isthmo-optic neurons normally send their axons to the wrong eye and subsequently appear to die, but the vast majority of these neurons send their axons to the correct eye and most of the cell death is found among appropriately connected neurons (Clarke *et al.*, 1976; Clarke and Cowan, 1976). As described above, some caudally situated spinal motor neurons in *Xenopus* project to inappropriate muscles early in development and are later eliminated leaving only appropriate connections with more rostrally situated cells. Once again, however, much more cell death is found among the appropriately connected population (Lamb, 1976; see also Landmesser, 1980).

Neuron death is also proposed to underly the formation of correct connections in systems where specificity may be more strictly controlled. For example, the elimination of inappropriately connected retinal ganglion cells is suggested to underly the normal topography of connections in the chick's retinotectal pathway (Rager, 1978; Rager and Rager, 1978). We have examined the distribution of degenerating cells in both the ganglion cell layer and the superior colliculus of the neonatal rat in order to determine the relationship between neuron death and topography in these connecting structures (Cunningham *et al.*, 1981). At postnatal ages when large numbers of degenerating profiles appear in both areas, the nasal part of the ganglion cell layer and the caudal part of the superior colliculus (which are normally connected) have about twice as many degenerating cells as the temporal ganglion cell layer and the rostral colliculus (which are also connected). It may be that the corresponding patterns of cell death serve to "fine tune" the normal topographic relationship between the two structures. On the other hand, it is more likely that the regional correspondence merely *reflects* the specificity of retinocollicular connections. In other words, the cell death may not remove inappropriate retinocollicular connections but simply provide a mechanism for adjustments of neuron number in regions which are or will be *appropriately* connected.

C. Competition and Quantitative Balance: A Hypothesis

Given the well-documented relationship between neuron survival and target contact, and the increasing evidence that a similar relationship exists between afferents and neuron survival, it now seems appropriate to introduce some new ideas about the role neuron death plays in assuring appropriate development of the nervous system. The suitability of this hypothesis lies in the fact that it does not discard the idea of simple competition for afferent or target contact but extends this idea to define more precisely the relationship between afferents, targets, neuron maturation, and neuron survival.

The hypothesis suggests that just as there are two morphological types associated with degenerating neurons, there are two reasons why neurons die. The first reason is for present purposes, virtually identical to the simple competition hypothesis, i.e., some neurons die because they fail to receive adequate signals (presumably, adequate amounts of trophic factors) from connecting cell populations during early stages of their maturation. This failure arrests the development of the neuron and in most cases results in the death of the cell. The second reason neurons die applies to those cells which compete successfully for initial contact with afferent sources or target structures and therefore are able to continue their maturation. However, these cells fail in a second important aspect of neuronal differentiation; they are either unable to gain target contact which is quantitatively appropriate to their afferent input, or they are unable to gain adequate afferent input relative to the amount of target contact. The function of cell death during development is then twofold: first, to eliminate cells which do not find adequate sources of input or adequate amounts of target, and second, to eliminate cells which do not have the proper balance of input and target contact.

I have already discussed the importance both of afferents and targets in neuronal maturation and presented evidence which suggests that either of these connecting cell populations can trigger the development of mature protein synthetic machinery in a particular neuron. For some neurons, target removal arrests maturation (producing the "nuclear" or immature form of cell death described above) and for other neurons afferent removal produces this effect. These experiments show an exaggerated version of the first cause of cell death, i.e., a failure by the cell to gain signals necessary for further maturation. As a result development is arrested. It is reasonable to assume that the neuron dies because this arrest precludes any further attempts by the cell to gain other inputs or targets.

Other neurons however receive the appropriate signals and are allowed to continue their maturation yet still degenerate (showing the "cytoplasmic" or more mature form of cell death described above). These neurons, like those which survive, are suggested to have an intrinsic requirement for a quantitative balance of afferent and target contacts. During development, the attainment of this balance is predicted to be strictly controlled, and within limits, more strictly controlled than the absolute numbers of either type of contact. If target structures or their associated trophic factors are contacted before substantial contact with afferents (or their trophic factors), then the afferent contact must be adjusted quantitatively to balance the target contact. The reverse would be true if the afferent contact is achieved first. Thus, part of the development of any neural pathway includes balancing of input and output, a process which could occur in either a stepwise or continuous fashion. Neurons which fail in this process are suggested to be incompatible with normal neural functioning and therefore are eliminated.

Starting with the simplest situation, the hypothesis predicts that depriving this population of more mature neurons of their afferents or targets will produce a gross mismatch between input and target contact and the cells will die. However, the extent of cell loss in any particular neuron population will depend on the ability of that population to achieve additional inputs or find additional targets. Spinal motor neurons for example have few options after removal of a limb bud and therefore neuron loss is greatly exaggerated. On the other hand, rat retinal ganglion cells have several options after removal of the dorsal lateral geniculate nucleus (Cunningham *et al*., 1979) and so there is little or no loss of ganglion cells (Perry and Cowey, 1979).

In the more complicated reverse situation, the hypothesis predicts that provision of extra input or targets will have different effects in different systems depending on the limitations imposed on the neuron because of prior target contact (in the case of extra input) or prior afferent contact (in the case of extra targets). For example, some neurons may have relatively little afferent contact at the time they invade the extra target and therefore are able to extend their target contacts to the upper limits of the cell (thereby gaining access to more life-sustaining trophic substances). Other neurons might not ramify as freely in an additional target because of prior influence by afferents and the necessity only to balance this afferent contact with an appropriate (and not necessarily excessive) amount of target contact. In amphibia with supernumerary limbs, the low percentage of rescued cells is explained by assuming that most motor neurons already have been contacted by afferents or afferent trophic factors at the time the limb becomes available for innervation. In birds, where the rescue rates are higher, fewer neurons are expected to have this prior contact. Amphibian motor neurons therefore appear to be less competitive than bird motor neurons because the extent of their target requirements has already been defined.[3]

The hypothesis predicts also that in cases of afferent or target enhancement the neurons involved will, respectively, adjust the amount of target contacted or the amount of afferent innervation in order to achieve the necessary balance of input and output. For example, neurons which form excessive

[3] These "preset" target requirements may explain the recent results of Lamb (1980). He finds that when both sides of the spinal cord innervate a single limb transplanted to the back of *Xenopus*, the amount of motor neuron death is considerably less than is found when the limb is removed and there is no opportunity for innervation. In this case, the motor neurons appear to gain their target contact even though the target is reduced.

target contacts after experimental manipulations may require additional afferent innervation. Unfortunately, there is little information on the ability of cells in such situations to increase their normal number of afferent connections. For example, in the experiments of Pilar *et al.* (1980) where targets were enhanced for some ciliary ganglion neurons by destruction of other competing neurons, the synaptic complement of the surviving cells was not determined. However, a similar experiment on the superior cervical ganglion of the neonatal rat shows that ganglionic neurons are able to increase their afferent connections (Smolen and Lindley, 1980). After cutting one of the two main postganglionic nerves, careful quantification of the input to the surviving neurons shows a 30% increase in the number of synapses per neuron. Although it is not known if these neurons actually expanded their innervation territory as did the ganglion cells in the Pilar *et al.* (1980) study, it is notable that neurons, at least to some extent, are able to adjust their input after such experimental manipulations.

The hypothesis also explains the elimination of the small percentage of neurons which represent developmental errors. Cells which migrate to or project to totally inappropriate positions should be at a severe disadvantage, not only in gaining sufficient afferent or target contact for continued maturation, but also in achieving the appropriate balance of these contacts. A mismigrated but appropriately projecting neuron might be expected to sometimes fail on the afferent side because of its bad location; the misprojected cell which had successfully migrated might be expected to sometimes fail at the inappropriate target.

It is not difficult to speculate on the intrinsic factors which govern the proper balance of input and target contact for a particular neuron. On one level, the anatomy of the neuron will determine the extent of its contact with other cell populations. For example, the ultimate form of the dendritic tree relative to the branching pattern of the cell's axon should impose some limits, at least to the extent that these are intrinsic properties of the neuron. On another level, the determining factors may be related to intrinsic metabolic capabilities, such as the synthesis of receptor molecules relative to the synthesis of transmitter related enzymes. In any case, it appears that an understanding of the progress of development at both ends of the nerve cell is important in studies of neuron viability during development.

Acknowledgments

I wish to thank Linda Wright for many valuable discussions. Arnold Smolen, Debra Giordano, Leonard Ross, and Michael Goldberger made several helpful suggestions. I also thank Irene Mohler and Betti Goren for photography and illustrations, and I am especially grateful to Kathy Golden for preparing the manuscript. During the writing of this article, I was supported by Grant No. 1-RO1-NS-16487 from N.I.N.C.D.S.

References

Arees, E. A., and Ångström, K. E. (1977). *Anat. Embryol.* **151,** 29–34.

Armstrong, R. C., and Clarke, P. G. H. (1979). *Neuroscience* **4,** 1635–1647.

Berg, A. T., and Finlay, B. L. (1979). *Neurosci. Abstr.* **5,** 153.

Boydston, W. R., and Sohal, G. S. (1979). *Brain Res.* **178,** 403–410.

Cantino, D., and Daneo, L. (1972). *Brain Res.* **38,** 13–25.

Chiappinelli, V., Giacobini, E., Pilar, G., and Uchimura, H. (1976). *J. Physiol.* **257,** 746–766.

Chu-Wang, I. W., and Oppenheim, R. W. (1978a). *J. Comp. Neurol.* **177,** 33–58.

Chu-Wang, I. W., and Oppenheim, R. W. (1978b). *J. Comp. Neurol.* **177,** 59–86.

Clarke, P. G. H., and Cowan, W. M. (1976). *J. Comp. Neurol.* **167,** 143–164.

Clarke, P. G. H., Rogers, L. A., and Cowan, W. M. (1976). *J. Comp. Neurol.* **167,** 125–142.

Cowan, W. M. (1970). *In* "Contemporary Research Methods in Neuroanatomy" (W. J. H. Nanta and S. O. E. Ebbesson, eds.), pp. 217–251. Springer-Verlag, Berlin and New York.

Cowan, W. M. (1973). *In* "Development and Aging in the Nervous System" (M. Rockstein, ed.), pp. 19–41. Academic Press, New York.

Creazzo, T. L., and Sohal, G. S. (1979). *Exp. Neurol.* **66,** 135–145.

Cunningham, T. J. (1972). *Anat. Rec.* **172,** 298.

Cunningham, T. J., and Murphy, E. H. (1979). *In* "Handbook of Behavioral Neurobiology" (R. B. Masterton, ed.), Vol. 1, pp. 39–72. Plenum, New York.

Cunningham, T. J., Huddelston, C., and Murray, M. (1979). *J. Comp. Neurol.* **184,** 423–434.

Cunningham, T. J., Mohler, I. M., and Giordano, D. L. (1980). *Anat. Rec.* **196,** 39A.

Cunningham, T. J., Mohler, I. M., and Giordano, D. L. (1981). *Dev. Brain Res.* **2,** 203–215.

DeLong, G. R., and Sidman, R. L. (1962). *J. Comp. Neurol.* **118,** 205–224.

Dribin, L. B., and Barrett, J. N. (1980). *Dev. Biol.* **74,** 184–195.

Giordano, D. L., and Cunningham, T. J. (1981). *Anat. Rec.* **199,** 95A.

Giordano, D. L., Murray, M., and Cunningham, T. J. (1980). *J. Neurocytol.* **9,** 603–614.

Gordon, T., Tuffery, A. R., and Vrbova, G. (1975). *In* "Recent Advances in Myology" (W. G. Bradley, D. Gardner-Medwin, and J. N. Walton, eds.), pp. 22–26. Elsevier, Amsterdam.

Hamburger, V., and Levi-Montalcini, R. (1949). *J. Exp. Zool.* **111,** 457–502.

Hamburger, V., Brunso-Bechtold, J. K., and Yip, J. W. (1981). *J. Neurosci.* **1,** 60–71.

Harkmark, W. (1954). *J. Comp. Neurol.* **100,** 333–371.

Harper, G. P., and Thoenen, H. (1980). *J. Neurochem.* **34,** 5–16.

Hendry, I. A., and Campbell, J. (1976). *J. Neurocytol.* **5,** 351–360.

Hollyday, M., and Hamburger, V. (1976). *J. Comp. Neurol.* **170,** 311–319.

Hollyday, M., and Mendell, L. (1976). *Exp. Neurol.* **51,** 316–329.

Hughes, A. (1973). *J. Embryol. Exp. Morphol.* **30,** 359–376.

Hughes, W. F., and LaVelle, A. (1975). *J. Comp. Neurol.* **163,** 265–284.

Jacobson, M. (1970). "Developmental Neurobiology," pp. 226–270. Holt, New York.

Jacobson, M. (1978). "Developmental Neurobiology." Plenum, New York.

Kessler, J. A., and Black, I. B. (1980). *Brain Res.* **189,** 157–168.

Knyihar, E., Csillik, B., and Rakic, P. (1978). *Science* **202,** 1206–1208.

Laing, N. G., and Prestige, M. C. (1978). *J. Physiol. (London)* **282,** 33–34.

Lamb, A. H. (1976). *Dev. Biol.* **54,** 82–89.

Lamb, A. H. (1979a). *J. Embryol. Exp. Morphol.* **49,** 13–16.

Lamb, A. H. (1979b). *Dev. Biol.* **71,** 8–21.

Lamb, A. H. (1980). *Nature (London)* **284,** 347–350.

Landmesser, L. (1980). *Annu. Rev. Neurosci.* **3,** 279–302.

Landmesser, L., and Pilar, G. (1974a). *J. Physiol. (London)* **241,** 715–736.

Landmesser, L., and Pilar, G. (1974b). *J. Physiol. (London)* **241,** 737–749.

Levi-Montalcini, R. (1950). *J. Morphol.* **86,** 253–283.

Levi-Montalcini, R., and Hamburger, V. (1951). *J. Exp. Zool.* **116,** 321–351.

Lund, R. D., and Bunt, A. H. (1976). *J. Comp. Neurol.* **165,** 247–264.

Lund, R. D., and Lund, J. S. (1972). *Brain Res.* **42,** 1–20.

Narayanan, C. H., and Narayanan, Y. (1978). *J. Embryol. Exp. Morphol.* **44,** 53–70.

Nishi, R., and Berg, D. K. (1979). *Nature (London)* **277,** 232–234.

O'Connor, T., and Wyttenbach, C. R. (1974). *J. Cell Biol.* **60,** 418–459.

Oppenheim, R. W. (1981). *In* "Studies in Developmental Neurobiology: Essays in Honor of Viktor Hamburger" (W. M. Cowan, ed.). Oxford Univ. Press, London and New York.

Oppenheim, R. W., Chu-Wang, I. W., and Maderdrut, J. L. (1978). *J. Comp. Neurol.* **177,** 87–112.

Panesse, E. (1976). *Neuropathol. Appl. Neurobiol.* **2,** 247–267.

Parks, T. N. (1979). *J. Comp. Neurol.* **183,** 665–678.

Perry, V. H., and Cowey, A. (1979). *Exp. Brain Res.* **35,** 85–95.

Pilar, G., and Landmesser, L. (1976). *J. Cell Biol.* **68,** 339–356.

Pilar, G., Landmesser, L., and Burstein, L. (1980). *J. Neurophysiol.* **43,** 233–254.

Pittman, R., and Oppenheim, R. W. (1978). *Nature (London)* **271,** 364–366.

Pittman, R., and Oppenheim, R. W. (1979). *J. Comp. Neurol.* **187,** 425–446.

Pollack, E. D. (1969). *Teratology* **2,** 159–162.

Pollack, E. D. (1980). *Neurosci. Lett.* **16,** 269–274.

Prestige, M. C. (1970). *In* "The Neurosciences: Second Study Program" (F. O. Schmitt, ed.), pp. 73–82. Rockefeller Univ. Press, New York.

Rager, G. (1978). *Exp. Brain Res.* **33,** 79–90.

Rager, G., and Rager, V. (1978). *Exp. Brain Res.* **33,** 65–78.

Saunders, J. W., Jr. (1966). *Science* **154,** 604–612.

Schneider, G. (1973). *Brain Behav. Evol.* **8,** 73–109.

Silver, J., and Hughes, A. F. (1973). *J. Morphol.* **140,** 159–170.

Smolen, A. J., and Lindley, T. (1980). *Abstr. Soc. Neurosci.* **6,** 683.

Sohal, G. S., and Weidman, T. (1978). *Exp. Neurol.* **61,** 53–64.

Stoeckel, K., and Thoenen, H. (1975). *Brain Res.* **85,** 337–341.

Tennyson, V. M., and Mytilineous, C. (1979). *In* "SIF Cells. Structure and Function of the Small Intensely Fluorescent Sympathetic Cells" (O. Eranko, ed.), pp. 35–53. DHEW Publ. No. (NIH) 76–942.

Wechsler, W., and Meller, K. (1967). *In* "Progress in Brain Research, Developmental Neurology" (C. G. Bernhard and J. P. Shade, eds.), Vol. 26, pp. 93–144. Elsevier, Amsterdam.

INTERNATIONAL REVIEW OF CYTOLOGY, VOL. 74

The Brown Fat Cell

JAN NEDERGAARD AND OLOV LINDBERG

The Wenner-Gren Institute, University of Stockholm, Stockholm, Sweden

ISBN 0-12-364474-7

I. Introduction

A. An Energy-Dissipating Cell

Living cells are fighting against ever increasing entropy. They have developed the capability of maintaining a high degree of organization against devastating tendencies toward disorder and purposeless dissipation of energy as heat. The existence of cells which function solely to produce heat at the price of chemically stored energy must therefore focus our interest and attention. The brown fat cell is such a cell, and within brown adipose tissue, "each cell . . . acts as a regulated energy converter" (Plant and Horowitz, 1978), and the regulation of the heat production will be the main scope of this article.

The function of the brown fat cell as such has not been extensively reviewed, but the metabolism of brown adipose tissue has been reviewed earlier by Smith and Horwitz (1969) and in the book "Brown Adipose Tissue" (Lindberg, 1970).

B. The Brown Fat Cell at Work

From being understood as a cell mainly working in hibernators and thus only attracting the interest of an esoteric group of hibernation researchers, the brown fat cell has within less than two decades acquired a position in physiological research which is a reflection of the understanding that the brown fat cell is active in a number of physiological states. Today these include—besides arousal from hibernation—physiological adaptation to cold in homeotherms, defense against hypothermia in the newborn mammal, and prevention of obesity in the mammal living on an unbalanced diet.

1. That the brown fat cell is actively taking part in the rewarming of the hibernator during *arousal* was the observation which allowed Smith and Hock (1963) to postulate that the "brown adipose tissue is the prime thermogenic effector organ for hibernator arousal." The relationship between hibernation, arousal, and brown fat metabolism has been reviewed by Hayward and Lyman (1967).
2. The significance of brown fat for *the newborn mammal* was first realized by Dawkins and Hull (1964). Recently, the physiological role of brown fat for the newborn has been reviewed by Cannon and Johansson (1980) and Cannon and Nedergaard (1982), and the ontogenetic development of the tissue has been reviewed by Hahn and Novak (1975) and Barnard *et al.* (1980).
3. The significance of brown fat for *adaptation to cold* was first realized by Smith (1961) and has been reviewed by Smith and Horwitz (1969) in a very comprehensive way. In order to summarize the development since then it is sufficient to cite the conclusion of Foster and Frydman (1978): "nonshivering thermogenesis occurs principally in brown adipose tissue."
4. The relationship between the activity of brown fat and the development of *obesity* was first realized by Himms-Hagen (1979) and by Rothwell and Stock (1979). Presently it is suggested that in a variety of states where animals become obese this is due to a decreased activity of brown fat and that brown

fat is responsible for the phenomenon of diet-induced nonshivering thermogenesis.

Thus a concensus is developing, according to which the brown fat cell is responsible for all "facultative" or "norepinephrine" (or "nonthyroid") nonshivering thermogenesis, whether that be hibernation induced, ontogenetically induced, cold induced, or diet induced. The assumption throughout this article is that in all these states the brown fat cell is working according to the patterns described here.

Man being a mammal, there is a priori no reason to think that he cannot display the same types of nonshivering thermogenesis, mediated by brown fat cells, and induced in the same ways (except hibernation!). Evidence for this has been summarized by Brück (1978) (ontogenically induced), by Cannon *et al.* (1978) (cold induced), and by Himms-Hagen (1979) (diet induced).

However, we shall here concentrate on the function of the brown fat cell, adhering to the statement of Angel (1969) in one of the early investigations with isolated brown fat cells: "it is possible that of all the available preparations *in vitro*, the isolated cell system may best reflect the physiological properties of the parent tissue."

C. Overview

In order to facilitate the detailed discussion, we shall start by giving an overview of the properties of brown fat cells. Brown fat cells (Fig. 1) produce heat by the oxidation of fatty acids. Isolated cells are prepared by collagenase treatment of brown fat from hamsters and rats. The cells are normally incubated in a Krebs–Ringer bicarbonate buffer, and in this medium they show a low basal respiration. The addition of catecholamines, normally norepinephrine (NE), results in a more than 10-fold increase in the rate of oxygen consumption, up to a value close to 1000 nmoles O/minute/million cells (representing a heat production of 3 mW/million cells). This value is in good agreement with the values which would be expected, based on calculations from experiments *in vivo*. Except for free fatty acids, no substrates added to the medium will induce their own oxidation within the cells proper. The action of catecholamines is mainly through β_1-receptors; evidence of α-adrenergic involvement is still meager but α-adrenergic stimulation may proceed via Ca^{2+}-dependent mechanisms. Catecholamine stimulation leads to cell membrane depolarization; the functional significance of this is unknown. β-Adrenergic stimulation leads to an increased formation of cAMP; the cAMP stimulates protein kinases in the cell; one of these probably activates a hormone-sensitive lipase and free fatty acids are released. Some of these

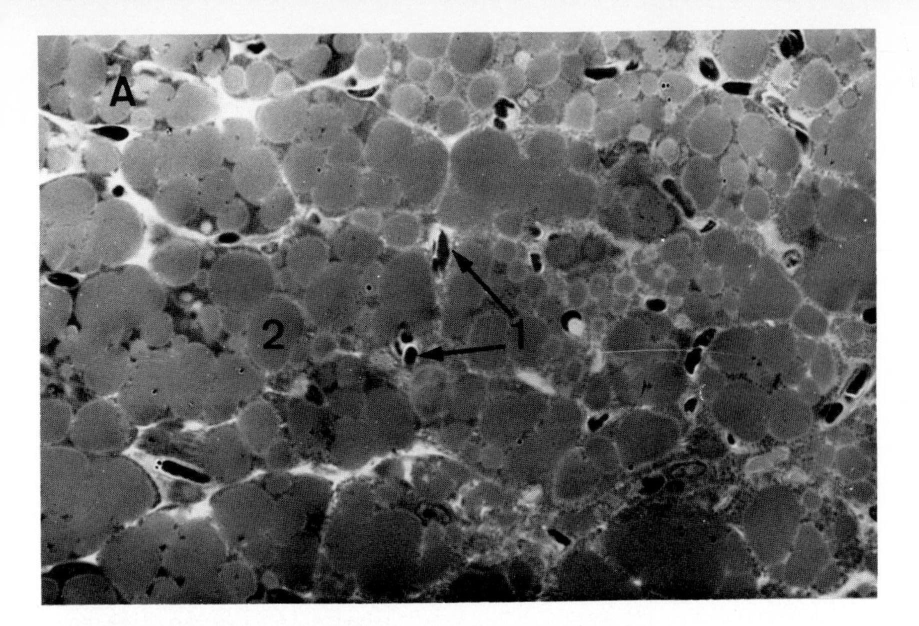

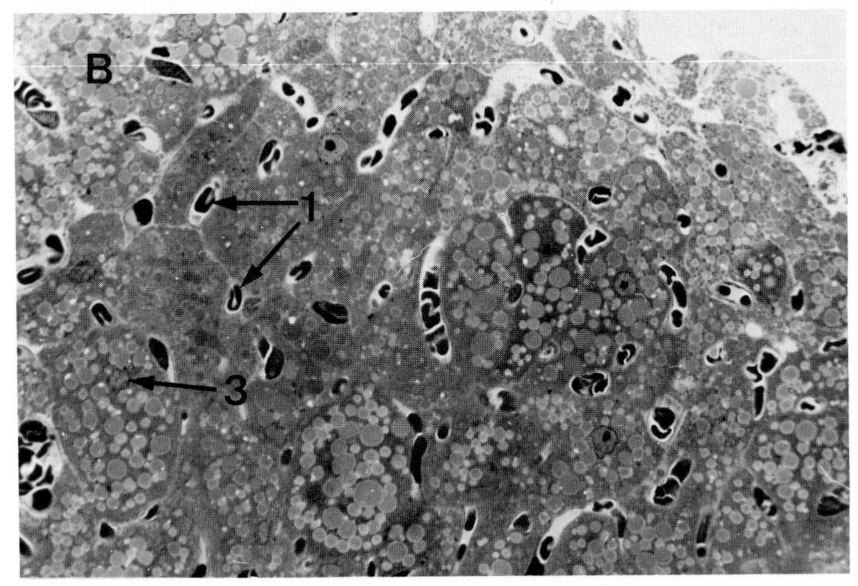

FIG. 1. The brown fat cell. (A) and (B): the brown fat cell *in situ.* Light micrographs of brown adipose tissue from a control (A) and a cold-adapted (B) rat (same magnification). Note the abundant blood capillaries with red blood cells (1) and the large lipid droplets (2) which have become much smaller in the cold-adapted rat. The cells have a central nucleus (3). (Courtesy of Myriam Nechad.) (C) Light microscopic picture of isolated hamster brown fat cells. Note the presence of large and small fat droplets (1) together with intact (2) and possibly damaged (3) brown fat cells. The suspension contained about 200,000 cells/ml and is shown for counting in a Bürker chamber. (Courtesy of Kent Nelson.) (D) Electron micrograph of an isolated brown fat cell. Note the central nucleus (1), the lipid droplets (2), and the abundant mitochondria (3). (Courtesy of Tudor Barnard.)

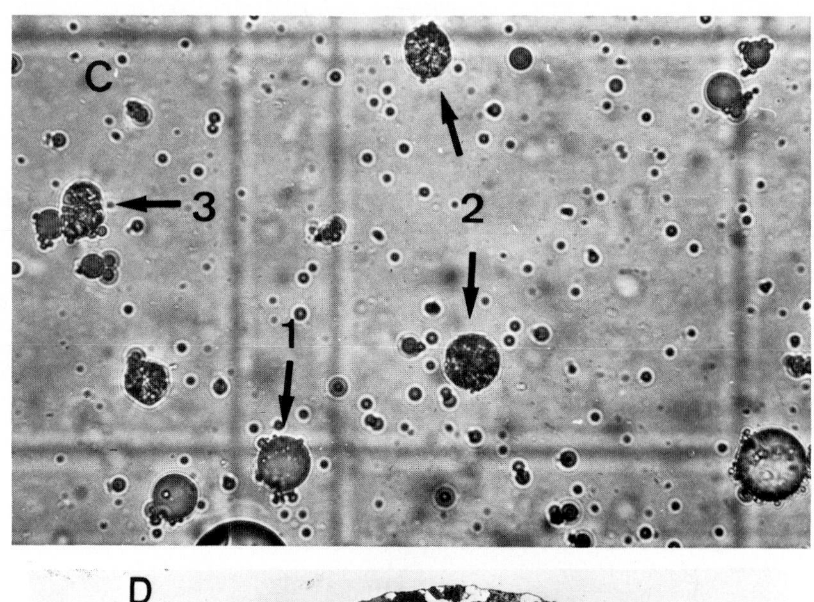

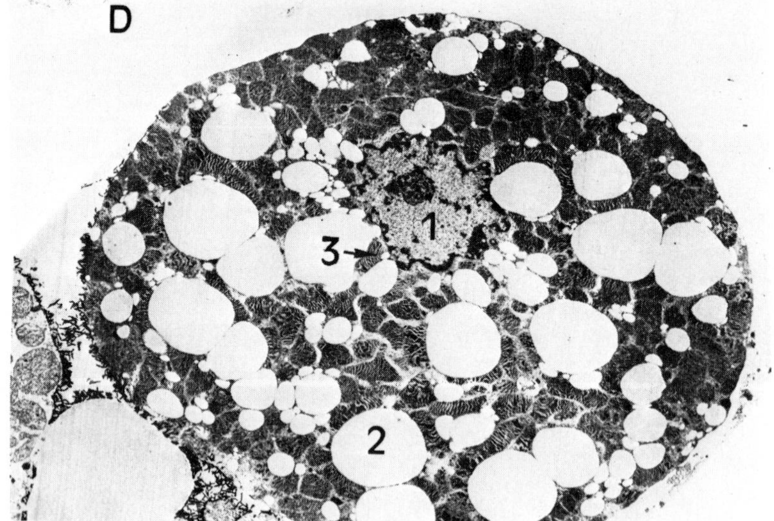

FIG. 1. (*continued*)

are exported from the cell, some are activated to longchain acyl-CoAs. These acyl-CoAs are the substrate for mitochondrial β-oxidation; further they may partially uncouple the mitochondria within the cell in a specific way by competing with purine nucleotides bound to *thermogenin*, a polypeptide of molecular weight 32,000 in the mitochondrial membrane. The acyl-CoAs, degraded in β-oxidation to acetyl-CoA, enter the citric acid cycle and are ultimately combusted to CO_2 and H_2O.

The isolated cells rapidly synthesize fatty acids from added 2, 3, or 6 carbon precursors, and added fatty acids are rapidly esterified to triglycerides. Some of these anabolic processes may be stimulated by insulin but the anabolic processes of the cells have not been thoroughly investigated.

The factors controlling the formation of brown fat cells and their differentiation are at present poorly understood.

II. The Isolation of Brown Fat Cells

Martin Rodbell, who introduced the collagenase method for the isolation of white fat cells, also tried his method on interscapular brown fat, but found that brown fat was not dispersed by this collagenase treatment, probably because of a higher collagen content in brown fat than in white fat (Rodbell, 1965). Simply by using a more vigorous procedure, Fain *et al.* (1967) and Prusiner *et al.* (1968a) were able to obtain cells from rats and hamsters, respectively. Their procedure, with smaller modifications, is still the one used, and cells obtained from rats and hamsters are the most investigated.

A. The Sources of Cells

1. *Cells from Different Animals*

Cells have been prepared from a number of different animals. The list in Table I gives for each animal the first (or only) reports of experiments with such cells and a few characteristics: particularly whether catecholamine-stimulated respiration, the most critical characteristic of a functional brown fat cell preparation, has been observed.

As can be seen from the table, only cells prepared from adult rats and hamsters, living under normal animal house conditions, have been well characterized and studied, and in the following "rat cells" and "hamster cells" will refer to these preparations. Nedergaard (1982) has directly compared rat cells and hamster cells but did not observe any principle differences in the variables measured.

2. *Cells from Different Anatomical Deposits*

Brown adipose tissue is distributed in different deposits in the body. There are no studies discussing any differences between cells prepared from these different sites, and the choice of tissue for cell preparation is governed only by practical aspects. *Interscapular* fat is the easiest accessible deposit; for rats it is often the only deposit used, and it is used for hamsters and rabbits

TABLE I

Brown Fat Cells, Isolated from Different Animals

Animal	Comment
Rat	Prepared from adult rats (130–160 gm) living at ambient temperatures around 20°C, first by Fain *et al.* (1967); the preparation has remained the one most used by Fain and co-workers at Brown University. Sprague–Dawley rats from Charles River are used (Fain *et al.*, 1973). The cells show catecholamine-stimulated respiration. The cell preparation has been very much studied. However, adult rats, living at around 20°C, contain only little brown fat, and the deposits are difficult to distinguish from deposits of white fat
Hamster	Prepared from adult golden (Syrian) hamsters (80–110 gm), living at ambient temperatures around 20°C, first by Prusiner *et al.* (1968a). This preparation is still the one most used by Lindberg and co-workers at the Wenner-Gren Institute, Stockholm. The cells show catecholamine-stimulated respiration. The cell preparation has been much studied; the advantage of using hamsters lies in the fact that hamsters retain brown fat as adults even when living at temperatures close to their thermoneutral zone.
Ground squirrel	Prepared from adult squirrels, hibernating and aroused, only by Burlington *et al.* (1969). The cells have only been used for determinations of lipid and DNA
Guinea pig	Prepared from adult guinea pigs (300–350 gm), living under normal conditions, only by Forn *et al.* (1970). Catecholamine-stimulated respiration was not reported, but NE-stimulated glycerol release and adenylate cyclase was found
Gerbil	Only mentioned by Fain and Reed (1970)
Rabbit	Prepared from *young* rabbits (400 gm), only by Chan and Fain (1970) and by Reed and Fain (1970) who state that "rabbit cells should be particularly useful in studies requiring large numbers of brown fat cells since a young rabbit has as much brown adipose tissue as three to six rats." Catecholamine-stimulated respiration was not reported, but epinephrine-stimulated fatty acid and glycerol release (Reed and Fain, 1970). The cells have been used to investigate some insulin effects (Chan and Fain, 1970). Prepared from *adult* rabbits (3 kg) by Kumon *et al.* (1976). These cells were not considered by the authors to be brown fat cells, but they were prepared from typical brown fat deposits (which look like white fat in the adult rabbit). Catecholamine-stimulated respiration was not studied, but catecholamines stimulated glycerol and fatty acid release in these cells, whereas catecholamines were without effect on cells prepared from typical white fat deposits (epididymis and omentum)
Lamb	Prepared from 3- to 5-day-old lambs, only by Cannon *et al.* (1977a). NE did not stimulate respiration in these cells, but oleate induced its own oxidation. Succinate and glycerol 3-phosphate, which often elicit respiration in other cell preparations, did not do this in the lamb cell preparation. The cells were "sinking cells" (i.e., they were collected as a pellet by centrifugation)

but the brown fat is there infiltrated within fat of white appearance. *Cervical* fat is the most "clean" anatomical site; it can be used in rats and it is, together with *axillary*, *scapular*, and *periaortic* fat, routinely used for hamster cell preparations. *Perirenal* fat was used in the preparation from lamb.

3. *Cells from the Two Sexes*

Fain *et al.* (1967) compared cells from female and male rats. They measured the basal and the epinephrine-stimulated rate of glycerol and fatty acid release; they did not observe any differences in these respects between the sexes. No other reports exist.

4. *Cells from Animals of Different Ages*

Cells which show catecholamine-stimulated respiration have been prepared from 10-day-old rats (Kopecký and Drahota, 1978), from young rats (130–160 gm) (Fain *et al.*, 1967), and from older rats (230–250 gm) (Herd *et al.*, 1973). In general the use of younger animals, both rats and hamsters (Nedergaard and Lindberg, 1979), seems to be the most convenient, in being the best compromise between the lower total yield in small animals and the tendency to a decreased hormonal sensitivity seen in older animals, not only in brown fat, but also, e.g., in white (Olefsky, 1977).

5. *Cells from Animals Living at Different Temperatures*

Adaptation to cold (5°C) increases the amount of brown fat, particularly in rats (Pagé and Babineau, 1950); this is the anatomical parallel to the development of cold-induced nonshivering thermogenesis (see Section I,B,3). A series of morphological and biochemical changes can be observed (e.g., Suter, 1969; and Thomson *et al.*, 1969). As the brown fat is most active in cold-acclimated animals, and as the total mass of brown fat is so much higher in the cold (in rat a threefold increase between 20 and 5°C (e.g., Nedergaard *et al.*, 1980)), the preparation of cells from cold-adapted animals looks very promising. Alas, in reality this is not so. The problems with cells from cold-adapted animals can be summarized in four points.

a. *Difficulties in Collecting Cells.* When an animal is exposed to cold, it will use much of its stored fat during the first days, especially if it is not fed. If any cells are obtained, they do not float (Fain, 1975), but rather sink during centrifugation and can be collected only in the pellet ("sinking cells," Fig. 2) (Pettersson, 1976). The same is true for cells from newborn lambs (Cannon *et al.*, 1977a).

b. *Difficulties in Obtaining Cells.* When the animals have become fully cold-adapted, the fat stores within the cells are partially replenished, and the cells will float again. It is however still often difficult to obtain any cells from animals living in the cold (Himms-Hagen, 1972); the cells seem to break

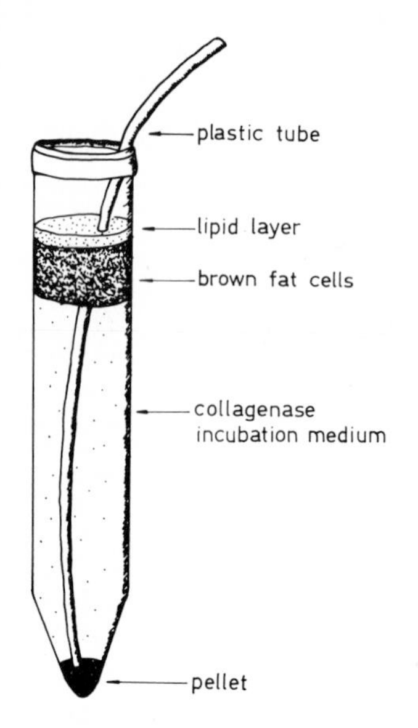

FIG. 2. Flotation of freshly prepared brown fat cells. The cells lie as a brownish-white layer between a thin film of transparent lipid and the collagenase incubation medium. Through a plastic tube the collagenase incubation medium (and later the wash medium) can easily be aspirated. The pellet contains debris from the tissue, epithelial cells, preadipocytes, red blood cells, "sinking cells" (Section II,A,V,a).

during the isolation procedure. This fragility may be related to the fact that the cells are isolated when they are very active, and the yield may be increased by cessation of the immediate cold stress. This is done simply by allowing the animals to stay in warmer conditions for some hours before the cells are prepared. Such a cessation of acute cold stress, which within minutes decreases blood flow through the tissue, probably does not to any significant extent counteract effects of cold adaptation which in the intact animal requires weeks to recede (Jansky *et al.*, 1967). Cells from animals living under daily temperature fluctuations often show better catecholamine sensitivity than cells from animals living under constant cold stress (Bertin, 1976; Bertin and Portet, 1976). In such cycles the coldest temperatures are often set during the night, and the cells are thus really prepared from animals which are not acutely maximally cold stressed at the moment of sacrifice, i.e., a situation equivalent to the removal of the animal from the cold room some hours before cell preparation.

c. *Lack of Response to Catecholamines.* Even when cells can be obtained from cold-adapted animals, such cells often lack sensitivity to catecholamines [when measured as increase in cAMP (Hittelman *et al.*, 1974; Bertin and Portet, 1976), glycerol release (Bertin, 1976), or respiration (Pettersson, 1969, internal report)]. In general the rates or levels measured are not higher than those found in unstimulated cells from control animals; this indicates that the lack of effect of catecholamines is not due to the cells being already maximally stimulated, but rather to a lack of catecholamine sensitivity. The cells may be said to be "absolutely refractory." If, however, such cells are allowed to recover for about 1 hour in a bicarbonate buffer in the presence of glucose (and pyruvate), the sensitivity to NE reoccurs, and these cells show NE-stimulated respiration and fatty acid release not significantly different from those of control cells, prepared and treated in the same way (Nedergaard, 1982).

d. *Lower Sensitivity to Catecholamines.* However, the EC_{50} for NE is about 10 times higher in cells from cold-adapted animals than in cells from control animals; the cells are "relatively refractory." This can be stated in terms of a desensitizing process, but the molecular nature of this desensitization—as well as of the cellular processes occurring during the recovery phase—is so far unknown (Nedergaard, 1982).

In practice, cells from cold-adapted animals are thus not in common use, and our knowledge of brown fat cell metabolism has largely been obtained with cells from animals living only slightly below thermoneutral temperatures.

Cells from *hibernating* animals have not been used for metabolic studies, although there are some indications that such cells would behave differently from those from control animals (e.g., Grodums, 1977, found differences in the electron microscopic appearance of the mitochondria in brown fat from hibernating, arousing, and active golden-mantled squirrels); nor have cells been prepared from *heat-adapted* animals (i.e., animals living at temperatures above thermoneutral).

6. *Food Conditions*

The first preparations of cells were made from rats which had been *deprived of food* for 18 hours (Fain *et al.*, 1967) or from hamsters which had been starved for 48 hours (Prusiner *et al.*, 1968a), in order "to reduce the size of the lipid globules within the cells in order to prevent rupture of the cellular membranes during centrifugation" (Prusiner *et al.*, 1968a). However, although starvation of rats which are not particularly cold stressed does lead to a decrease in fat content in the brown fat [from 60 to 35% in 2 days (Chalvardjian, 1964)], starvation of hamsters at 25°C does *not* lead to significant changes in the fat content (35%) even after 3 days (Lindberg *et al.*, 1976). Probably, in order to see effects of starvation in brown fat, the animals must be starved at an ambient temperature somewhat below their thermoneutral zone; there are no reports of a decrease in fat content under thermoneutral conditions. As the effect of starvation in hamster is minimal, this pretreatment has been discontinued, and it is most often not used in rats either.

The *kind of food* given, e.g., the fat content, may also influence the experimental results. Thus higher lipogenesis is expected if animals are fed a high-carbohydrate-containing diet than if they get sufficient fat, and differences in measured lipogenic capacity of isolated cells have been ascribed to differences in diet (Angel, 1969). The pattern of fatty acids in brown fat is changed by feeding rats different diets, thus the inclusion of sunflower seeds leads to an increased level of linoleic acid in the fat (our unpublished observations).

Cells have not yet been isolated from rats fed a "cafeteria" diet. Such rats, according to Rothwell and Stock (1979), obtain as much brown fat as cold-adapted rats, together with a similar increase *in vivo* of NE-stimulated thermogenesis, indicating that the brown fat cell is stimulated by this palatable diet. However, hamster cells are routinely obtained from animals living on a "palatable" diet (sunflower seeds, dried carrots, maize, etc.) and their high brown fat content may well be a reflection of this; also the concentration of thermogenin in their mitochondria is high, compared to control rats (Sundin *et al.*, 1981b). Thus it is possible that the routine hamster cell preparation is in reality obtained from animals showing diet-induced non-shivering thermogenesis (see Section I,B,4); this may also explain the lack of difference between cells from "control" (i.e., then diet-adapted) and cold-adapted hamsters (Nedergaard, 1982).

Thus, brown fat cells are routinely prepared from adult rats and hamsters, living at normal ambient temperatures (20°C), and eating "normal diets." There is, however, no reason at present to suspect that results obtained with such cells are not representative for cells from animals in which the brown fat is more activated.

B. The Isolation Procedure

Although many procedures have been described, the one presented here is the most commonly used. Some modifications of and comments to the procedure can be studied in Table II.

The animals are killed by decapitation and the brown fat is dissected out in Krebs–Ringer phosphate (Table II, 1) buffer with albumin (2) and glucose/fructose (3) at room temperature. The tissue is cut into small pieces with a pair of scissors, and the mince is washed with some extra buffer to get rid of blood. The mince is placed in plastic vials (4), e.g., scintillation vials, approximately at a concentration of 1 gm/ml (i.e., brown fat from one to two hamsters), still in the same buffer (5). Collagenase (6, 7) is added to a concentration of 2 mg/ml and the vials are incubated (8) at 37°C for approximately 30 minutes (9), until most of the tissue seems to have disintegrated. The incubation is performed in a slowly shaking water bath; every fifth minute the vial is removed and shaken for 10 seconds on a Vortex shaker at lowest speed or vigorously by hand (10). The suspension is then filtered through silk cloth (11). The filtrate (12) is collected in a plastic test tube,

TABLE II

NOTES AND COMMENTS TO THE CELL ISOLATION PROCEDURE[a]

Note number	Comments
1.	Fain *et al.* (1967) in the original method used Krebs–Ringer *bicarbonate* buffer; this buffer needs constant bubbling with 5% CO_2 to keep the pH, a procedure which is rather cumbersome. Reed and Fain (1968a) then changed to phosphate buffer for the isolation of cells and this buffer is now in common use. A "minimal essential medium" was used by Feldman (1978) in his studies on glucocorticoid hormone effects on brown fat cells
2.	4% albumin is currently used; although generally fatty acid-free albumin is to be preferred (in investigatory incubations), crude albumin (fraction V) is routinely used in the preparation procedure, during the collagenase incubation, for economical reasons
3.	10 m*M* fructose and 10 m*M* glucose may be added; if, however, effects of additions of glycolytic intermediates are studied, these additions should be avoided (Cannon and Nedergaard, 1979a)
4.	As is the case with white fat cells, glass vials are deleterious to the cells. Fain *et al.* (1967) measured a lower rate of glucose conversion to glyceride-glycerol in glass vials than in plastic vials. Instead of plastic tubes, siliconized glass can be used
5.	For preparation of brown fat cells from neonatal rats for cell culture, Dyer (1968) used a Ca^{2+},Mg^{2+}-free buffer system (and trypsin instead of collagenase); no statement on catecholamine-stimulated reactions in these cells was given. For preparation of *liver* cells a procedure with an initial perfusion with Ca^{2+}-free medium (with the chelators EGTA or EDTA present) before the collagenase treatment is routinely used (Seglen, 1976). Perfusion of brown fat is very difficult because of the diffuse localization of the tissue, but it is possible to start the incubation by 5–10 minute treatment with a Ca^{2+}-free EGTA-containing buffer, and then to continue as described. In some preliminary observations (unpublished) we did see an increase in cell yield of 25–50% in tissue treated in this way, compared with paired controls, but the extra yield did not seem to be worth the procedure
6.	The collagenase used is commonly the "crude" fraction from either Worthington or Sigma
7.	Soybean trypsin inhibitor (0.3 mg/ml) may be added (Horwitz, 1973), especially as the insulin receptor is very sensitive to trypsin-like proteases
8.	During the collagenase incubation period, the preparation was originally bubbled either with 95% O_2 and 5% CO_2 (for the bicarbonate buffer) (Fain *et al.*, 1967) or with 100% O_2 (Prusiner *et al.*, 1968b), but Bernson (1971, internal report) found that this bubbling with O_2 did not give more cells or cells showing higher NE-stimulated respiration than untreated controls, and the bubbling (or blowing of O_2 on the surface of the incubation) has been discontinued (Houštěk *et al.*, 1975a). However, Feldman (1978) could not measure glucocorticoid receptors in brown fat cells, unless the incubation medium was constantly gassed with 95% O_2 and 5% CO_2 during the preparation

(*continued*)

TABLE II (*continued*)

Note number	Comments
9.	Pettersson and Vallin (1976) introduced a modification in which the incubation is first allowed to proceed for only 5 min; the incubation is then filtered as described, but the filtrate is discarded and the nondigested tissue on the silk cloth is returned to a medium with the same composition as before, including the collagenase, and the incubation is continued as described. They state that "this procedure removes a lot of triglyceride droplets, broken cells and possibly white fat cells and contributes significantly to giving a cleaner final suspension of brown adipocytes"
10.	This step is undoubtedly the most critical in the procedure; if shaken too much the cells break, if too little the tissue does not disintegrate
11.	Silk cloth is obtained from Joymar Scientific, New York. It is possible to increase the yield somewhat by collecting the nondigested tissue from the silk cloth and continuing the incubation
12.	Alternatively, Horwitz (1973) and Hamilton and Horwitz (1979) discard this filtrate and used what is left on the filter to make their cell suspension, simply by expressing the remaining mince into a test tube
13.	Originally a centrifugation step was performed here, in a table-top centrifuge at low speed for a few minutes (Fain *et al.*, 1967: 400 *g*, 1 minute; Prusiner *et al.*, 1968b: 50 *g*, 2 minutes). Horwitz (1973) collected the cells simply by floatation, as did Pettersson (1977), who stated that centrifugation reduced the number of intact cells collected
14.	Here only one wash is described. More washes seem to remove the mitochondrial fragments which are responsible for the observed exogenous glycerol 3-phosphate and succinate respiration (Herd *et al.*, 1973; Bernson *et al.*, 1979). However, the yield decreases and in practice a compromise between purity and yield must be made
15.	The final amount of triglyceride droplets is decreased if not only the infranatant but also the supernatant clear fat and the white layer is removed here (Williamson, 1970)
16.	We have observed (unpublished) that cells are best stored on ice; only a small change in respiratory capacity is then observed during a working day. If cells are kept at room temperature, the rate of basal respiration increases with time and NE-stimulated respiration decreases. Pettersson (1976) found that cells could be stored for up to 48 hours in this way and he pointed out that it is feasible to prepare cells on one day and perform experiments on the next. It may be argued that storage of cells at 0°C leads to loss of the intracellular ions and that the lipids solidify, but the cells seem to be able to recover quickly from such effects, when brought to higher temperatures

[a] Numbers relate to the discussion of the isolation procedure in Section II,B.

and the cells are allowed to float (13) to the surface for approximately 30 minutes, giving an appearance as in Fig. 2. The infranatant is sucked off and new buffer added, and the cells are again allowed to float and the infranatant is removed (14, 15). A cell suspension with approximately 2 million cells per milliliter will be the result. This cell suspension can be kept on ice (16) or in the cold for hours or even days; if it is used after storage it is of advantage again to wash and filter it as above, as a high basal respiration and threadlike conglomerates develop upon storage. The yield of cells is very variable; in a routine preparation from three hamsters between 3 and 30 million cells can be obtained.

C. Properties of the Isolated Cell Suspension

The method described in the preceding section results in a suspension of isolated cells. Here we shall discuss the qualities of this preparation, i.e. (1) the purity of the suspension (and how impurities may affect parameters to be studied), (2) its quantification, (3) how the isolated cells compare with the parent tissue, and the (4) advantages and (5) disadvantages of the use of isolated cells.

1. *The Purity of the Isolated Cell Suspension*

Figure 3 illustrates those impurities in the cell suspension which have so far been described. They include:

a. *Broken Brown Fat Cells.* Houštěk *et al.* (1975a) added 0.1% Alcian blue to the cell preparation; this dye stains the nuclei if it can permeate into the cells, and cells with a broken cell membrane thus have a blue nucleus, seen in the light microscope. In this way the number of broken cells can be estimated; routinely 5–10% of the cells stain with Alcian blue. Unfortunately, it is not known how these cells influence the properties of the cell suspension; they probably lack response to catecholamines, but they may in part be

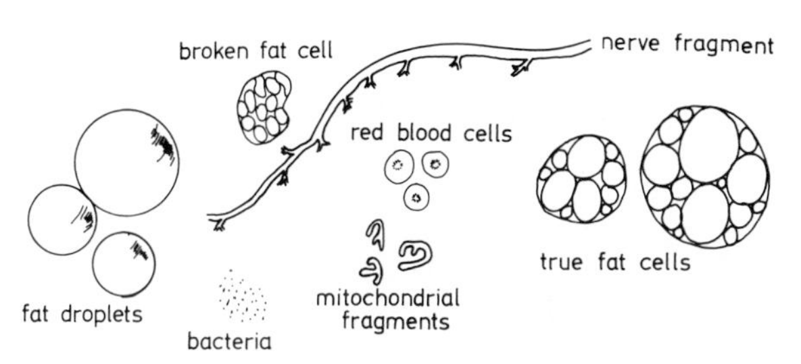

FIG. 3. A visualization of the impurities encountered in a preparation of brown fat cells. The impurities and their effects on the results obtained are discussed in Section II,C,1.

responsible for, e.g., the oxidation of substrates such as glycerol-3-phosphate which are poorly permeable over the intact cell membrane.

b. *Fat Droplets.* These circular droplets often outnumber the number of true brown fat cells (see, e.g., Fig. 1C), and their presence in principle invalidates determinations of activity, expressed per milligram dry weight or per millimole triglyceride. Further, in automatic cell counting systems (Coulter Counters) they may count as true cells, also invalidating this system. They easily absorb organic-soluble additions to the suspension, in this way disturbing dose–response calculations for such agents. Their number can be diminished (Table II, 15). In the light microscope, they are indistinguishable from white fat cells, but electron microscopic pictures show that they lack a cytoplasmatic rim around the fat droplet (T. Barnard, personal communication). They are probably formed by coalescence of fat droplets, liberated from broken cells during the isolation procedure.

c. *White Fat Cells.* The question of the existence of white fat cells in a brown fat cell preparation is more semantic than practical. It may be possible to find cells with a unilocular appearance in a preparation, but there is no accepted way to distinguish between a brown fat cell disguised in unilocular form and a "true" white fat cell. Hittelman *et al.* (1974) calculated that 10–15% of a hamster brown fat cell preparation consisted of white fat cells, but the assumptions behind this calculation may be discussed, and in any case, as the authors themselves point out, "it is possible that this capacity [i.e., for a true brown fat cell to respond as a white fat cell] is lost during adaptation to cold." The brown fat cell isolation procedure is so harsh that the bigger and more fragile white fat cells—if they are present—will probably break during isolation. In principle, we favor the "anatomical" view, i.e., that cell preparations made from fat pads from typical brown fat loci consist solely of *brown* fat cells; these may in turn, however, be in differently differentiated states.

d. *Nerve Terminals.* A great advantage of the isolated cell preparation compared to tissue slices should be the absence of nerve terminals; however due to the dense innervation of brown fat [in the tissue every single cell is innervated (Derry *et al.*, 1969)] it is difficult to ascertain that nerves no longer cling to some cells after the isolation. Such nerves will influence hormonal studies in two ways: by being able to release stored NE and by being able to take up added catecholamines. *Release* of NE can, it seems, be caused by a multitude of factors. Indeed, we suggest that all observed stimulatory effects of serotonin (and perhaps some other noncatecholamine hormones), high K^+ and ascorbic acid are solely due to these agents causing release of stored NE from the vesicles, and we suggest that before any noncatecholamine agent is accepted to be a stimulator of thermogenesis, the following controls should be performed: it should be shown that the cell preparation cannot be stimulated by ascorbic acid (around 10 μM) and it should be shown that a purportive thermogenesis stimulator cannot be inhibited by the β-adrenergic inhibitor propranolol [rather high doses of propranolol (up to 50 μM) must be tested as a release of NE from cell-attached nerve fibers directly into the synaptic cleft must be competed away]. *Uptake* of added catecholamines into the nerve vesicles may make the interpretation of hormonal experiments difficult; e.g., an added α-inhibitor may inhibit the uptake of NE into the vesicles; if small amounts of NE are added, the presence of the inhibitor may thus in fact lead to higher concentrations of NE at the fat cell receptor and such experiments may be interpreted as a demonstration of effects of α-adrenergic receptors on the brown fat cells. It may be advantageous to include an uptake inhibitor, e.g., cocaine, under such circumstances.

e. *Mitochondrial Fragments.* The substrate-induced respiration of added succinate or glycerol-3-phosphate, which can be observed in most cell preparations, is extracellular and may take place in mitochondrial fragments (see Section III,D,2,3). The mitochondrial contaminations are considered to be fragmentary, as only flavoprotein-linked respiration (i.e., of glycerol-3-phosphate and succinate) is observed; other examples of substrate-induced respiration are not known (except fatty accids, see Section III,D,1), although a series of other substrates have been tested. This substrate-induced respiration can be greatly diminished by successive washings of the cell suspension (Herd *et al.*, 1973; Bernson *et al.*, 1979). The presence of mitochondrial fragments has blurred the discussion on the coupling state of the fat cells, but these problems now are clarified.

f. *Red Blood Cells or Hemoglobin.* Hemoglobin absorbs light around 400 nm, and as this absorption is influenced by the oxygen tension in the cuvette it may disturb spectrophotometric measurements. Most of this hemoglobin can be washed away in an additional wash (Prusiner, 1970).

g. *Bacteria.* Bacteria do not normally cause problems in respiratory studies, but in metabolic or anabolic studies they may. They can be kept at bay by penicillin and streptomycin (Horwitz, 1973).

Thus most of the impurities in the cellular preparation can be washed away in an extra series of washings; the cost is naturally a decreased total amount of cells.

2. *Quantification of the Cell Preparation*

Several methods have been used to quantify the cell preparation (Table III) making comparisons between results from different laboratories difficult. As seen from the table, most of the methods do not distinguish between metabolically active and inert fat, and the method which does best in this regard (manual cell counting) is unfortunately prone to some subjectivity. In order to be able to compare results from different laboratories we have in this article often expressed the results as nanomoles per million cells, using the conversion factors stated in Table III (indicated in the text by "calc.").

TABLE III
METHODS FOR QUANTIFICATION OF THE CELL PREPARATION

Method	Description and comments
mmole triglyceride	An aliquot of the cell suspension is extracted with Dole's mixture and the organic phase is analyzed for ester content with a colorometric method (Rodbell, 1964). This has been the preferred method of Fain and co-workers (Fain *et al.*, 1967). The limitation of this method is that it does not discriminate between metabolically active and inert fat (i.e., fat droplets in the suspension), and it thus underestimates the activity, compared with the parent tissue. Based on a fat content of around 40% in the tissue and 100 million cells per gram tissue, 1 mmole triglyceride is approximately equivalent to 200 million cells. Activities are often in this case expressed as micromoles/millimole triglyceride; this is then equivalent to 5 μmole/million cells. Respiration rates may be measured in microliters O_2/micromole triglyceride; this is then equivalent to 415 nmole O/million cells
"Microhematocrit" (or better microadipocrit)	An aliquot of the cell suspension is allowed to enter a microhematocrit tube and centrifuged and the percentage cells determined (Prusiner *et al.*, 1968a). This method does not distinguish between metabolic active and inert fat, but it is very quick to perform. According to Prusiner *et al.* (1968b), 1% cells is approximately equal to 100,000 cells/ml
Dry weight	An aliquot of the cell suspension is pipetted onto a preweighed Millipore filter paper (0.8 μm) and the buffer filtered away under vacuum. The filter is dried overnight and weighed (Butcher *et al.*, 1968). Used by Hittelman *et al.* (1974). This method mainly measures fat weight, and it does not distinguish between metabolically active and inert fat. Based on a water content of 50% in the tissue and 100 million cells/gm tissue, 1 mg dry weight is equivalent to 200,000 cells
DNA determination	In an aliquot of the cell suspension, DNA is measured fluorometrically (Kissane and Robins, 1958). Used by Herd *et al.* (1973). This method does not distinguish between DNA in broken and intact cells. Based on a value of 6 pg DNA per mammalian cell, 1 μg DNA is equivalent to 130,000 cells
Hemacytometer	Used by Czeck *et al.* (1974), but this instrument seems to give very high counts; as a result Czeck *et al.* state that they incubate 20 million cells in 0.1 ml buffer but such a concentration of cells is probably not even seen in the tissue itself
Cell counter (Coulter Counter)	This automatic device was recommended by Fain (1975) and used by Feldman (1978). Cell counters also count fat droplets as cells, and the number obtained is thus an overestimate of the number of true cells

(*continued*)

TABLE III (*continued*)

Method	Description and comment
Manual cell counts	An aliquot of the cell suspension is diluted 10 times with preparation buffer, and 0.1% Alcian blue is added. The suspension is placed in a Bürker chamber, and only profiles considered to be true cells (i.e., not fat droplets) are counted; cells with a blue nucleus are considered broken (Houštěk *et al.*, 1975). (A similar method was used by Forn *et al.*, 1970.) The advantages of this method are that it is quick (can be performed prior to experiments), that it counts only metabolically active cells, and that results are not dependent on whether cells are fat-depleted or have a changed protein content. The disadvantages are the need of a subjective analysis of the cell preparation (see Fig. 1C) and the inherent uncertainty due to the limited amount of cells which can in practice be counted. If a total of around 100 cells is counted the estimated standard error of the determination is 10% and a 95% confidence interval is ±20%. Thus if results from different cell preparations are to be compared, it is often advantageous to express the results in a relative manner (as percentage of, e.g., maximal NE response)

TABLE IV

OXYGEN CONSUMPTION CAPACITY AND NUMBER OF BROWN FAT CELLS IN BROWN FAT TISSUE

A. Measurements of the oxygen consumption capacity of brown fat tissue

Heim and Hull (1966) measured the blood flow through newborn rabbit intrascapular brown fat and the arteriovenous difference in oxygen content over the tissue. They found a basal rate of 10 μmole O min^{-}1 gm^{-1} and a NE-stimulated rate of 50 μmole O min^{-1} gm^{-1}

Hayward and Davies (1972) found that ligature of the artery to brown fat in mice leads to a 40% decrease of NE-stimulated oxygen consumption. From their values it can be calculated that if brown fat respiration is solely responsible for the difference, it must use 400 μmole O min^{-1} gm^{-1}

Foster and Frydman (1978), by measuring blood flow to and arteriovenous difference over brown fat, calculated that in NE-injected cold-adapted rats, brown fat used 140 μmole O min^{-1} gm^{-1}

B. Estimates of number of brown fat cells in brown fat tissue

Sheldon (1924) found that the cell size in well-fed rats varied between 25 and 40 μm; this corresponds to between 20 million and 60 million cells cm^3 (approx. 1 gm)

Schierer (1956) calculated that in normal-fed rats there are 320 million cells/cm^3, whereas in rats starved to death, as many as 1500 million (!)

Teodorn and Grishman (1959) found that the cell size in Syrian hamsters in different physiological states varied between 12 and 23 μm; this corresponds to between 80 and 600 million cells/cm^3

Williamson (1970) measured cytochrome c spectrophotometrically (540–550 nm) in a rat cell suspension and found 0.2 nmole/million cells. By relating to cytochome c content in the tissue, he estimated that there are 200 million cells/cm^3

T. Barnard (personal communication) estimated by morphometric techniques that there are 100 million cells in 1 cm^3 hamster tissue and 80 million cells in 1 cm^3 rat tissue

3. *Comparison between Cells and Parent Tissue*

In order to compare the respiratory (i.e., thermogenic) capacity of isolated cells with that of the cells in the parent tissue the number of cells per gram tissue and the respiratory capacity of the tissue must be known (Table IV). As Table IVA shows, it is reasonable to approximate the oxygen consuming capacity of brown fat to a value of 100 μmoles O/minute/gm wet weight. However (as also reviewed by Afzelius, 1970), the cell diameter of brown fat cells varies much (Table IVB), and the estimated content of cells in 1 cm^3 (approximately 1 gm) naturally much more. A value of 100 million cells/gram tissue is only an acceptable rule of thumb, which nevertheless may be used for calculations of expected activities of isolated cells. From these two approximations, an expected oxygen consuming activity of close to 1000 nmoles O/minute/million cells can be calculated; comfortingly, under optimal conditions, isolated brown fat cells show respiratory efficiencies approaching this value (Nedergaard and Lindberg, 1979). This good agreement between data obtained *in vivo* and *in vitro* has earlier been pointed out (Lindberg *et al.*, 1976; Nedergaard *et al.*, 1977; Cannon *et al.*, 1978). Conversely, we consider that in cell preparations which show respiratory capacities one order of magnitude or more below this, the reactions measured may not be those of greatest quantitative importance *in vivo*.

The amount of *fat per cell* and the *proportion of brown fat cells* in brown fat tissue have been estimated by Fain *et al.* (1967) in rat and by Burlington *et al.* (1969) in ground squirrel. The authors agree that only 25% of the cells in the tissue are brown fat cells, but this estimate is based on comparisons between the fat/DNA ratio in tissue and in isolated cells [in both reports equivalent to around 5 mg (10 μmoles) triglyceride per million isolated cells (calc.)] and is thus dependent on a fat droplet-free cell preparation.

4. *Advantages of the Isolated Cell Preparation*

In comparison with tissue slices the isolated cell preparation has the following advantages:

Oxygen supply does not readily become rate limiting. Due to their extremely high rate of oxygen consumption, stimulated brown fat cells, if not greatly diluted, immediately use all available oxygen; e.g., if concentrated as much as in the intact tissue, they use all oxygen dissolved in tissue water (50% of tissue) in less than one-tenth of a second! Accordingly, Friedli *et al.* (1978) comment on their experiments on tissue slices that already at a NE concentration of 0.1 μM, "oxygen diffusion was probably rate limiting at such high respiratory rates" [80 nmoles O/minute/million cells (calc.)]. Anoxia occurring unnoticed during measurements other than respiratory may lead to a series of secondary effects, affecting the variable under study.

Only one cell type is studied, whereas in tissue slices the majority of cells are probably not brown fat cells, but other cell types which may interact with brown fat cells in an unpredictable manner. This is particularly important for nerve cells or nerve endings.

Compared to *in vivo* studies, the *lack of secondary effects* may be specifically noticed. Especially in hormonal studies, *in vivo* effects on other organs and feedback mechanisms may influence the results; e.g., the presence of α-adrenergic receptors on the blood vessels leading to brown fat may lead to vasoconstriction and thus to a non-fat cell-mediated inhibition of the tissue activity.

5. *Disadvantages of the Isolated Cell Preparation*

a. *Low Yield.* Only 3–30 million cells are obtained from three hamsters, i.e., from about 3 × 600 mg brown fat. This means that the yield is only 2–10%. However, in many experiments, e.g., measurements with an oxygen electrode, 50,000 cells may be enough for one trace; a routine preparation is thus more than sufficient for one series of experiments.

b. *Selected Class.* The low yield may in turn be due to the possibility that only cells in a narrow zone between being too fat depleted to float and too fat filled to withstand vortexing during cell preparation are harvested. The surprising similarity between cells isolated from different species under different physiological conditions (Nedergaard, 1982) may bear on this point.

c. *Lack of Contact between Cells.* Gap junctions are the only intercellular contacts observed in electron microscopic studies; there are no desmosomes and no tight junctions between the cells (Revel *et al.*, 1971). When isolated, cells obviously cannot communicate with each other through gap junctions, and possible important intercellular regulation systems cannot be studied. Further one may wonder whether at the place of the broken gap junction an unnatural permeability region has been created.

d. *Changes during Isolation.* The existence of such changes cannot be excluded; e.g., Feldman (1978) found only one-fifth the number of glucocorticoid receptors per milligram cytosolic protein if cytosol was prepared from isolated cells, compared to cytosol prepared directly from tissue. It is further possible that trypsin-like contaminations in collagenase may degrade insulin receptors.

III. Incubation Conditions

Nearly all studies of brown fat cells have been performed in a—sometimes significantly—modified Krebs–Ringer buffer, in principle similar to the one described by Krebs and Henseleit (1932). In the following we shall discuss

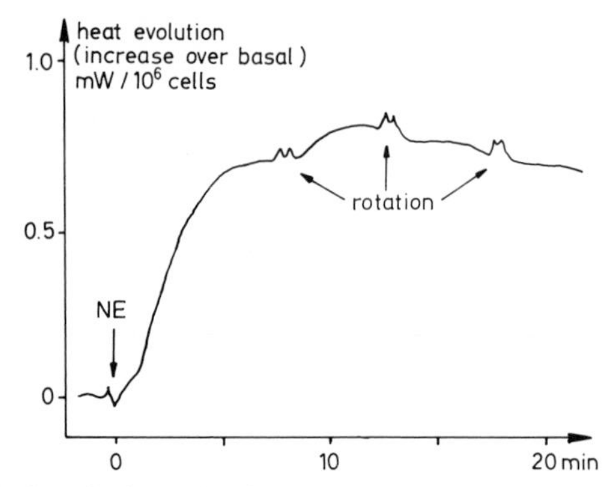

FIG. 4. Microcalorimetric determination of NE-stimulated heat production in brown fat cells. Heat production was monitored in an LKB 2107 batch microcalorimeter. At NE the compartments containing brown fat cells and NE were mixed by rotation; in order to ensure unlimiting oxygen supply to the cells the vessels were rotated every fifth minute (rotation). Adapted from Nedergaard *et al.* (1977).

the effect of the incubation conditions on those parameters which are traditionally studied in brown fat cells. They include the *membrane potential*, *lipolysis*, and *respiration* as an indirect measure of *thermogenesis*.

The membrane potential is always measured with a microelectrode over the plasma membrane of a single cell. Practically these studies cannot yet be performed in isolated cell suspensions; instead tissue slices are routinely used.

Lipolysis is measured as glycerol or fatty acid release in incubations.

Thermogenesis. Only by microcalorimetry can heat, the main product of the brown fat cell, be directly measured (Fig. 4). Although microcalorimetry is the method closest to the real function of brown fat cells, "indirect calorimetry," i.e., measurement of oxygen consumption, has been the most popular experimental procedure, because of the possibility of immediately seeing the influence of various agents on thermogenesis. In parallel oxygen electrode and microcalorimetric experiments, Nedergaard *et al.* (1977) determined the ratio between heat production and oxygen consumption to be 250 kJ/mole O, this value being so close to the theoretical one that it could be concluded that no other major heat-producing processes other than oxidation processes take place in the brown fat cell. The rate of oxygen consumption is most easily followed with an oxygen electrode in a sealed chamber, as first described by Prusiner *et al.* (1968a,b). Manometric techniques may be used (Fain *et al.*, 1967) but these demand prolonged incubations (1–2 hours) and a defined atmosphere, normally carbon dioxide free. Furthermore, it is not possible to follow minute-by-minute changes in thermogenesis. A typical oxygen electrode tracing (Fig. 5) shows first the

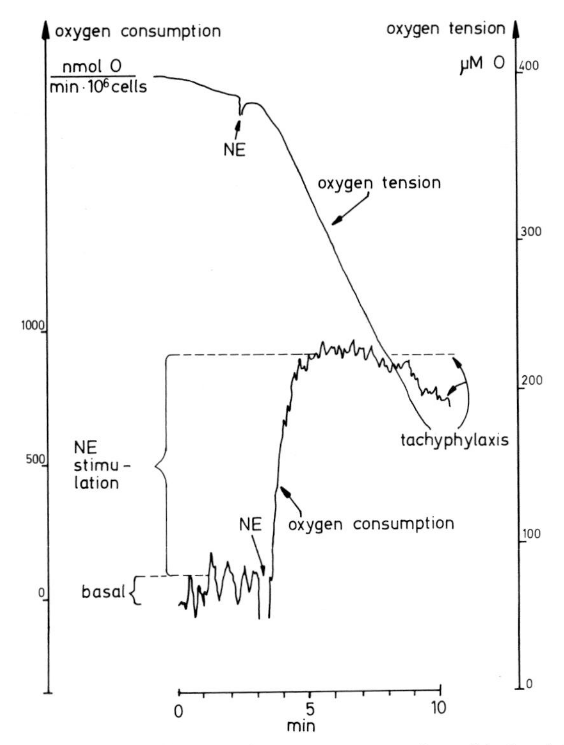

FIG. 5. Polarographic determination of oxygen consumption of isolated brown fat cells. Brown fat cells (50,000) isolated from adult hamsters were incubated in a Krebs–Ringer bicarbonate buffer in the presence of albumin. Both the oxygen tension in the electrode chamber and the rate of oxygen consumption of the cells are followed. The basal rate of oxygen consumption is stimulated about 10 times by NE. A maximal rate is obtained after 2–3 minutes after which a slow decrease is observed (tachyphylaxis). This decrease may be of biological origin or may be an incubation artifact (see Section III).

basal respiration of the diluted cell suspension. Second, as a response to the addition of NE, the respiration rate is drastically increased and a maximum is obtained. After this maximum, a decline—a tachyphylaxis—may be observed. Changes in both basal respiration, maximally stimulated respiration, and tachyphylaxis may be observed, depending on incubation conditions.

A. The Ionic Environment

The Krebs–Ringer buffer includes a high concentration of Na^+ and Cl^-, and low amounts of K^+, Mg^{2+}, and Ca^{2+}, and phosphate and SO_4^{2-}. In the "phosphate buffer," pH is stabilized with rather high extra amounts of phosphate; in the "bicarbonate buffer" pH is stabilized by the equilibrium between HCO_3^- and CO_2.

We shall here review the significance of the extracellular cations and anions for cellular metabolism. This is experimentally normally investigated either by *decreasing* (possibly omitting) or by *increasing* the amount of the ion, and measuring the effect on *membrane potential*, *lipolysis*, and *respiration* (=thermogenesis).

1. *Na^+*

The recommended concentration of Na^+ in Krebs–Ringer bicarbonate buffer is 145 m*M*.

If Na^+ is *decreased* to 5 m*M* (osmotically substituted with choline$^+$ or sucrose), the cell *membrane potential* depolarizes 20 mV, down to about the value normally seen after NE stimulation. Under these conditions, NE cannot depolarize the cell further (Williams and Matthews, 1974b). Due to the obligatory coupling of Na^+ efflux to K^+ influx of the Na^+,K^+-ATPase, a low extracellular Na^+ concentration may secondarily, at least in prolonged incubations, lead to a decreased intracellular K^+ concentration. As the membrane potential is largely governed by the K^+ gradient, this may explain the observed depolarization (although not the subsequent lack of NE effect).

NE-stimulated *fatty acid* and *glycerol release* are largely independent of extracellular Na^+ (Williams and Matthews, 1974b; Nedergaard *et al.*, 1979b; Nedergaard, 1981), but the rates of NE-stimulated *respiration* are very much influenced by Na^+; indeed this respiration is proportional to the logarithm of external Na^+. Below 6 m*M* Na^+ no further decrease is seen and this may be interpreted as indicating that the intracellular Na^+ level is of this order (Nedergaard *et al.*, 1979b). Also Herd *et al.* (1973) observed a lower respiration in a low-Na^+ buffer. Al-Shaikhaly *et al.* (1979) demonstrated that under these conditions (i.e., when substrate is produced but not oxidized), addition of an artificial uncoupler (FCCP) leads to an increased respiration, implying that lack of Na^+ interferes with the physiological uncoupling mechanism.

Due to the high standard concentration of Na^+ in the buffer, any significant *increase* of Na^+ leads to hyperosmosis, which in itself diminishes NE-stimulated respiration (our unpublished observations). However, in an indirect way, Al-Shaikhaly *et al.* (1979) studied effects of increased Na^+. By addition of the Na^+ ionophore monensin they increased Na^+ permeability over the cell membrane, probably by this increasing intracellular Na^+. This had no effect in itself; but if an exogenous substrate (pyruvate) was also added, a respiratory stimulation occurred. The authors implied that this respiration is due to Na^+/Ca^{2+}-futile cycling in the cell, but there is no direct evidence for this.

There is good evidence that Na^+ is necessary for thermogenesis but not for lipolysis; its action is perhaps connected with the action of the mitochondrial Na^+/Ca^{2+} exchanger (see Section V,D) and is thus really related to changes in intracellular Ca^{2+}.

2. *K^+*

The recommended concentration of K^+ in Krebs–Ringer bicarbonate buffer is 6 m*M*.

Decrease of extracellular $[K^+]$ below 6 m*M* depolarizes the cell *membrane potential* successively with time (Girardier *et al.*, 1968; Williams and Matthews, 1974a). Girardier *et al.* suggested that this is due to an inhibition of the Na^+,K^+-ATPase, which does not function in the absence of extracellular K^+. As intracellular K^+ leaks out and is diluted in the medium, the K^+ gradient decreases and the measured potential falls. Lowering of extracellular K^+ decreases catecholamine-stimulated *glycerol and fatty acid release* drastically (greater than two-thirds) (Reed and Fain, 1968b; Herd *et al.*, 1973; Williams and Matthews, 1974b). NE-stimulated cAMP accumulation is not inhibited (Williams and Matthews, 1974b); this points to a direct effect of lowered K^+ on the activity of the hormone-sensitive lipase, and is in principle in agreement with earlier observations on white fat cells (Mosinger and Vaughan, 1967). Cs^+ and Rb^+ can, if added to a K^+-free buffer, partially restore catecholamine sensitivity; NH_4^+ cannot (Reed and Fain, 1968b); this ion has other, inhibiting, effects (Cannon and Nedergaard, 1979). In light of the inhibitory effect of lack of K^+ on lipolysis, it is no surprise that catecholamine-stimulated *respiration* is even more inhibited, simply due to lack of substrate (Reed and Fain, 1968b; Herd *et al.*, 1973; Prusiner *et al.*, 1970). Fain and Reed (1970) also observed that K^+ is necessary for fatty acid-induced respiration (the biochemical mechanism for this is unknown), and, based on this, they have discussed K^+ cycling as a thermogenic system (see Section XIII,C).

If K^+ is *increased* above 6 m*M*, the cell *membrane potential* is decreased 60 mV per 10 times increase in K^+, as expected if the K^+ gradient is largely responsible for the magnitude of the membrane potential (Girardier *et al.*, 1968). Zero potential is measured with 135 m*M* external K^+; this can be taken as an indication of the intracellular K^+ level. These results could alternatively have been secondary to K^+-induced release of NE from nerve terminals within the tissue, but Barde *et al.* (1975) could measure a depolarization by increasing external K^+ even in reserpinized (NE-depleted) rats; thus extracellular K^+ directly affects the membrane potential. In tissues from young rats, increased K^+ ($\times 10$) led to a stimulation of *glycerol release*, but not in tissues from older rats (Williams and Matthews, 1974b). This age difference probably reflects differences in catecholamine content, as the stimulated glycerol release here is probably secondary to release of endogenous catecholamines. Barde *et al.* (1975) confirmed earlier observations that incubations of brown fat tissue in high K^+ (40 mM) showed an increased *respiration.* They convincingly demonstrated that this was secondary to a high-K^+-induced release of stored catecholamines within the tissue, e.g., the stimulation was inhibitable with propranolol, and tissue from reserpine-treated rats was not stimulated. In agreement with this, we have not observed respiratory stimulation by increasing K^+ in the incubation medium for isolated hamster cells (Nedergaard, 1981).

Thus K^+ is necessary for the stimulation of hormone-sensitive lipase, and, as a consequence, for thermogenesis. The cellular membrane potential responds as expected to changes in external K^+, but the significance, if any, of such changes is not known.

3. *Mg^{2+}*

The recommended concentration of Mg^{2+} in Krebs–Ringer bicarbonate buffer is 1 mM.

Removal of all Mg^{2+} by addition of the Mg^{2+} chelator EDTA to a Mg^{2+}-free buffer leads to only a slight inhibition of NE-stimulated *respiration* (Nedergaard, 1981), and lack of Mg^{2+} does not influence fatty acid (octanoate)-stimulated respiration (Fain and Reed, 1970).

Addition of as high as 20 mM Mg^{2+} does not alter the *membrane potential* (Williams and Matthews, 1974a).

Thus, extracellular Mg^{2+} is of no or little importance for the short-term responses of the brown fat cell.

4. *Ca^{2+}*

The recommended concentration of Ca^{2+} in Krebs–Ringer bicarbonate buffer is 2.5 mM.

Removal of Ca^{2+} in itself does not alter the *membrane potential;* only if the Ca^{2+} chelator EGTA is also added is depolarization observed. During such depolarization, NE can still induce further depolarization (Williams and Matthews, 1974a). No explanation is offered for these effects. Omission of Ca^{2+} does not influence NE-stimulated *glycerol release* from tissue pieces (Williams and Matthews, 1974b), and even the presence of EGTA in a Ca^{2+}-free buffer leads to only a slight inhibition of NE-stimulated *respiration* (Nedergaard, 1981). However, if cells are isolated and incubated without Ca^{2+}, their catecholamine-stimulated respiration is sensitive to Ca^{2+}: addition of 1 mM Ca^{2+} then leads to a doubling of the respiratory rate, and addition of EGTA abolishes the response (Williamson, 1970). Furthermore, lack of Ca^{2+} diminishes fatty acid (octanoate)-stimulated respiration (Fain and Reed, 1970).

There are no reports on effects of *increased* Ca^{2+}; in fact the recommended Ca^{2+} concentration is at the level of solubility, and often in bicarbonate buffers, and always in phosphate buffers, the Ca^{2+} is halved.

Thus, extracellular Ca^{2+} is of little importance for the short-term responses of the cell.

5. *pH*

The recommended pH of Krebs–Ringer bicarbonate buffer is 7.4.

At pH 6.9, compared to pH 7.4, the rate of NE-stimulated *fatty acid release* is decreased (Nedergaard and Lindberg, 1979), the basal *respiration* rate is lower, the maximal rate is slightly lower, and tachyphylaxis is reduced (Nedergaard and Lindberg, 1979). The EC_{50} for NE-stimulated respiration is increased (i.e., lower sensitivity) (Pettersson, 1977; Nedergaard and Lindberg, 1979).

Within a close interval above 7.4, the effects of increased pH are the opposite of a decreased pH; a general inhibition occurs above pH 8 (our unpublished observation).

The effects of pH on brown fat cells are closely similar to the effects on lipolysis in white fat cells (Hjelmdahl and Fredholm, 1977). The molecular nature of the effects is not understood. As increased pH seemingly leads to an increased thermogenesis, change in intracellular pH has been discussed as a mediator of catecholamine stimulation (see Section XIII,B).

6. *Cl^-*

The recommended concentration of Cl^- in Krebs–Ringer bicarbonate buffer is 130 m*M*.

Substitution of Cl^- with the (from Cl^-) very dissimilar anions isethionate, benzensulfonate, or sulfate depolarizes the cell *membrane potential* by 30 mV. The more similar anions NO_3^- and Br^- only depolarize 15 and 4 mV, respectively (Williams and Matthews, 1974a). The large effects of lack of Cl^- may indicate that the membrane potential includes an active Cl^- part. Williams and Matthews (1974b) did not observe any effect on NE-induced *glycerol release* of substituting isethionate for Cl^- in brown fat slices.

There are no reports on the effects of *increased* Cl^-.

7. *HCO_3^-/CO_2*

The recommended HCO_3^- concentration (addition) in Krebs–Ringer bicarbonate buffer is 25 m*M*. The buffer is bubbled with 5% CO_2 in air (or 5% CO_2/95% O_2). The commonly used Krebs–Ringer *phosphate* buffer does not contain any added HCO_3^-; thus differences between results obtained in this buffer and the bicarbonate buffer may be due to lack of bicarbonate, and/or may be due to higher phosphate.

Removal of CO_2/HCO_3^- leads with time to depolarization of cells; this depolarization also occurs in the presence of the buffer Tris, and is thus not due to changes in extracellular pH (Williams and Matthews, 1974a). Lack of CO_2/HCO_3^- does not influence the NE-stimulated rise in cAMP level (Pettersson, 1977), and there are no reports on effects on *lipolysis*.

In phosphate buffer, the maximal *respiration* rate is lower than in bicarbonate buffer (Williamson *et al.*, 1970; Pettersson, 1977). If, however, CO_2/HCO_3^- is added to the phosphate buffer (by bubbling the buffer before the experiment with 5% CO_2 in air) the respiration rate observed is higher than in normal (air-bubbled) phosphate buffer (Pettersson, 1977). [As a secondary effect of this CO_2 bubbling the pH decreases from 7.4 to 6.9; the effect of CO_2 bubbling on maximal respiration is not secondary to a change in pH (Pettersson, 1977), but other differences (e.g., change in NE sensitivity) between CO_2-bubbled and air-bubbled buffer may so be.] In the phosphate buffer, a high degree of tachyphylaxis is seen (Prusiner *et al.*, 1968b). If the buffer is CO_2 bubbled, an "almost constant high respiration" is seen after NE stimulation (Pettersson, 1977). This absence of tachyphylaxis is not seen at optimal HCO_3^- (i.e., in a bicarbonate buffer) (Pettersson, 1977); but this is probably related to the very high rate of fatty acid release in the bicarbonate buffer and can be counteracted by addition of albumin (Nedergaard and Lindberg, 1979). Under conditions of lack of HCO_3^-, i.e., in phosphate buffer, a decrease in *cellular energy potential* (or ATP/ADP ratio) occurs after NE stimulation (Prusiner *et al.*, 1968c; Pettersson and Vallin, 1975). If 5% CO_2 is bubbled into the buffer, this decrease is much less pronounced (Pettersson, 1977).

There are no reports of effects of *increased* HCO_3^-.

The use of an incubation medium without CO_2/HCO_3^- is a relic from manometric experiments, where these components interfere with the measurement; generally all CO_2 produced in such experiments is absorbed in a KOH vial in the center of the manometer, although Horwitz (1973) performed manometric experiments in a medium, which contained 5 m*M* HCO_3^- and was in equilibrium with 1% CO_2. In oxygen electrode measurements, these problems do not arise. Pettersson (1977) discussed the possibility that CO_2 directed more released fatty acids toward oxidation than to export; this is probably combined with a higher rate of lipolysis (Nedergaard and Lindberg, 1979). The metabolic action of HCO_3^- is linked to its reaction with pyruvate to form oxaloacetate, an anaplerotic reaction (Cannon and Nedergaard, 1978, 1979a) (see Section VIII,C). The effect of lack of CO_2 can then be understood in the following terms: acetyl-CoA, formed from fatty acids released by NE stimulation, needs oxaloacetate to enter the citric acid cycle.

If, in the absence of CO_2, pyruvate carboxylase is inactive and oxaloacetate is not formed, a pile-up of unmetabolizable compounds occurs, and ATP-producing reactions are inhibited. The lack of ATP then makes the Na^+,K^+-ATPase unfunctional, and a successive depolarization occurs.

8. PO_4^{3-}

The recommended concentration of inorganic phosphate in Krebs–Ringer bicarbonate buffer is 1 m*M*.

Prusiner *et al.* (1968c) *removed* phosphate from the medium; they found a normal respiratory response of the cells to NE under these conditions.

The commonly used Krebs–Ringer phosphate buffer contains 20 m*M* phosphate; thus differences between results obtained in this buffer and in bicarbonate buffer may as stated above be due to high phosphate *or* to lack of CO_2.

Addition of 20 m*M* phosphate to the bicarbonate buffer diminishes NE-stimulated *fatty acid release* (Nedergaard and Lindberg, 1979). Similarly, lower stimulation of fatty acid and glycerol release occurs in hamster brown fat slices in phosphate buffer than in bicarbonate buffer (Rabi *et al.*, 1977); in rabbit cells, catecholamine stimulation of fatty acid and glycerol release is abolished in phosphate buffer (Kumon *et al.*, 1976). Probably as a consequence of the effects on lipolysis above, NE-stimulated *respiration* is diminished if 20 m*M* phosphate is added to a bicarbonate buffer (Nedergaard and Lindberg, 1979).

Thus, although Fain (1975) states that there is no difference in hormonal response of fat cells in phosphate and bicarbonate buffers, it seems clear that the high phosphate concentration encountered in a phosphate buffer is not beneficial to the cells. The high phosphate is added solely to buffer the pH, and it may be more advantageous to use a buffering system which in itself is not a metabolite in the cells.

9. SO_4^{2-}

The recommended concentration of inorganic sulfate in Krebs–Ringer bicarbonate buffer is 1 m*M*. Fain *et al.* (1967) did not use sulfate in their buffer system; the existence of a response to NE under these conditions suggests that extracellular sulfate is not essential for the short-term response of the cells.

B. Other Additions

1. *Albumin*

Albumin acts as an acceptor of fatty acids. When used, it is normally added to the concentration found in blood 4% (0.6 m*M*). Commercially, albumin can be obtained as crude albumin (unpurified, "fraction V") and as the more expensive fatty acid-free albumin, which is also dialyzed. The discussion of effects of albumin is made more difficult by these two albumins having been used interchangeably.

Albumin in the incubation medium does not influence the resting *membrane potential* or the NE-induced depolarization (Williams and Matthews, 1974a). Albumin increases *fatty acid release*, both at maximal and submaximal NE additions; at low NE concentrations no fatty acid release may be seen without added albumin (Bieber *et al.*, 1975). When the release of fatty acids ceased after an NE addition, a ratio fatty acids/albumin of on average 2.3 was measured (Bieber *et al.*, 1975), probably a ratio related to the number of high-affinity binding sites for fatty acids on an albumin molecule, which is normally stated as 3.

Albumin is necessary for a *sustained respiration* in bicarbonate buffer (Nedergaard and Lindberg, 1979). The lack of albumin in the incubation medium in manometric measurements (Horwitz, 1973) may then explain the low NE stimulations observed (only 3.5 × basal rate). Under certain conditions, albumin may inhibit NE-stimulated respiration (Bieber *et al.*, 1975; Kopecký and Drahota, 1978); it is not known whether this is due to NE binding to albumin, inhibitory impurities still present in albumin, or to a true competition between oxidation and a withdrawal of fatty acids by albumin.

In addition to the effects described above, the impurities in *crude albumin* may interfere with the measurements. Some batches of crude albumin contain an (insulin-like?) inhibitor of lipolysis (Fain *et al.*, 1967), and the problems with such inhibitors have been discussed by Fain (1975). At suboptimal NE concentrations, crude albumin may thus inhibit respiratory stimulation more then fatty acid-free albumin does (Nedergaard and Lindberg, 1979). Also substrates (citrate, lactate) present as impurities in crude albumin may influence respiration (Nedergaard and Lindberg, 1979).

2. *Carnitine*

Carnitine takes part in the shuttle mechanism for transfer of fatty acids from cytosol to mitochondria. It is lost from liver cells during preparation (Christiansen and Bremer, 1976), and Zuk *et al.* (1978) found that during 30-minute incubation of brown fat slices from rats, which had been injected with labeled carnitine, 80% of the label was released. One would therefore expect that the isolated cells may be in a carnitine-depleted state, and that addition of carnitine should increase respiration. This is, however, not the case. Carnitine addition has insignificant effect on fatty acid-stimulated respiration (Reed and Fain, 1968b), and addition of 10 m*M* carnitine [or of the stereoisomer (D)-carnitine or deoxycarnitine which both would be expected to compete with the natural, endogenous carnitine and thus to inhibit respiration] does not influence NE-induced respiration (our unpublished observations).

3. *pH buffer*

The pH of the bicarbonate buffer is very dependent upon the partial pressure of CO_2, and is not very stable. As the addition of high amounts of phosphate (as a pH buffer) has an inhibitory effect on the cells, other buffers may be used to stabilize pH. We would suggest H-Tes; if the medium is then adjusted with Tris–OH to the required pH, there will be no change of the ionic environment due to pH adjustment. Tes has been used in brown fat cell incubations by Prusiner *et al.* (1968c) and Hepes by Al-Shaikhaly *et al.* (1979).

4. *Antibiotics*

Horwitz (1973) added 400 μg streptomycin and 400 U penicillin/ml incubation; she noted that omission led to high endogenous respiration rates and 10^3–10^6 bacteria being present per ml medium.

C. Other Conditions

1. *Temperature*

A series of different temperatures have been used for studies of brown fat cells, leading to differences in quantitative measurements. It would be helpful if all studies were performed at 37°C—except those which specifically examine brown fat cell metabolism at lower temperatures. Such studies are of physiological interest, as cells from hibernators even *in vivo* have on occasion to work under temperatures much below normal mammalian body temperature. However, brown fat cells from hibernators seem to be as temperature sensitive as other cell types. Thus, Nedergaard *et al.* (1977) compared the rate of NE-stimulated *respiration* at different temperatures. The rate at 37°C was set to 100%; it was then 67% at 27°C, 10% at 15°C, and immeasurable at 7°C. Bernson (1979) obtained similar results.

The resting plasma membrane potential is also very temperature sensitive (see Section V,A,1).

2. *Oxygen*

Respiratory studies on cells are normally performed in buffers which have been equilibrated with air, whereas slices and tissue pieces are often oxygenated with 95% O_2.

Change of the partial pressure O_2 from 95 to 25% leads to a cell *membrane depolarization* of at least 25 mV (Girardier *et al.*, 1968). The cause of this depolarization is unknown, but the observed effect is of considerable importance, as it complicates the interpretation of changes of membrane potential observed after catecholamine stimulation. It is thus not possible at present to exclude the possibility that all observed catecholamine-induced depolarizations are secondary to the fall in oxygen tension caused by the combustion of substrate within the tissue.

Prusiner *et al.* (1968b), who incubated cells at 23°C, did not observe any inhibition of *respiration* before the oxygen tension fell below 5 mm Hg (approx. 5% of air-saturated or 20 μM O). Kopecký and Drahota (1978) found, when incubating at 30°C, a decrease in respiratory rate occurring at a higher oxygen tension. Working at 37°C, we observe a respiratory inhibition already when the oxygen pressure has dropped to 25 mm Hg (25% of air-saturated or 100 μM O). The higher rates of combustion at higher temperatures, together with nearly unchanged diffusion rates, may explain this relationship between limiting oxygen tension and incubation temperature.

D. Added Substrates

Brown fat cells respire well on endogenous substrate, i.e., mainly fat stores. Thus addition of substrate is normally unnecessary, and, if done, should not a priori lead to substrate-induced respiration or change NE-stimulated respiration, as the cells cannot be considered to be substrate-limited. However, a series of exogenous substrates do influence the respiration of the cellular preparation; they do so in at least three different ways: inducing their own respiration, being extracellularly oxidized, or (possibly synergistically) increasing catecholamine-induced respiration.

1. *Fatty Acids*

Although fatty acids are potentially beneficial for the cell, addition of free fatty acids has some inhibitory effects. Thus, octanoate slowly *depolarizes* the cell (Williams and Matthews, 1974b), and if the ratio of fatty acids added/albumin present is 5/1 or above, NE-stimulated *glycerol release* is inhibited (Reed and Fain, 1968a), but there is no effect on the basal rate of glycerol release (Reed and Fain, 1968b). Under these conditions where lipolysis is fatty acid inhibited, NE-stimulated *respiration* is also inhibited by fatty acids.

However, it is the *stimulation* of respiration by fatty acids which is most interesting for the understanding of thermogenesis. Thus, if added in amounts clearly exceeding albumin-binding capacity, octanoate, palmitate, and oleate all stimulate their own oxidation in brown fat cells (Prusiner *et al.*, 1968b; Reed and Fain, 1968b) [whereas propionate and butyrate do not (Prusiner *et al.*, 1968b,c)]. This fatty acid-induced oxidation is in all measured respects indistinguishable from hormonal-stimulated respiration, and the sum of NE-stimulated respiration and fatty acid-stimulated respiration remains the same whichever NE concentration is used (Prusiner *et al.*, 1968b). This parallelism between fatty acid and hormone-stimulated respiration caused both Fain *et al.* (1968b) and Prusiner *et al.* (1968b) to use the expression that fatty acids *mimic* the effects of hormone. Prusiner *et al.* (1968b) further state that the "main action of catecholamine . . . is the release of endogenous fatty acids," giving rise to the "minimal theory" (Bieber *et al*, 1975) that an increase in intracellular fatty acids is both the necessary and the sufficient initiator of thermogenesis. This fatty acid-induced increase in respiration is probably more specific than a classical uncoupler-like effect and may be understood in light of the experiments of Cannon *et al.* (1977b, 1980) on isolated brown fat mitochondria. It was there shown that fatty acyl-CoAs can compete with GDP for the binding site on thermogenin and thus uncouple the mitochondrion in a specific way (see Section XIII,E). Thus, at low concentrations of added fatty acid (10–100 μM), the fatty acids are activated within the cell, and the triglyceride-synthesizing machinery takes care of all added fatty acids, as measured by Lindberg *et al.* (1976). However, at higher fatty acid concentrations (400 μM) this pathway—although still functioning—does not have the capacity to take care of all activated fatty acids; the intracellular acyl-CoA concentration increases, and uncoupling and respiratory stimulation occur, as measured by Lindberg *et al.* (1976). Fatty acid uncoupling also increases the conversion of glucose to CO_2 (Fain *et al.*, 1970), and the added fatty acids do not (solely) act as uncouplers of the oxidation of endogenous substrates but are themselves oxidized (Prusiner *et al.*, 1968c; Fain *et al.*, 1970).

As the effects of added fatty acids are indistinguishable from those of added hormone, the decrease in cellular *energy potential* is also the same in both cases (Williamson *et al.*, 1970; Pettersson and Vallin, 1976).

2. *Succinate*

Succinate induces a high rate of respiration in most brown fat cell preparations [but not in lamb cells (Cannon *et al.*, 1977a)] (Prusiner *et al.*, 1968a; Williamson *et al.*, 1970).

a. *Extracellular Effects.* That the oxidation of succinate really is of *extra*cellular nature was only successively realized, based on, and explaining a long series of observations:

Succinate oxidation does not interfere with NE-stimulated respiration (Prusiner *et al.*, 1968a); succinate-induced respiration is much more easily inhibited by malonate than is NE-stimulated respiration (Prusiner *et al.*, 1968b); and "[in cell suspensions, it seems that] only the region of the respiratory chain before cytochrome-*b* is coupled to phosphorylation. However, such a postulation is contradictory to results obtained from experiments with isolated mitochondria" (Prusiner *et al.*, 1968c). Furthermore, Pettersson (1969, internal report) commented that "it is very difficult to prepare so bad cells that the succinate response disappears." Prusiner (1970) followed the redox state of cytochrome b in cell suspensions as caused by succinate; he concluded when comparing them with results of NE stimulation that "the interpretation of these experiments remains unclear." However, extra washings of the cell preparation led to a loss of the response to added tricarboxylic acid cycle intermediates (Herd *et al.*, 1973) and succinate added to "intact" cell suspensions can reduce *added* cytochrome c (Pettersson *et al.*, 1978).

Further, the –SH reagent 5,5′-dithiobis(2-nitrobenzoate) (DTNB) totally blocks succinate oxidation, without affecting NE-stimulated respiration; and successive cell washes diminish succinate oxidation without affecting NE-stimulated respiration (per million cells) (Bernson *et al.*, 1979).

Thus, there is ample reason to conclude that succinate-induced respiration is of extracellular origin and probably occurs in mitochondrial fragments (Bernson *et al.*, 1979).

b. *True Intracellular Effects.* In the presence of DTNB it is possible to investigate intracellular effects of succinate. Under these conditions, little effect is seen of addition of succinate itself (induced respiration less than 10 nmole O/minute/million cells), but if the cells are NE-stimulated, the respiration in a phosphate buffer with succinate is 20% higher than in a buffer without (Bernson *et al.*, 1979). This is probably due to an anaplerotic effect of succinate (see Section VIII,C). Similarly, Williamson *et al.* (1970) observed that 45 m*M* succinate was able to somewhat restore the ATP status of cells which were arsenite inhibited. The authors claimed this to show a replenishment of the citric acid cycle after arsenite block of 2-oxoglutarate dehydrogenase.

3. *Glycerol 3-phosphate*

When glycerol 3-phosphate is added to a cell preparation, a high rate of respiration is observed (Prusiner *et al.*, 1968a; Williamson, 1970). Maximal effect is obtained with 10 m*M* glycerol-3-phosphate. The evidence that this respiration is of extracellular nature is not as compelling as that for succinate. Certainly this stimulation can also be much diminished by successive washes of the preparation (Bernson *et al.*, 1979) but the best evidence is obtained by combining observations on mitochondria and cells. Bukowiecki and Lindberg (1974) found that mitochondrial glycerol 3-phosphate dehydrogenase was inhibited by the Ca^{2+} chelator EDTA. This chelator cannot penetrate the cell membrane and is practically devoid of action on NE-stimulated respiration (Nedergaard, 1981). However, addition of EDTA (or

EGTA) to a cell preparation, actively respiring on glycerol 3-phosphate, inhibits this respiration (Williamson, 1970). Thus, the respiration is probably of extracellular origin; it can also be inhibited if the extracellular concentration of fatty acids is increased by NE stimulation of the cells (Houštěk *et al.*, 1975a).

There is no evidence for a true intracellular effect of added glycerol 3-phosphate, and this compound, as other phosphorylated compounds, must have difficulties in penetrating the cellular membrane.

4. *Pyruvate*

There is no direct effect of addition of pyruvate to a cell preparation (Prusiner *et al.*, 1968c).

a. *Oxidation of Pyruvate Proper.* If an uncoupler is also added, respiration is stimulated if pyruvate is present (M. Czech, unpublished observations according to Reed and Fain, 1970). This rise in respiration is quite large (more than 100 nmole O/minute/million cells) (Bernson *et al.*, 1979), and the control of respiration indicates that it takes place in intact mitochondria, probably within the cells. All these experiments were performed in media containing CO_2, and the pyruvate carboxylase reaction (see Section VIII,C) probably facilitated the high respiratory rates.

b. *Synergistic Effects on NE-Stimulated Respiration.* Pyruvate and CO_2 have synergistic effects on NE-stimulated respiration in phosphate buffer, i.e., this rate is increased further when both of these agents are present than the sum of both would predict. This effect is caused by the action of pyruvate carboxylase which leads to an increased level of citric acid cycle intermediates in the cells (Cannon and Nedergaard, 1979a). Furthermore, especially in bicarbonate buffer, the NE-stimulated respiratory rate is elevated if pyruvate is added (our unpublished observations).

Note that most of the pyruvate effects are observed in the presence of 10 m*M* glucose/fructose; thus it must be concluded that glycolysis does not proceed fast enough to produce optimal amounts of pyruvate.

Pyruvate is of great interest as a model compound, as its oxidation is under respiratory control. Thus any metabolic sequence or regulatory compound, postulated to uncouple respiration *in vivo* and through this to cause thermogenesis, must also uncouple cellular pyruvate oxidation.

5. *Lactate*

A higher NE-stimulated respiration occurs if lactate is added to a bicarbonate buffer, although the increase is only half of that observed with pyruvate (our unpublished observations). This difference must be due to a limited capacity of lactate dehydrogenase; this dehydrogenase has the isozymic pattern which is normally found in organs utilizing (not synthesizing) lactate (A. Bass, personal communication) and it is probable that brown fat cells *in vivo* may combust circulatory lactate.

6. *2-Oxoglutarate*

Lindberg *et al.* (1967), working with brown fat slices, observed that 2-oxoglutarate, although not metabolized in itself, enhanced NE-stimulated respiration. We have observed (unpublished observations) that this is also the case with isolated cells.

7. *Glucose and Fructose*

Traditionally, 10–20 m*M* glucose and/or fructose is often present in the incubation medium (e.g., Reed and Fain, 1968a; Pettersson and Vallin, 1975). This is thought to increase the survival of the cells. Although the presence of glucose is in principle a physiological condition, glucose may change metabolic patterns. Thus, e.g., in the absence of glucose, uncoupler-stimulated respiration is of a transient nature (Prusiner *et al.*, 1968a), but this is not the case when glucose is present. Furthermore, the presence of glucose makes the cells less sensitive to oligomycin, and it increases octanoate-stimulated respiration (Fain *et al.*, 1970), perhaps via pyruvate formation. Glucose may also provide the glycerol backbone for triglyceride formation; thus glucose doubles the rate of esterification of added palmitic acid (Angel, 1969).

8. *Other Substrates*

Citrate, isocitrate, malate, acetoacetate, 2-OH-butyrate (Prusiner *et al.*, 1968c), and palmitoyl-carnitine (Reed and Fain, 1968b) do not induce their own oxidation.

E. Summary of Incubation Conditions

In the *ionic* environment of the cells, K^+ is necessary for lipolysis (and thus for thermogenesis) whereas Na^+ is necessary for combustion of released fatty acids. The importance of any other extracellular ion for the short-term regulation of thermogenesis has not been demonstrated.

Added *substrates* fall into several groups: those which can induce their own intracellular oxidation (i.e., only middle- and long-chain fatty acids); those which are oxidized by impurities in the cell preparation (i.e., only flavoprotein-linked substrates: succinate and glycerol 3-phosphate); those of which the oxidation can be induced by artificial uncouplers (pyruvate + CO_2); and those which increase the rate of NE-stimulated respiration (pyruvate + CO_2, lactate + CO_2, succinate, 2-oxoglutarate).

IV. Hormonal Effects

Brown fat thermogenesis *in vivo* is stimulated by NE, liberated from sympathetic nerve endings in the tissue.

Metabolically competent brown fat cells must therefore also be catecholamine sensitive. Besides this, there is ample evidence that they are insulin sensitive, and some evidence that they may be glucocorticoid hormone sensitive. Cells from some species may be glucagon sensitive *in vitro*. As we shall see, no other hormones have been convincingly demonstrated to affect the cells proper.

We shall discuss here only short-term effects, as only these effects can be studied with isolated, mature cells. The question of the hormonal control of proliferation and differentiation of brown fat is however also of great interest (see Section XIV).

A. Catecholamines

Ahlqvist (1948) divided catecholamine receptors into α and β receptors. This division was based upon the sensitivity of the system in question to the synthetic agonist isoprenaline compared to the sensitivity to NE and epinephrine: for β, isoprenaline was a better agonist than NE and epinephrine; for α, it was inferior to those. Currently the following relationships are accepted:

For α: phenylephrine is as good (but not a better) an agonist as NE or epinephrine; isoprenaline is much weaker. Phentolamine is a better antagonist than propranolol.

For β: isoprenaline is as good (or a better) an agonist as NE or epinephrine; phenylephrine is much weaker and is only a partial agonist. Propranolol is a better antagonist than phentolamine.

It must be emphasized that these differences are of a relative, quantitative nature. Thus the use of only one agonist or inhibitor to demonstrate that an effect is α- or β-adrenergic is of limited value, and the common use of both α- and β-(ant)agonists in one experimental series, but the two in different concentrations, does little to clarify reasoning in this area. For simplicity, we have in this article divided observed adrenergic effects into β- and α-adrenergic effects. Whereas the β-adrenergic effects are in general well examined, this is certainly not, as we shall see, the case with α-adrenergic effects.

1. *β-Adrenergic Effects*

We suggest (see Fig. 6) that β-adrenergic effects consist of binding to β_1 receptors, stimulation of adenyl cyclase, protein phosphorylation, increased

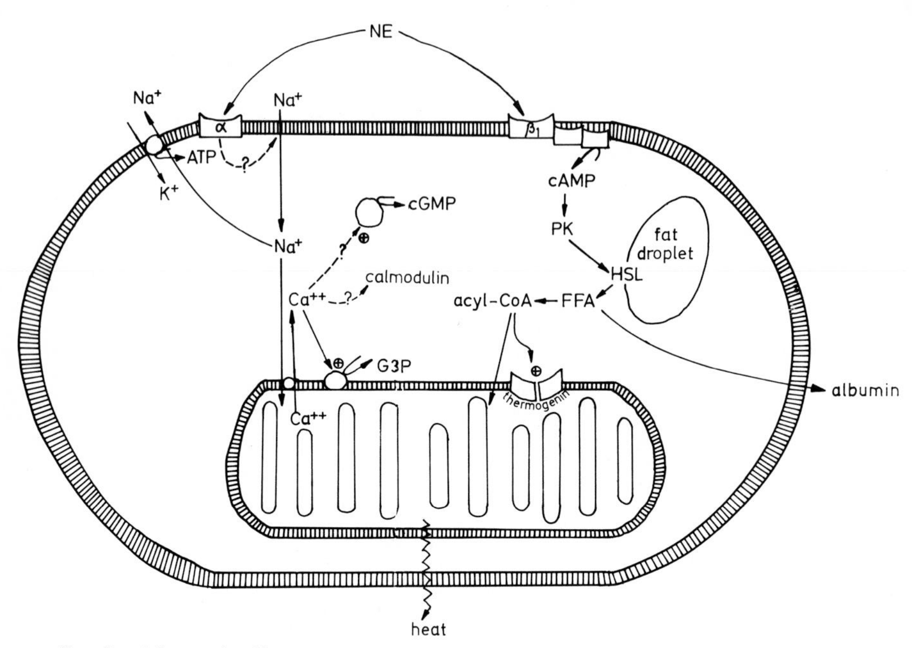

FIG. 6. Adrenergic effects on brown fat cells. The figure is divided into an α-adrenergic (hypothetical) and a β-adrenergic (well substantiated) side. G3P, Glycerol 3-phosphate; PK, protein kinase; HSL, hormone sensitive lipase; FFA, free fatty acid (see Section IV,A). Adapted from Cannon and Nedergaard (1982).

lipolysis and (the main part of) thermogenesis. Some antianabolic effects may also occur.

a. *β_1 vs β_2 Effects.* Lands *et al.* (1967) subdivided β-adrenergic effects into β_1 and β_2 effects. For β_1, isoprenaline is more efficient than NE, which in its turn is about as efficient as epinephrine. For β_2, isoprenaline and NE are about equally efficient, and better than epinephrine. Both from receptor-binding studies (see below) and from stimulation of respiration (Pettersson and Vallin, 1975; Mohell *et al.*, 1980; and Fig. 7) it is clear that the criteria for the receptor being a β_1 receptor are fulfilled.

b. *The β_1-Receptor.* Figure 8 summarizes what is known about the β_1 receptor. Most of this information has been collected from experiments with isolated cells, where the labeled β-adrenergic antagonist dihydroalprenolol has been used to examine the binding site (Cannon *et al.*, 1978; Svoboda *et al.*, 1979), in principle according to the methods developed by Lefkowitz and co-workers. Most of these results are in good agreement with those of Bukowiecki *et al.* (1978a) who studied the β receptor in homogenates of brown fat, and for competition studies Bukowiecki *et al.* and Svoboda *et al.* used very similar concentrations of dihydroalprenolol (13 resp. 5 n*M*). How-

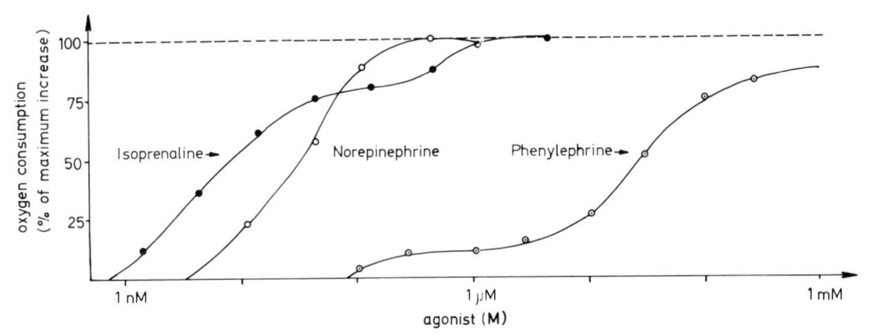

FIG. 7. The effect of different adrenergic agonists on oxygen consumption of isolated hamster brown fat cells. The responses are related to the maximal response obtained with NE. The response can be analyzed to consist of a β-adrenergic effect (about 80% of maximum) and an α-adrenergic effect (about 20%) (see Section IV,A). For further conditions see Mohell *et al.* (1980). Adapted from Mohell *et al.* (1980).

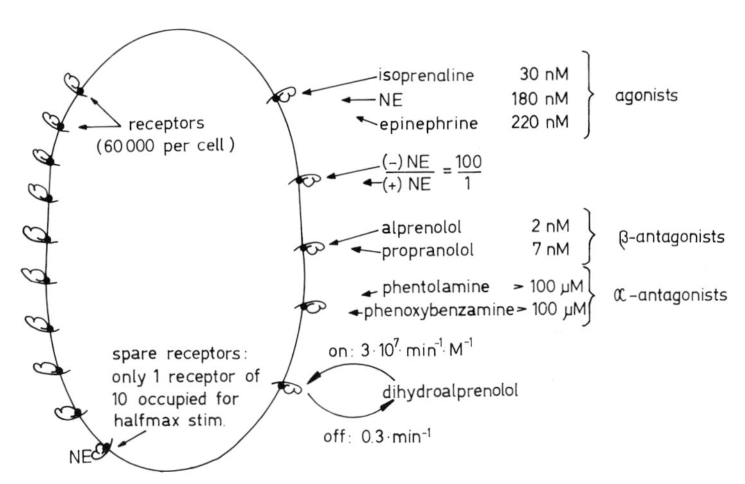

FIG. 8. The β_1 receptor of the brown fat cell. The figure summarizes results by Cannon *et al.* (1978) and Svoboda *et al.* (1979) from studies on brown fat cells. The α-antagonist data are from Bukowiecki *et al.* (1978a). The values given for the different agonists and antagonists are their binding affinity determined from competition studies with dihydryalprenolol (see Section IV,A,1,b).

ever, for the determination of the affinity of the β receptor for dihydroalprenolol, Svoboda *et al.* examined the interval 1–20 nM and Bukowiecki *et al.* 5–240 nM; thus the site studied by Svoboda *et al.* was in principle saturated by the lowest dihydroalprenolol concentration used by Bukowiecki *et al.* As a consequence of this, Scatchard plots of the data of Bukowiecki *et al.* (1978b) show two binding sites, and the Hill plot analysis indicates negative cooperativity ($n = 0.85$) and a K_D for dihydroalprenolol of 100 nM (100 times higher than that of Svoboda *et al.*) and a large amount of binding sites (Bukowiecki *et al.*, 1978a,b).

As pointed out by Svoboda *et al.* (1979) the presence of "spare receptors" (Fig. 8) endows the cell with an apparent sensitivity for agonists greater than the affinity of the receptor itself.

c. *Stimulation of Adenyl Cyclase.* If cells are studied when theophylline or caffeine is added to the incubation medium and phosphodiesterase thus is inhibited, the activity of adenyl cyclase is in reality studied by measuring cAMP levels. Under these conditions, epinephrine stimulates the cyclase about five times (Loken and Fain, 1971). Half-maximal stimulation occurs at 0.1 μM and maximal at 1 μM NE (Moskowitz and Krishna, 1973). The stimulation by 30 μM NE can be totally inhibited by 2 μM propranolol (Bertin and Portet, 1976) and this inhibition is clearly of a competitive nature (Skala *et al.*, 1970).

d. *Stimulation of Lipolysis.* Epinephrine stimulates glycerol release (15 times increase) and fatty acid release (Fain *et al.*, 1967); maximal stimulation occurs at 1 μM. Propranolol blocks this release (without affecting release caused by dibutyryl-cAMP) (Fain, 1970). Also the β_2-agonist salbutamol stimulates glycerol release, but 6 μM is needed for maximal stimulation (Fain *et al.*, 1973). In guinea pig cells (Forn *et al.*, 1970) and in hamster cells (Lindberg *et al.*, 1976) maximal glycerol release is obtained with 1 μM NE.

e. *Stimulation of Respiration.* Epinephrine or NE stimulates respiration in rat and hamster cells (Fain *et al.*, 1967; Prusiner *et al.*, 1968a,b,c). Maximal effect is obtained at about 1 μM (Prusiner *et al.*, 1968b; Horwitz, 1973). This respiration is blocked by propranolol (Prusiner *et al.*, 1968a) in a competitive way (Pettersson and Vallin, 1975). The relative efficiency of different agonists (Pettersson and Vallin, 1975; Mohell *et al.*, 1980) (Fig. 7) clearly points to respiratory stimulation—and by this to thermogenesis—as being β_1-adrenergic receptor mediated; α-specific agonists are nearly 10,000 times less potent than β-specific agonists.

f. *Other Effects of a β-Adrenergic Nature.* Isoprenaline stimulates the *catabolism* of glucose (Fain and Loken, 1969); lactate is the end product but this may be due to anaerobiosis occurring under the incubation conditions.

An injection of NE into rats leads to a doubling of *lipoprotein lipase* in brown fat (Radomski and Orme, 1971); the regulatory mechanism is unknown.

NE inhibits the *anabolic* incorporation of acetate into fatty acids in rat cells, but rather high NE concentrations are necessary (Steiner and Evans, 1976).

Results from some studies can be interpreted as showing a β-adrenergic stimulated *depolarization* of the plasma membrane (Krishna *et al.*, 1970; Fink and Williams, 1976). We suggest that these effects in brown fat cells may result from the lack of oxygen within the tissue (as seen by Girardier *et al.* (1968)), which results from β-adrenergic respiratory stimulation.

2. *α-Adrenergic Effects*

The study of possible α-adrenergic effects has been complicated by a series of factors. Thus, in *in vivo* experiments, an α stimulation may lead to a decreased blood flow to brown fat due to vasoconstriction; the resulting lack of oxygen may cause inhibition of thermogenesis (Flaim *et al.*, 1977) and be misinterpreted as an α effect on the brown fat cell proper. In experiments on slices and with impure cell preparations, uptake of added catecholamines into nerve vesicles may take place; and if (ant)agonists are added at high doses, they may have a fat cell receptor-unspecific or a fat cell receptor-unrelated action.

In spite of these complications we suggest that brown fat cells are endowed with α-adrenergic receptors, and we propose a scheme (Fig. 6) which is only a hypothesis but which does link together a series of studies. In this hypothesis, α receptor stimulation results in an increased Na^+ permeability over the plasma membrane (measured as a depolarization); the increased intracellular Na^+ (besides indirectly stimulating Na^+,K^+-ATPase) leads to an efflux of Ca^{2+} from the mitochondria; this Ca^{2+} may to some extent take part in a futile cycle, and the increased cytosolic (and perhaps even the decreased mitochondrial) Ca^{2+} concentration may have a series of regulatory functions, e.g., on adenyl cyclase and glycerol-3-phosphate dehydrogenase. We shall now look into these processes in greater detail.

a. *α-Adrenergic Receptors.* In a preliminary communication, Svartengren *et al.* (1980a) demonstrate the existence of α-adrenergic receptors in brown fat homogenates by finding a reversible, high affinity (0.5 n*M*), binding site for the α-adrenergic antagonist dihydroergocryptine. These receptors have also been demonstrated on isolated cells (Svartengren *et al.*, 1980b).

b. *Stimulation of Membrane Depolarization.* Catecholamines depolarize the cellular membrane (Girardier *et al.*, 1968). This is an early effect, occurring before oxygen consumption increases (Seydoux and Girardier, 1978), and the depolarization can thus not be secondary to lack of oxygen. There is a series of studies indicating that this depolarization may be α receptor mediated.

Horwitz *et al.* (1969) found that in intact rats, 10 ppm phentolamine blocks 73% of the depolarization caused by nerve stimulation, but this treatment does not prevent the increase in brown fat temperature (indicating that β effects are unimpaired). Furthermore, Fink and Williams (1976) found that phenylephrine is as potent as NE in eliciting depolarization, that the α-antagonists phentolamine and phenoxybenzamine in nanomolar concentrations block catecholamine-induced depolarization, and that 0.1 μM phentolamine is a better inhibitor of NE-induced depolarization than is 3.0 μM propranolol (Fink and Williams, 1976). Finally, Flaim *et al.* (1977) found that in intact rats, 0.1 ppm phenylephrine depolarized more than did the same amount of isoprenaline.

c. *Indirect Stimulation of* Na^+,K^+*-ATPase.* If α receptor stimulation leads to membrane depolarization, caused by increased permeability to Na^+ and therefore by an increased influx of Na^+, the plasma membrane Na^+,K^+-ATPase is indirectly stimulated as it has to restore the ion balance. This may in its turn cause some increase in oxygen consumption, e.g., some of the ~20% shown in Fig. 7. Herd *et al.* (1970) and Horwitz and Eaton (1975) have discussed the possibility that the Na^+,K^+-ATPase is *directly* stimulated by catecholamines. We feel that such a stimulation should lead to a *hyper*polarization of the membrane. Furthermore, the experimental evidence presented shows a series of unusual qualities of the receptor regarding the NE concentration giving maximal effect (6 mM) and the effects of antagonists. Before these experiments are taken as evidence for a direct physiological catecholamine stimulation of Na^+,K^+-ATPase, more precise investigations are needed, e.g., concerning stereospecificity of effects and competitivity between agonists and antagonists. We cannot presently exclude the possibility that the measured effects are secondary to a chelating action of catecholamines (cf. Svoboda *et al.*, 1981a,b).

d. *Stimulation of* Ca^{2+} *Release.* See Section V,D. for a discussion of Na^+-induced Ca^{2+} release from mitochondria.

e. *Possible Inhibition of Adenyl Cyclase and Stimulation of Guanyl Cyclase.* Under conditions where phosphodiesterase is inhibited by caffeine, an α-adrenergic inhibition of adenyl cyclase can be observed, both in hamster cells (Hittelman *et al.*, 1974) and in rat cells (Bertin and Portet, 1976). In both cases, 500 μM phentolamine doubles the adenyl cyclase activity, stimulated by 30 μM NE. In several tissues, adenyl cyclase has been shown to be inhibitable by Ca^{2+} in very low amounts (Rasmussen and Goodman, 1977). We suggest that an α-stimulated increase in cytosolic Ca^{2+} leads to an inhibition of adenyl cyclase also in brown fat cells; such an indirect way of inhibition also explains why no inhibitory effect of α stimulation can be seen in membrane preparations (Skala *et al.*, 1970). One would expect, as a consequence of such an inhibition, that lipolysis and respiration were also inhibited by α stimulation. No evidence for this exists; it is possible that the effects are evident only under conditions where phosphodiesterase is inhibited.

Phenylephrine at high doses (2–4 mM) inhibits 2 μM isoprenaline-stimulated glycerol release by 63% (Horwitz, 1977; Hamilton and Horwitz,

1977). This is probably not due to α-mediated inhibition. Rather the 1000 times excess of phenylephrine competes isoprenaline away from the β-receptor; phenylephrine is only a partial agonist on the β receptor (cf. Fig. 7) and thus in practice an inhibition is observed.

In intact rats, α stimulation causes an increase in brown fat cGMP levels (Skala and Knight, 1979). This may be a Ca^{2+}-mediated effect (Fig. 6), but it may also be secondary to vasoconstriction and subsequent lack of oxygen, known to increase cGMP levels in other tissues (Busuttil *et al.*, 1976).

B. INSULIN

Insulin, which can interact with white fat cells, has in many ways similar effects on brown fat cells. There are no direct studies on the insulin receptor itself in brown fat cells, but insulin stimulates glucose uptake and anabolic reactions, and it inhibits catabolic reactions. The nature of the intracellular "second messenger" for insulin is not known. The actions of insulin, in order to be considered physiologically meaningful, should be elicited by less than 1 mU/ml (less than 10 n*M* concentrations). A series of compounds have "insulin-like" effects in stimulating glucose oxidation. Figure 9 summarizes the effects of insulin.

1. *The Insulin Receptor*

Incubation of cells for 1 hour with 1 mg trypsin/ml abolishes all effects of insulin, probably because of digestion of the receptor (Fain and Loken, 1969). Cells treated in this way do not recover insulin sensitivity, even when

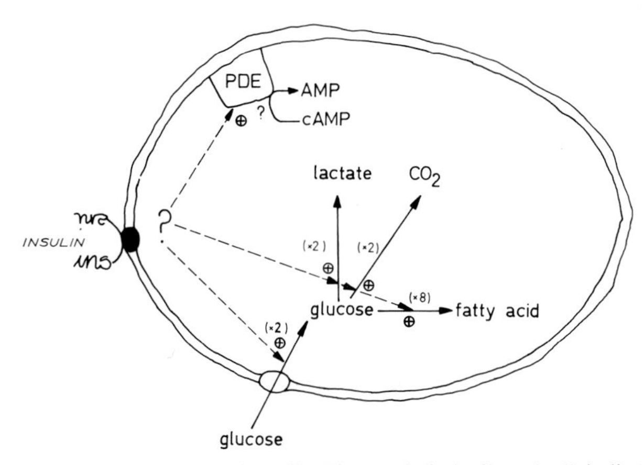

FIG. 9. Insulin effects on brown fat cells. The symbols (×2) or (×8) indicate the insulin stimulation of the conversions. PDE, Phosphodiesterase (see Section IV,B).

incubated for 4 hours without trypsin (Rosenthal and Fain, 1971); thus cells treated with trypsin can be used to ensure that insulin cannot have any effect (Fain *et al.*, 1970). The possible presence of trypsin-like contaminations in collagenase preparations has made several investigators use a trypsin inhibitor during cell preparation (Murthy and Steiner, 1972; Horwitz, 1973). Murthy and Steiner (1972) did not obtain any insulin effect if this precaution was not taken, but they used high amounts of collagenase (5 mg/ml) and long incubations (90 minutes); by this treatment the cells were 10 times more exposed to collagenase than in the standard cell preparation procedure (Section II,B).

2. *Stimulation of Glucose Uptake*

Traditionally, the action of insulin is said to be to increase the glucose uptake. This is, however, routinely measured as an increased rate of glucose metabolism, but Czech *et al.* (1974a) did successfully demonstrate that insulin doubled the rate of 3-*O*-methyl-glucose uptake into brown fat cells. As this uptake occurs as a "facilitated diffusion," there is no change in the amount of total uptake. However, the effects are clearly seen only at very low (micromolar) extracellular glucose concentrations; at millimolar concentrations the non-carrier-mediated diffusion has a much higher "capacity" than the glucose carrier, and, in consequence, it is perhaps doubtful that the insulin stimulation of glucose metabolism is caused by stimulation of uptake.

3. *Anabolic Actions of Insulin*

Insulin doubles the rate of acetate incorporation into fatty acids (Angel, 1969; Murthy and Steiner, 1972) and it increases 8-fold the rate of fatty acid synthesis from glucose (Fain *et al.*, 1967). These effects are probably caused by a doubling of the degree-of-activation of the enzymes pyruvate dehydrogenase and acetyl-CoA-carboxylase (McCormack and Denton, 1977).

Insulin also doubles the incorporation of glucose into glyceride-glycerol (Fain *et al.*, 1967; Angel, 1969).

4. *Stimulation of Glucose Oxidation*

Insulin doubles the rate of glucose conversion to CO_2 (Fain *et al.*, 1967) and to lactate (Fain and Loken, 1969), and glycolysis and the pentose–phosphate shunt are equally stimulated (Czech *et al.*, 1974a). The cause of these effects is unknown; brown fat cells are not thought to be in a substrate-depleted state, and an insulin-stimulated increase in glucose uptake cannot in itself explain the increased glucose oxidation, unless insulin makes glucose a preferred source of energy for basal metabolism.

5. *Anticatabolic Effects*

Insulin decreases the NE-induced increase in cAMP in brown fat, and this effect is abolished if the phosphodiesterase inhibitor theophylline is present (Knight, 1974). Likewise, insulin has in itself no effect on cAMP levels; nor has it, if caffeine is present as a phosphodiesterase inhibitor, any effect on cAMP levels in isolated cells (Bertin and Portet, 1976). Thus, insulin probably exerts its anticatabolic effect by increasing the activity of a phosphodiesterase, as is the case in white fat (Lovell-Smith *et al.*, 1977). As a consequence of this, insulin diminishes epinephrine-stimulated glycerol and fatty acid release (Fain *et al.*, 1967). Furthermore, it is in itself without effect on respiration (Fain and Reed, 1970) and membrane potential (Krishna *et al.*, 1970), but it counteracts NE-induced depolarization (Krishna *et al.*, 1970) and it diminishes submaximally NE-stimulated respiration, without having any effect on maximally stimulated respiration (Nedergaard and Lindberg, 1979). Furthermore, it diminishes the cold-induced (NE-mediated?) increase in lipoprotein-lipase (Radomski and Orme, 1971).

6. "*Insulin-Like Effects*"

Any substance which increases glucose conversion to CO_2 is said to have "insulin-like effects." Among such substances are several "bacterial factors," cystein (+ a heavy metal ion) (Rosenthal and Fain, 1971), phospholipase C (Fain, 1975), and concanavalin A (Czech *et al.*, 1974b). For most of these substances, trypsin has been shown *not* to inhibit the "insulin-like effects," indicating that the effects are *not* insulin receptor-mediated. Fain (1975) concludes, that "there are a wide variety of agents which stimulate glucose oxidation by fat cells, and many of these agents at higher concentrations lyse fat cells. Apparently mild membrane damage results in an increased rather than decreased basal glucose metabolism of fat cells." We would add to this that mild membrane damage leads to an increased Na^+ influx. The increased action of the Na^+,K^+-ATPase then demands extra energy which may lead to higher glucose metabolism.

C. Corticosteroids

Glucocorticoids in general, like other steroid hormones, are slowly acting hormones, which probably interact with the synthesis of certain enzymes, especially enzymes concerned with carbohydrate metabolism. Often the synthetic glucocorticoid hormone dexamethasone is used in experimental studies, instead of the natural hormones corticosterone and cortisone. Figure 10 summarizes glucocorticoid effects on brown fat cells.

Brown fat has about the same amount of glucocorticoid receptors as other fat tissues (Feldman and Loose, 1977): 75 fmole/mg cytosolic protein. These

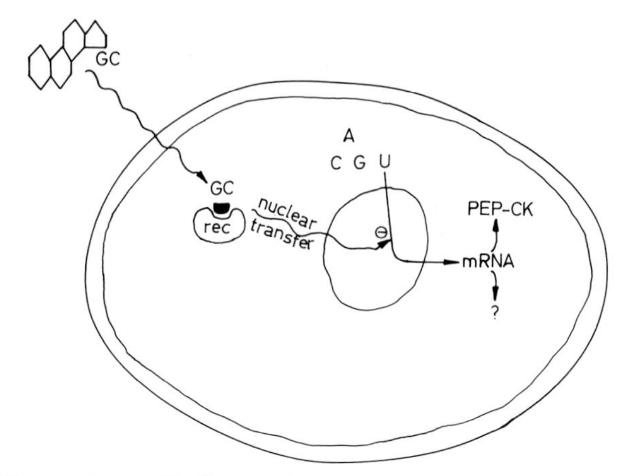

FIG. 10. Glucocorticosteroid effects on brown fat cells. GC, Glucocorticoid; A, C, G, U, the nucleotides for mRNA synthesis; PEP-CK, phospho*enol*pyruvate-carboxykinase (see Section IV,C).

receptors are most easily studied in adrenalectomized rats. Dexamethasone, bound to the receptor, is competed out easily by corticosterone, but sex hormones and cortexolone (a minerocorticoid) are weak competitors, indicating that the receptor is glucocorticoid-specific (Feldman, 1978).

When dexamethasone is bound to the receptor, it is transferred to the nucleus (Feldman,1978), and here dexamethasone inhibits uridine incorporation into RNA in a dose-dependent, time-dependent way. Maximal effect is obtained with 100 n*M* after 4 hours; this results in a 30% inhibition (Feldman, 1978).

When administered *in vivo*, dexamethasone (1 ppm) abolishes the increase in phosphoenolpyruvate-carboxykinase induced by a high-fat diet or fasting; it slightly reduces nonstimulated kinase activity (Feldman and Hirst, 1978). The physiological significance of this is unknown.

Dexamethasone does not change glucose metabolism in brown fat slices (Fain, 1965) or cells (Fain *et al.*, 1967). Some, very age-dependent, effects have been observed *in vivo* (Hahn *et al.*, 1969; Skala and Hahn, 1971), and Wünnenberg *et al.* (1974) found that the major part of nonshivering thermogenesis in hedgehogs could be stimulated by 2 ppm corticosterone. The authors suggested that one part of the brown adipose tissue was glucocorticoid-sensitive. None of these effects has been studied further.

D. SEROTONIN

The observed stimulatory effects of serotonin (5-HT) on brown fat and brown fat cells are secondary to release of NE from nerve filaments within the tissue or from nerve vesicles occurring as impurities in cell preparations. This conclusion is based on the observations that serotonin does not stimulate adenyl cyclase (Skala *et al.*, 1970), that cell preparations do not normally respond to serotonin addition (Dryer *et al.*, 1971), and that some cell preparations respond to 5–100 μM serotonin by a slight stimulation of cAMP accumulation and some glycerol release. Respiration is not stimulated. These serotonin effects can be blocked by 33 μM propranolol (Fain *et al.*, 1973). Further, in reserpine-treated rats, there is no effect of serotonin on the cell membrane potential (Fink and Williams, 1976), and serotonin does not decrease acetate incorporation in cells or increase fatty acid release from cells, and there is no NE-like effect on brown fat slices prepared from reserpine-treated rats. However, serotonin efficiently releases NE, preloaded into nerve endings in tissue slices (Steiner and Evans, 1976).

Thus, observed effects of serotonin are not due to a serotonin effect on the fat cell itself.

E. Corticotropin

The observed stimulatory effects of corticotropin (ACTH) on brown fat cell preparations or tissue are probably also secondary to release of NE from nerves. This is based on the following evidence.

Corticotropin does not stimulate adenyl cyclase (Skala *et al.*, 1970), and corticotropin does not stimulate respiration in isolated cells (Dryer *et al.*, 1971). Although corticotropin increases the level of cAMP in some rat cell preparations (Bertin and Portet, 1976), the increase is tiny (4.0 ± 0.5 to 5.9 ± 1.0), and corticotropin does not stimulate glycerol or fatty acid release from brown fat cells (Fain *et al.*, 1967). Corticotropin may increase cAMP levels in some hamster cell preparations (Hittelman *et al.*, 1974). However, Rosak and Hittelman (1977) demonstrated that at least in white fat cells, the corticotropin-stimulated glycerol release is inhibitable by propranolol (albeit at very high concentrations) and, similarly, the effects of corticotropin on brown fat slices can be blocked by propranolol (Beviz *et al.*, 1968). Thus, Skala *et al.* (1970) conclude that "the metabolic effects [of corticotropin] in intact tissue are exerted through catecholamine release from storage sites."

We conclude that in impure cell preparations the same occurs, and that the brown fat cell itself is probably insensitive to corticotropin.

F. Glucagon

There are a few studies on glucagon effects on brown fat. Although glucagon does not stimulate the respiration of isolated hamster cells (Dryer *et al.*, 1971), glucagon somewhat stimulates glycerol release from tissue in rats in some age groups of rats (Williams and Matthews, 1974b), and glucagon infusion into rats leads to a slight stimulation of fatty acid release from the tissue *in situ* (Kuroshima *et al.*, 1977). It is also possible to measure glucagon-stimulated thermogenesis in isolated rat cells (Kuroshima and Yahata, 1979) but the concentrations needed to observe effects *in vitro* are very high (about 1000 ng/ml) compared to the glucagon levels which have been measured *in vivo* by the same group (maximally 0.3 ng/ml) (Kuroshima and Dai, 1976). Thus the physiological significance of the glucagon response of rat cells has not been elucidated; we would like to draw attention to the fact that glucagon is a member of a polypeptide–hormone family, which also includes secretin, VIP, and GIP. Thus the response to glucagon may result from an artificial stimulation of similar receptors, and we have observed (unpublished) that rat cells also respond to VIP addition.

The glucagon response in rat cells has as yet not been further investigated; it is, e.g., not known if it is mediated by cAMP (as it is in liver cells) and if the response to catecholamines is additive or complementary to the glucagon response.

G. Thyroid Hormones

Thyroxine and T_3 exert their effects on the synthesis of enzymes; no acute effects are expected. In agreement with this, T_3 does not stimulate cell respiration (Dryer *et al.*, 1971).

Cells prepared from hyperthyroid rats do not show an NE stimulation of oxygen consumption different from that of euthyroid controls (Herd *et al.*, 1973). However, hyperthyroid rats have more brown fat than euthyroid controls, but the difference is due to an increased amount of fat; no increase in protein occurs (Heick *et al.*, 1973). Furthermore, under conditions of equal cold stress, hyperthyroid rats have less thermogenin in their mitochondria (and totally in brown fat) than controls (Sundin, 1981).

These results are best understood in the following way: hyperthyroid rats have an increased metabolism and an increased body temperature. At equal environmental conditions they have a higher heat production (basal metabolic rate) than controls, and their brown fat is consequently less activated, leading to fat deposition instead of combustion, and to a reduction in thermogenic capacity, i.e., amount of thermogenin.

Thus, thyroid hormones are in principle anti-brown fat hormones, and they affect the tissue only in an indirect way.

H. Other Hormones

Thyroid-stimulating hormone (TSH) has been discussed by Doniach (1975) as being responsible for the cold-induced increase in brown fat protein. The evidence is only circumstantial, and there are no reports on effects on cells. *Melatonin* leads to an increase in wet weight of brown fat in the djungarian hamster (Heldmaier and Hoffmann, 1974) and in the white-footed mouse (Lynch and Epstein, 1976). There are no reports on cellular effects. *Oxytocin* does not influence glucose metabolism in brown fat cells (Fain and Loken, 1969). *Growth hormone* does not stimulate adenyl cyclase (Skala *et al.*, 1970). *Histamine* does not stimulate adenyl cyclase (Skala *et al.*, 1970). *Adenosine*, which is presently discussed as a feedback regulator of fat cell metabolism, does in the form of phenylisopropyl-adenosine inhibit the isoprenaline-stimulated increase of cAMP (Hittelman *et al.*, 1974). *Dopamine* in high concentrations inhibits NE-stimulated respiration in cells, probably in a competitive manner (Bergström and Brynolf, 1969, internal report). It slightly depolarizes the cell membrane, but as this depolarization is inhibited by phentolamine and not by the specific dopaminergic antagonist pinozide, Fink and Williams (1976) concluded that dopamine was a weak agonist on the α receptor.

I. Conclusion: Hormones

The hormones which have been thoroughly studied on brown fat cells can be divided into three major groups: those which exert their effect on the brown fat cell itself (only catecholamines, insulin, and glucocorticoids; perhaps glucagon but only in unphysiological concentrations); those which act solely by releasing stored catecholamines, within the tissue or in impure cell preparations (at least serotonin and corticotropin); and those which only affect the tissue indirectly in the intact animal (at least thyroid hormones, perhaps melatonin).

V. Second Messengers

Of the hormones which can affect the brown fat cell itself, only glucocorticoids penetrate the cellular membrane, and the glucocorticoid–receptor complex is in itself the "second messenger." Insulin and NE bind to receptors on the cell surface and the information about their presence must be transferred into the cell by a "second messenger." The membrane potential, cAMP and cGMP, Ca^{2+} and prostaglandins have all been discussed in this context. Here we shall review which agents influence these possible messengers, and how changes in the messengers may affect cellular metabolism.

A. The Membrane Potential

As visualized in Fig. 11, the membrane potential is measured with a glass capillary electrode within the cell. If no ionic permeability exists over the cell membrane, no electric circuit can be established, and no membrane

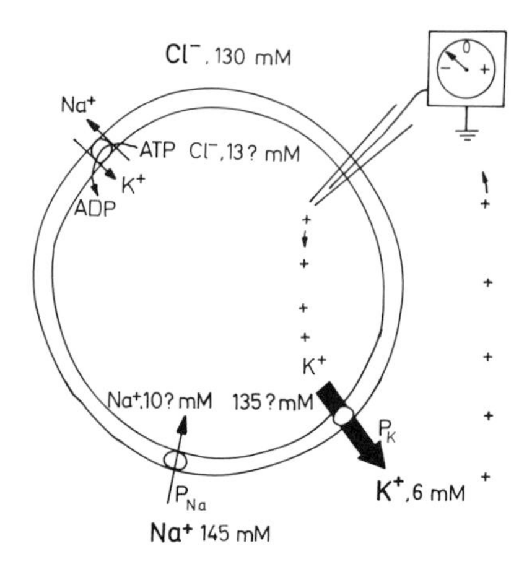

FIG. 11. The plasma membrane potential of the brown fat cells. Inside ionic concentrations are estimates. P_{Na} and P_K are the permeabilities for Na^+ and K^+ (see Section V,A).

potential is measured. In reality, a low current (a few pA) passes through the voltmeter; this is measured as a membrane potential of about 60 mV.

Traditionally the discussion is restricted to Na^+ and K^+. These ions are, by virtue of the Na^+,K^+-ATPase, not in equilibrium over the cell membrane, and if the indicated ionic concentrations (taken from Girardier *et al.*, 1968) are used for calculations, the energy equivalent for the K^+ gradient according to the Nernst equation is 60 mV · log $(K_i^+/K_o^+) = -81$ mV (inside negative), i.e., the energy gain for K^+ efflux is 8 kJ/mole. Similarly, the Na^+ potential is 60 mV · log $(Na_i^+/Na_o^+) = +59$ mV, i.e., the energy cost for Na^+ pumped out is 6 kJ/mole. The resulting current over the cell membrane is governed by these potentials and the permeability of the ions (P_K and P_{Na}). As the measured membrane potential is 65 mV, negative inside, this must imply that P_K is much greater than P_{Na}. Using the Goldman equation, Girardier *et al.* (1968) calculated the ratio P_{Na}/P_K in resting brown fat cells to be 1/25. If P_{Na} is increased (P_K being unchanged), the resultant current is less outwardly directed, and the membrane potential is measured as lower: a depolarization has occurred.

Chloride ions are expected to equilibrate over the membrane; therefore the expected intracellular Cl^- concentration is about 13 m*M*.

1. *The Resting Membrane Potential*

The value of the resting membrane potential has been measured by different authors. In Fig. 12 these results have been plotted against incubation temperature. It is clear from the figure that the greater part of the difference between the values found is due to the different incubation temperatures; at 37°C, the resting membrane potential is close to 60 mV. When a series of

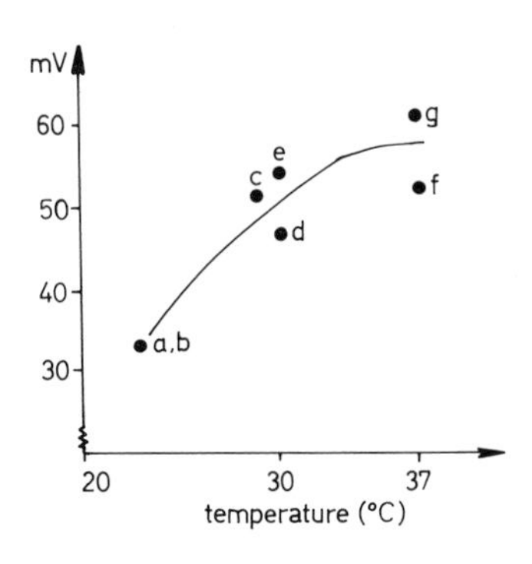

FIG 12. Temperature dependence of the resting membrane potential of brown fat cells. The graph is based on data obtained by (a) Krishna *et al.* (1970); (b) Moskowitz and Krishna (1973); (c) Girardier *et al.* (1968); (d) Horwitz *et al.* (1969); (e) Seydoux and Girardier (1978); (f) Williams and Matthews (1974a); (g) Fink and Williams (1976) (see Section V,A,1).

single cell potentials is measured, they form a Gaussian distribution about the mean; this indicates that only one cell type is involved (Girardier *et al.*, 1968).

2. *Changes of the Membrane Potential*

a. *Agents Causing Hyperpolarization.* As the resting membrane potential is close to, but not at, the K^+ potential, the membrane may be hyperpolarized, either by increasing the P_K or by increasing the K^+ gradient. P_K is increased by addition of the K^+ ionophore valinomycin; this hyperpolarizes the membrane by 40% (Krishna *et al.*, 1970). Increase of the K^+ gradient—by a lowering of extracellular K^+—does not yield the expected hyperpolarization; and in practice the result is a decreasing gradient and a depolarization which has been explained as due to the requirement of the Na^+,K^+-ATPase for extracellular K^+ (Girardier *et al.*, 1968).

b. *Agents without Effect on Resting Membrane Potential.* These include: the hormones acetylcholine, serotonin (Fink and Williams, 1976), and insulin (Krishna *et al.*, 1970); cAMP (added from the outside) and theophylline (Girardier *et al.*, 1968; Horwitz *et al.*, 1969); the inhibitors of oxidation CN^- and antimycin (Girardier *et al.*, 1968; Williams and Matthews, 1974a); increase in external Mg^{2+} or decrease in Ca^{2+} (Williams and Matthews, 1974a); and the Na^+ channel inhibitor tetrodotoxin (Krishna *et al.*, 1970).

c. *Agents Causing Depolarization.* As the resting membrane potential is mainly governed by the K^+ gradient, a depolarization may occur (i) by a decrease of this gradient, (ii) by a decrease of P_K, and, of most interest (iii) by an increase of P_{Na}.

i. *Depolarizations caused by a decrease of the K^+ gradient. Increased external K^+* leads to the expected decrease of membrane potential of 60 mV per decade change of K^+ (Girardier *et al.*, 1968); this is also observed in reserpinized rats and is thus not secondary to release of endogenous catecholamines (Barde *et al.*, 1975). *Decreased internal* K^+ is caused by any action which inhibits the Na^+,K^+-ATPase. Thus lowered temperature decreases the potential 1 mV/°C, much more than expected from the Nernst equation (0.2 mV/°C) (Williams and Matthews, 1974a; and Fig. 12). Furthermore, the specific Na^+,K^+-ATPase inhibitor ouabain (Girardier *et al.*, 1968; Williams and Matthews, 1974a), lack of external K^+ (Girardier *et al.*, 1968), lack of external Na^+ (Williams and Matthews, 1974a), iodoacetate (a glycolytic inhibitor) (Williams and Matthews, 1974a), and CCCP and fatty acids (uncouplers of oxidative phosphorylation) (Williams and Matthews, 1974a) all lead to depolarization.

All the depolarizations caused by inhibition of Na^+,K^+-ATPase occur rather slowly (a question of minutes) and are of a detrimental, nonphysiological nature. They can be considered as artifacts.

ii. Depolarizations caused by a decrease of P_K. No example known.

iii. Depolarizations caused by an increase of P_{Na}. These are the depolarizations which are of a physiological nature. We have already summarized the evidence that α-adrenergic stimulation causes membrane depolarization (Section IV,A,2); we suggest that this occurs by an increased P_{Na}. We would like to point out that this in practice means an increased inflow of Na^+ and an increased intracellular level of Na^+. We suggest that this increased level is the real significance of the stimulation of P_{Na} and that the change in membrane potential is of little importance per se.

3. *Effects of Changes in the Membrane Potential*

There are no reports on any effects of *hyperpolarization.*

Depolarization. In excitable tissue (nerves and muscle), depolarization is an event which in itself, electrically, leads to new events (propagated action potentials). However, in nonexcitable tissues like brown fat, the measured electric quality of depolarization is probably not important; instead the influx of Na^+ is perhaps the important event. This can be understood from the demonstration by Barde *et al.* (1975), that depolarization by increased K^+ in brown fat tissue does not lead to any increase in respiration; similarly, increased K^+ does not affect respiration in isolated cells (Nedergaard, 1981). Thus the suggestion by Moskowitz and Krishna (1973), that the "primary event leading to increased cyclic AMP in intact brown fat cells is depolarization of the fat cell membrane by NE," has not been substantiated.

Rather, the NE-induced entry of Na^+ seems to be a necessary step in thermogenesis, although the precise mechanism for the intracellular action of Na^+ is not known (Al-Shaikhaly *et al.*, 1979). This Na^+ entry may lead to an increased cytosolic Ca^{2+} level, which in itself—or through calmodulin—may regulate cellular metabolism.

4. *Electric Coupling between Cells*

Brown fat cells within the tissue are electrically coupled. The junctional maculae are seen in electron micrographs stained with lanthanum (Sheridan, 1971; Linck *et al.*, 1974), and in freeze-fracture studies (Revel *et al.*, 1971; Sheridan, 1971). Revel *et al.* calculated that the junctional maculae, which are 0.3–1 μm in diameter, cover 1–2% of the cell membrane area.

In electrophysiological studies, Sheridan (1971) obtained evidence for coupling by injecting a weak current (36 nA) into a cell. A small change in membrane potential (2 mV) could then be measured in neighboring cells, although not in all of them; there were cases of 5–10 coupled cells.

The physiological significance of coupling is not presently known. The junctions may allow ions and smaller organic compounds to equilibrate rapidly between a series of cells.

B. cAMP

cAMP is normally the "second messenger" of β-adrenergic stimulation. The level of cAMP in the cell is regulated by the balance between the activity of adenyl cyclase and phosphodiesterase (Fig. 13).

1. *Adenyl Cyclase*

By histochemical techniques, Rechardt and Hervonen (1976) visualized adenyl cyclase in brown adipose tissue. It is situated at the plasma membrane, easily within reach of hormones.

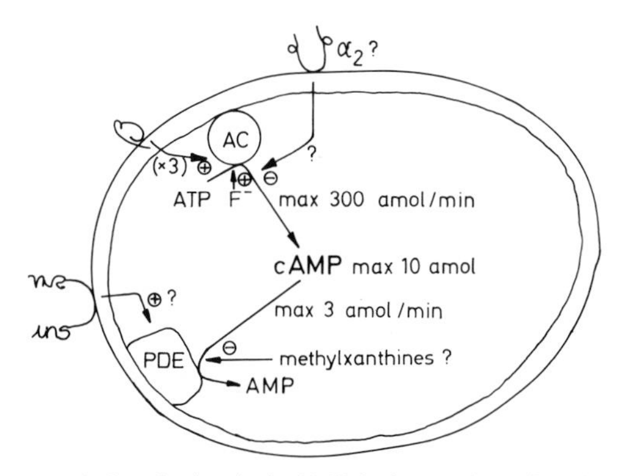

FIG. 13. Factors regulating the level of cAMP in brown fat cells. The activities indicated are calculated per cell. AC, Adenylate cyclase; PDE, phosphodiesterase; (×3) indicates degree of stimulation (see Section V,B).

In guinea pig cells, the cyclase activity is *stimulated* three times by NE; maximal activity (300 pmole/minute/million cells) needs 100 μM NE, although maximal glycerol release occurs already at 1 μM (Forn *et al.*, 1970). Similar results have been obtained by Muirhead and Himms-Hagen (1971) on homogenates, by Skala *et al.* (1970) and by Bertin and Portet (1976) and Bertin (1976). The cyclase is, as in other cell types, also stimulated by F^- in millimolar concentrations; the activity is then the double of the maximally NE-stimulated activity (Forn *et al.*, 1970; Muirhead and Himms-Hagen, 1971).

The adenyl cyclase activity is *not stimulated* by any noncatecholamine hormone tested (serotonin, histamine, corticotropin, glucagon, thyroid-stimulating hormone, thyroxine, dexamethasone, prostaglandin E_1) (Skala *et al.*, 1970).

There are no reports on *inhibition* of the basal activity, but NE-stimulated activity is competitively inhibited by β-antagonists (Skala *et al.*, 1970) and also inhibited by prostaglandin E_1 (Fain and Reed, 1970). Adenyl cyclase may be inhibited by α-adrenergic stimulation, perhaps in an indirect way (see Section V,D).

2. *Phosphodiesterase*

The phosphodiesterase activity in brown fat cells is about 2–3 nmole/minute/million cells (Forn *et al.*, 1970).

Stimulation of phosphodiesterase is probably the way in which insulin causes its anticatabolic actions (see Section IV,B). It has been suggested that in the relatively refractory cells found in cold-adapted animals, phosphodiesterase is in a stimulated state (Nedergaard, 1982).

Inhibition of phosphodiesterase activity is caused by methylxanthine derivatives; notably theophylline and caffeine have been used, at 1–3 mM concentrations. Methylxanthines increase the effect of suboptimal doses of NE (Reed and Fain, 1968b). We suggest that in all cases where theophylline alone has had NE-like effects (Reed and Fain, 1968a; Horwitz *et al.*, 1969; Williams and Matthews, 1974b) this is due to a low endogenous cAMP production, which in itself gives no observable effects but which is reinforced by theophylline.

3. *Changes in Cellular cAMP Levels*

The combined action of adenyl cyclase and phosphodiesterase may result in changes in cellular levels of cAMP. cAMP levels have been measured in brown fat cells; measurements in the presence of methylxanthines are excluded here, as these in reality mainly follow the activity of adenyl cyclase.

There are no reports on a *decreased* level. The level is *increased* by epinephrine; a doubling of the level is sufficient for maximal stimulation of glycerol release from cells, and a further fourfold increase of the level has no further effects on the rate of glycerol release irrespective of whether the level is changed by NE or theophylline (Fain *et al.*, 1972). Similarly, in brown fat slices, there is some correlation ($r = 0.37$) between the concentration of cAMP and the rate of lipolysis, but this is true only within a very narrow range of cAMP concentrations (0.6–0.8 nmole/gm) (Knight, 1974).

All these measurements have been performed at a specified time after NE stimulation. Pettersson and Vallin (1976) followed changes with time in cAMP after an NE stimulation which was just sufficient to give maximal respiration. The basal level (3 pmole/million cells) was increased fourfold within 3 minutes, and it then declined to basal levels after 10 minutes. At 10 minutes, lipolysis was still stimulated, as was respiration (Pettersson, 1977).

Thus, although changes in the level of cAMP can be observed after NE stimulation, these changes do not seem to be very well correlated with later effects (lipolysis and respiration).

We suggest that the levels of free cAMP are not of major significance for the regulation of the cellular metabolism. Rather, the protein kinases—and perhaps other cAMP-binding proteins—function as buffers for cAMP. They can well do so; if 1 pmole cAMP is produced in 1 million cells, this is equivalent to only a few thousand molecules per cell, and the number of kinases may well be in the same order. These may thus all be activated with only 1 pmole cAMP, but the free concentration of cAMP may remain virtually unchanged.

4. *Addition of cAMP*

It is evident from the figures given above that the effective concentrations of cAMP within the cells are in the micromolar range or below. Even addition of much larger (500–1000 μM) concentrations of cAMP outside the cells has normally no effect, and when it has (Hamilton and Horwitz, 1977; Horwitz and Eaton, 1975; Horwitz, 1973) it may be due to interaction with release of endogenous NE.

Rather, cAMP should be added in the form to the more membrane-permeable dibutyryl-cAMP; in millimolar concentrations, this agent mimics the effects of NE on fatty acid and glycerol release, and on respiration (Reed and Fain, 1968a). Added to the supernatant fraction from a homogenate, cAMP does not have permeability problems, and effects are to be expected in the micromolar range. Thus reported effects on fatty acid synthesis, which occurred with dibutyryl-cAMP with half-maximal effects at 500 μM (and which were not seen even at 5 mM cAMP) (Giacobini, 1971), are difficult to interpret, but may be due to released butyrate.

C. cGMP

Presently the regulation of cGMP values in any cell type is not understood. cGMP has not been measured in isolated brown fat cells, but only in tissue homogenates. A doubling of cGMP levels (from 50 to 100 pmole/gm wet weight) seen upon cold adaptation (Bertin *et al.*, 1978; Skala and Knight, 1979) may be secondary to changes in protein and water content.

Skala and Knight (1979) found changes in cGMP levels when intact rats were treated with α- and β-adrenergic (ant)agonists. The changes would indicate that cGMP levels were regulated by α-adrenergic stimulation. This stimulation may, however, not be on the fat cell proper; perhaps α stimulation may lead to vasoconstriction, and the observed effects on cGMP levels may be secondary to anaerobiosis, which in other cell types has been shown to cause an increase in cGMP (Busuttil *et al.*, 1976).

Changes in the level of cGMP do not seem to regulate lipoprotein lipase activity (Bertin *et al.*, 1978).

Skala and Knight (1979) suggest that an increase of cGMP may be related to "proliferative activity." This is based on a somewhat higher cGMP-dependent protein kinase activity in newborn and cold-adapted adult rats, compared to control adult rats (see Section VI,A,2,d).

Little is known about the regulation and the effects of cGMP; Fain and Butcher (1976) concluded an investigation on white fat cells with these remarks: "Our conclusion . . . is that the physiological significance of increased cyclic GMP levels in isolated white fat cells is minimal with respect to the regulation of lipolysis. Possibly there are other processes which are regulated by cyclic GMP in fat cells which have not been elucidated. Any postulated function of cyclic GMP in fat cells must account for the fact that it is elevated by a wide variety of agents which either accelerate, inhibit or have no effect on fat cell lipolysis."

D. Ca^{2+}

Intracellular levels of Ca^{2+} are difficult to monitor (Caswell, 1979), and there are no studies on such changes with brown fat cells.

If short-term changes in cytosolic Ca^{2+} are induced by NE, the Ca^{2+} is probably not entering from the extracellular fluid, as the Ca^{2+} channel inhibitor D600 is without effect on the NE-induced depolarization (Williams and Matthews, 1974a), and as extracellular Ca^{2+} concentrations are of little importance for the thermogenic response (Section III). Rather, cytosolic Ca^{2+} may be increased—and mitochondrial decreased—as an indirect effect of Na^+ entry (Section III). This may secondarily lead to an increase in cytosolic-free Ca^{2+} concentrations; certainly, in mitochondrial experiments, added

Na^+ increases the steady-state level of Ca^{2+} (Nedergaard and Cannon, 1980; Nedergaard, 1981).

The effects of increase in cytosolic Ca^{2+} (in the micromolar range) may include (see also Fig. 6) an increased activity of the glycerol-3-phosphate dehydrogenase (Bukowiecki and Lindberg, 1974). This may lead to less glycerol 3-phosphate being present as the precursor for glyceride-glycerol and by this to inhibition of fatty acid esterification (Nedergaard, 1981). Also a decreased activity of adenyl cyclase (Cheung, 1980) and changed affinity of thermogenin for nucleotides (Nicholls, 1979) may be an effect of increases in cytosolic Ca^{2+}. Furthermore, the accompanying change in mitochondrial Ca^{2+} may affect several intramitochondrial enzymes (pyruvate dehydrogenase, isocitrate dehydrogenase, and oxoglutarate dehydrogenase) (McCormack and Denton, 1979).

E. Prostaglandins

Changes in levels of *endogenous* prostaglandins in brown fat cells have not been studied; nor do any reports on effects of inhibitors of prostaglandin metabolism (acetylsalicylic acid, indomethacin) exist.

Addition of prostaglandin E_1 or E_2 (50–100 n*M*) does not affect unstimulated adenyl cyclase activity, but decreases or abolishes NE-stimulated activity (Bertin and Portet, 1976); higher doses certainly abolish the activity (Moskowitz and Krishna, 1973). As a consequence of this, catecholamine-stimulated lipolysis and respiration are diminished (Dryer *et al.*, 1971).

It is doubtful whether any of these effects has any physiological significance as endogenous levels of prostaglandins may well be in the picomolar range. The results reported may well be due to an inhibition of the interaction of NE with the receptor.

F. Conclusions: "Second Messengers"

Only for cAMP is there reason to consider it demonstrated that it is a second messenger. Changes may occur also in the levels of cGMP and Ca^{2+} but neither these changes nor their effects have so far clearly been demonstrated to be of physiological significance.

VI. The Catabolic Cascade

The effects of cAMP occur solely through activation of protein kinases. There are, e.g., no effects of cAMP on isolated mitochondria (Nicholls *et al.*, 1974).

A. Protein Kinases and Phosphoprotein Phosphatases

Kuo and Greengaard (1969) first reported the existence of cAMP-dependent protein kinase in brown fat. The kinase has been studied in greater detail, especially by Knight and Skala and their co-workers.

1. *The Regulation of the Activity of the Kinase*

Skala and Knight (1977) followed the degree of activation of the kinase(s). By homogenizing in 0.5 *M* NaCl they prevented the reassociation between the free catalytic subunit(s) and the regulatory subunit(s), and they could then measure the degree of activation as (the activity without added cAMP/the activity with cAMP added). As usual in investigations on protein kinase activity, a histone was used as an artificial substrate. Skala and Knight found that although the degree of activation varied only between 0.3 and 0.5, it was changed (by NE, theophylline, and insulin *in vitro*) in the directions expected from their actions on the cAMP level; even a cold stress of the animal manifested itself as an increased activation.

2. *The Molecular Nature of the Kinase(s)*

As the molecular nature (i.e., subunit composition) of the protein kinases has been investigated by three different techniques (ultracentrifugation, DEAE-sepharose columns, gel electrophoresis) and in two species (young rats and infant rabbits), the total picture is not quite clear; the interpretation is further complicated by the fact that autophosphorylation of the regulatory subunit may take place.

a. *Different Kinases.* In Fig. 14 we have constructed all possible combinations of protein kinases, based on the existence of the molecule as a tetramer of two catalytic and two regulatory subunits, this latter appearing both in a low-affinity (for cAMP) (=nonphosphorylated), and a high-affinity (=phosphorylated) form. As seen, 12 different kinases may theoretically be constructed, 10 of these showing cAMP dependency. Under some separation procedures, especially those where a high ionic strength is used, several of these forms may perhaps be created.

Thus it is no surprise that a lot of different kinases may be observed, as many as 10 have indeed been characterized in one preparation (Knight and Skala, 1977). Some of these may be different kinases; some may be the same kinase in different forms.

We have in Fig. 14 very tentatively tried to identify some of these 12 possible kinases. Although such a system can be extracted from the available data, other schemes, which include different protein kinases, can naturally be constructed.

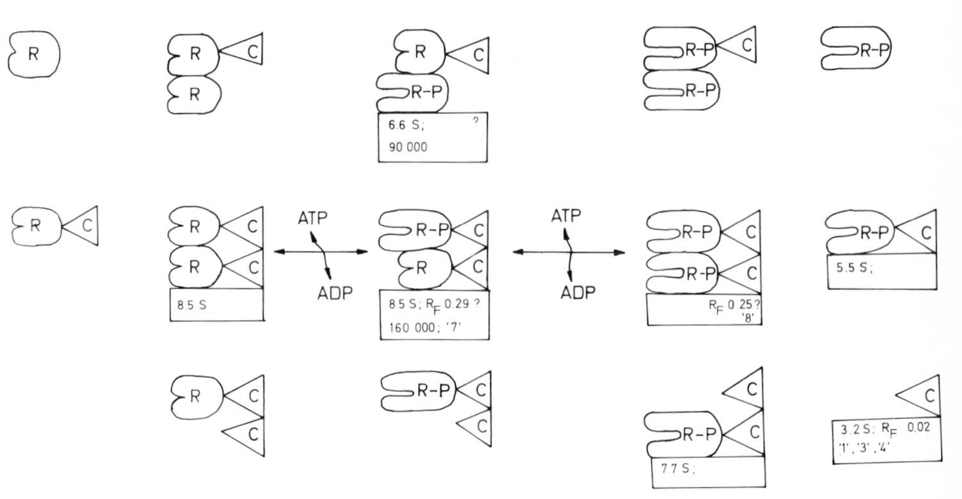

FIG. 14. The possible forms of a cAMP-dependent protein kinase complex. In this figure all the combinations of the protein kinase complex which can hypothetically be formed are presented. C, Catalytic subunit (only active when free); R, regulatory (cAMP-binding) subunit; R-P, phosphorylated regulatory subunit (autophosphorylation). Very tentatively, some of these complexes are identified with some of the kinases described in brown adipose tissue. The S values and the molecular weights refer to the determinations of sedimentation constants by Knight and Fordham (1975), and the R_f values and the numbers 1 to 8 to the peaks described by Knight and Skala (1977) (see Section VI,A,2).

b. *Autophosphorylation.* Incubation of brown adipose tissue with NE leads to autophosphorylation of the regulatory subunit of the protein kinase; this is associated with an increased affinity for cAMP (K_m changed from 100 to 10 n*M*) (Knight, 1975). Actually, the regulatory subunit is in itself the "endogenous protein" which can be phosphorylated by protein kinase activity (Knight, 1976), and is so far the only phosphorylated protein detected.

c. *ATP- and GTP-Utilizing Kinases.* Skala and Knight (1975) investigated the rate of histone IIA phosphorylation with either labeled ATP or GTP as phosphate donor. Although the activity was about the same with both donors, the authors concluded from a series of control experiments that only ATP was a true donor; the GTP activity was due to the combined action of a very highly active nucleotide phosphotransferase and the usual, ATP-utilizing protein kinase. Similarly, the increased activity seen by addition of 1 m*M* GTP to an incubation containing 200 μM ATP (Knight and Skala, 1975) is probably due to the ATP hydrolysis product ADP being rephosphorylated by GTP; the kinase activity must be very dependent upon the available ATP concentration as the K_m for ATP is 200 μM, the amount initially added. There is thus no reason to assume that the kinase can use any phosphate donor other than ATP.

d. *cGMP-Dependent Protein Kinases.* Skala and Knight (1979) found that protein kinase activity could be stimulated 25% over basal levels by 300 n*M* cGMP in homogenates from both newborn and cold-adapted adult rats, whereas no stimulation was obtained in homogenates from adult, control rats. In all cases, 300 n*M* cAMP led to a doubling of the rate. Whereas this experiment may indicate a cGMP-dependent kinase being present in brown fat and perhaps changing activity in parallel with the thermogenic activity of the tissue, the conclusion of the authors, that cGMP-dependent protein kinase activity is related to trophic stimulation of the tissue, may be discussed since cold-*adapted* rats (6 weeks in cold) are not normally considered to have a progressing trophic stimulation of their brown fat.

e. *The Kinase Inhibitor.* Similar to what is found in other tissues, the supernatant fraction from brown fat contains a heat-stable, nondialyzable, nonproteolytic, trypsin-sensitive inhibitor of protein kinase; the inhibition is noncompetitive with respect to ATP (Skala *et al.*, 1974). The regulatory function of this inhibitor is unknown.

f. *The Substrates for the Protein Kinase.* Besides phosphorylating the traditional substrate histone IIA, the kinase may phosphorylate a variety of other added proteins. It is, however, clear from the work of Knight and Skala (1977) that generally the activity of different kinase "peaks" from elution profiles versus different substrates tends to change in parallel, i.e., that "good" kinase peaks phosphorylate histones at a high rate and other added proteins (e.g., casein) at a lower rate, whereas "bad" peaks phosphorylate histones slowly and other proteins not at all. Furthermore, as stated by Skala and Knight (1978), all nonhistone–protein phosphorylations are cAMP *in*dependent.

Naturally, it is the activity versus endogenous proteins which is of physiological interest, but the only endogenous substrate of the protein kinase demonstrated so far is the kinase itself (autophosphorylation). However, Skala and Knight (1977) found a good correlation between the degree of activation of the kinase in the range 0.3–0.5 and glycerol release from tissue slices; this is at least an indication that the protein kinase studied does activate the hormone-sensitive lipase.

g. *Phosphoprotein Phosphatases.* Enzymes which dephosphorylate proteins phosphorylated by protein kinases must exist if the reactions are to be reversible, and Knight and Skala (1979) have investigated such enzyme(s) in brown fat.

B. Activation of Glycolysis

Activation of glycogen degradation and glycolysis in liver were the cAMP-mediated reactions to be first demonstrated. Also in brown fat these reactions may be stimulated. Glycogen granules are found in brown fat cells (Barnard and Skala, 1970), and all enzymes of glycolysis are present, although hexokinase activity is relativity low (Williamsson *et al.*, 1969). Glycolysis seems to be stimulated by NE; at least the pyruvate "content" of brown fat cells is doubled by NE (Williamsson *et al.*, 1970), and the high glycerol 3-phosphate-utilizing shuttle mechanism allows glycolysis to proceed, unhampered by NADH accumulation, during thermogenesis (Houštěk *et al.*, 1975a).

Activated glycolysis is beneficial to thermogenesis in at least two ways. It is a cytosolic ATP producer. Both the membrane potential (Section V,A,2,c) and NE-stimulated lipolysis are very sensitive to glycolytic inhibitors; iodoacetate and 2-deoxyglucose both inhibit lipolysis, whereas the uncoupler FCCP and the mitochondrial ATP-synthetase inhibitor oligomycin are virtually without effect (Houštěk *et al.*, 1975b). Thus it seems that mitochondrial ATP is not easily utilized for these reactions, but that they are dependent upon glycolysis. Further, glycolysis is a pyruvate producer. Indeed, Prusiner *et al.* (1968b) pointed out that "it could be visualized that the function of glycolysis might be to make pyruvate available for the formation of oxaloacetate necessary as a condensing partner for the acetyl-CoA formed from the β-oxidation of fatty acids." This hypothesis was later proven true by Cannon and Nedergaard (1979a).

C. Activation of Hormone-Sensitive Lipase

The most important action of cAMP-dependent protein kinase in brown fat must be the activation of hormone-sensitive lipase. Knight (1974) stated that "a stimulation of the lipase by cyclic AMP and protein kinase in a cell-free system has not yet been demonstrated. Experiments in this laboratory to show such an effect have so far failed." No other, more successful attempts have been reported.

The activation of lipase results in an increased rate of triglyceride breakdown. It may also lead to an increase in diglyceride breakdown, as Lindberg *et al.* (1976) found that the amount of labeled palmitate in diglycerides decreased after addition of NE to cells.

The end product of triglyceride breakdown is fatty acids, and, if the reaction proceeds, glycerol. However, resynthesis of triglycerides from monoglycerides or diglycerides may occur in brown fat (Schulz and Johnston, 1971), and such activities may make measurements of lipolysis by determination of glycerol release underestimates.

We consider glycerol and fatty acids the end products of lipase activity. *Glycerol* is mainly released from the cells; there is no evidence for glycerol kinase in rabbit brown fat (Dawkins and Hull, 1964), nor in hamster cells (Lindberg *et al.*, 1976), but there is some activity in rat cells (Bertin, 1976). The fate of *fatty acids* is discussed in the next section.

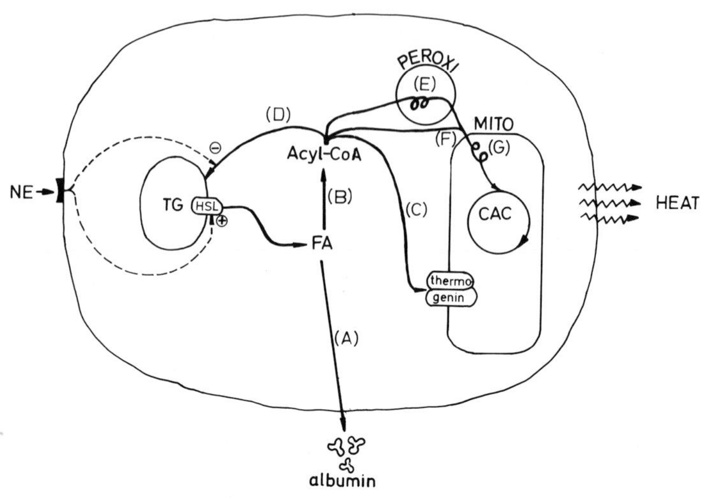

FIG. 15. Pathways of fatty acid utilization in brown fat cells. The pathways include (A) export, (B) activation, (C) regulatory functions, (D) reesterification, (E) peroxisomal β-oxidation, (F) mitochondrial transfer, and (G) mitochondrial β-oxidation. HSL, Hormone-sensitive lipase; CAC, citric acid cycle; TG, triglyceride droplet (see Section VII). Adapted from Nedergaard (1980).

VII. Fate of Released Fatty Acids

The fatty acids, which have been released by hormone-sensitive lipase activity, may either be exported as such or be activated to acyl-CoA in the cell. Besides having regulatory functions, acyl-CoA may be used for reesterification or be oxidized, in both peroxisomes and in mitochondria. In Fig. 15 we have visualized some of these possible pathways.

A. Export of Fatty Acids

Measured ratios of exported fatty acids/released glycerol are fairly close to 3 (Reed and Fain, 1968a; Nedergaard and Lindberg, 1979). This indicates that the major part of released fatty acids is exported from the cell, and only a fraction is reesterified or combusted. Also when the cell is not maximally NE stimulated with respect to respiration, more released fatty acids are actually exported than combusted (Cannon *et al.*, 1978). The capacity for fatty acid release is higher in brown fat than in white fat when expressed per millimole triglyceride (Fain and Reed, 1970) or per gram tissue; the values obtained are between 40 and 100 nmol/minute/million cells. In phosphate buffer, the release is "biphasic" after NE addition with a rapid release for 4–10 minutes, followed by a slower rate (Reed and Fain, 1968a; Bieber *et al.*,

1975; Pettersson and Vallin, 1975). A higher linearity and a higher rate are observed in bicarbonate buffer (Nedergaard and Lindberg, 1979).

The mechanism for regulation of the partition of released fatty acids between export and activation is not known.

B. Activation of Fatty Acids

Acyl-CoA synthetase, which is the activation enzyme system for the released fatty acids, follows mono-amine-oxidase activity in different fractionation procedures of a brown fat homogenate; Pedersen *et al.* (1975) concluded from this that the enzyme was probably located exclusively on the mitochondrial outer membrane in brown fat (in most other tissues the activating system is found both in mitochondria and in microsomes; due to the small amount of endoplasmic reticulum in brown fat, a microsomal fraction may not easily be detected).

The activity of the activating system is very high (100–150 nmole/minutes/mg mitochondrial protein) (Pedersen *et al.*, 1975; Norman and Flatmark, 1978); indeed this activity (with 2–3 mg mitochondrial protein per million cells) is probably higher than the maximal rate of fatty acid release.

C. Regulatory Functions of Acyl-CoAs

There is so far no direct evidence that cytosolic acyl-CoA levels are increased during thermogenesis. If they are—as may be expected—they may influence the following enzymes, perhaps fulfilling a regulatory role.

The mitochondrial glycerol-3-phosphate dehydrogenase is inhibited in a competitive manner by palmitoyl-CoA; K_i is 2 nmole/mg mitochondrial protein (Bukowiecki and Lindberg, 1974). The mitochondrial phosphate transfer system (Christiansen and Wojtczak, 1974) and the mitochondrial ADP/ATP exchange are both inhibited (Christiansen *et al.*, 1973); together these two effects may explain the dependency upon glycolysis of cytosolic reactions during thermogenesis. Thermogenin in isolated mitochondria is activated in a competitive way with GDP by palmitoyl-CoA; here 2 nmole/mg protein is the half-maximal level (Cannon *et al.*, 1977b). This regulatory effect is the one of most interest because it may be the way in which brown fat mitochondria are uncoupled *in vivo* during thermogenesis (see Section XIII,E).

D. Reesterification

Acyl-CoAs may be used for the resynthesis of triglycerides. In the NE-stimulated cell, this is not a very prominent way of utilization of acyl-CoA, and Fain *et al.* (1967) found that only in the simultaneous presence of both insulin and NE, may some released fatty acids be reesterified (about 3%). Esterification (of added, labeled fatty acids) does seem to take place even when cells are NE stimulated (Prusiner *et al.*, 1968c).

E. Peroxisomal β-Oxidation

In accordance with the situation in liver, brown fat peroxisomes are capable of β-oxidation of acyl-CoAs; this is evident from the existence of a palmitoyl-CoA-dependent, cyanide-insensitive NAD-reductase in the tissue (Cannon *et al.*, 1978; Kramar et al., 1978; Nedergaard *et al.*, 1979a, 1980). The peroxisomal localization of this enzyme in brown fat was demonstrated by Kramar *et al.* (1978) who described a fraction, obtained by isopycnic gradient centrifugation, which was enriched in the peroxisomal marker

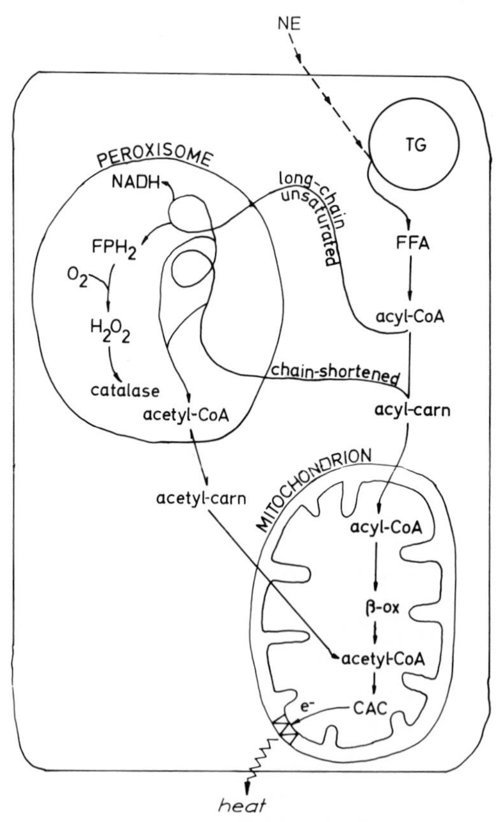

FIG. 16. Interaction between peroxisomal and mitochondrial β-oxidation. TG, Triglyceride droplet; carn, carnitine; FPH_2, reduced flavoprotein; β-ox, β-oxidation; CAC, citric acid cycle (see Section VII,E). Adapted from Alexson *et al.* (1980).

enzyme catalase and devoid of the mitochondrial marker cytochrome c–oxidase. In this fraction, palmitoyl-CoA-dependent NAD-reduction could be detected; this activity was not inhibitable by carbon monoxide. A similar activity in a mitochondrial fraction was carbon monoxide sensitive.

Thus the peroxisomes in brown fat are probably capable of β-oxidation. Nedergaard *et al.* (1980) found that the peroxisomal β-oxidation was increased more by cold adaptation than mitochondrial oxidases; furthermore, these authors suggested that the role of peroxisomal β-oxidation was mainly to standardize acyl-CoAs for mitochondrial β-oxidation, as the peroxisomal activity toward long-chain unsaturated fatty acids (which inhibit mitochondrial β-oxidation) generally is higher than the activity toward saturated fatty acids, both in liver (Osmundsen *et al.*, 1979), and in brown fat peroxisomes (Alexson, unpublished observations).

The need for such a system in brown fat can perhaps be illustrated by the fact that a rapeseed oil diet (which contains much erucic acid) leads to an increased fraction of the carnitine in brown fat from non-cold-treated rats being bound as acyl-carnitine (Gumpen and Norum, 1973), perhaps because the long-chain unsaturated acyl-CoAs are slowly oxidized in brown fat cells which have only few peroxisomes.

As illustrated in Fig. 16, β-oxidation in peroxisomes gives rise to a series of products: H_2O_2, NADH, acetyl-CoA, and an acyl-CoA with a shortened chain length. This acyl-CoA (as well as acyl-CoAs, which have not first been partially β-oxidized in peroxisomes) and acetyl-CoA, are transferred into the mitochondria by the action of the carnitine shuttle.

F. Mitochondrial Transfer of Acyl-CoA

The carnitine content in brown fat is high, about 0.5 μmole/gm wet weight (0.5–1 m*M*) (Hahn and Skala, 1972). The carnitine is mainly there to take part in the carnitine shuttle for the transfer of acyl units into the mitochondrial matrix (Fig. 17). The outer long-chain acyl-transferase has a maximal activity in excess of the rate of acyl-CoA production (300–400 nmole/minute/mg protein) (Hahn and Skala, 1972; Normann *et al.*, 1978). The inner acyl-transferase is about 1000 times more active (Normann *et al.*, 1978). The activity of the acyl-carnitine-permease itself has not been measured, but it must be very high.

The effective system for the transfer of carnitine esters over the mitochondrial membrane also explains why 2-hydroxy-butyryl-carnitine is oxidized six times faster than free 2-hydroxy-butyrate (Pedersen *et al.*, 1968) and why acetyl-carnitine is so easily oxidized (Cannon, 1971).

Addition of carnitine, together with the substrates for fatty acid activation, ATP and CoA, to isolated mitochondria results in transient respiration, probably due to oxidation of fatty acids, which have perhaps associated with

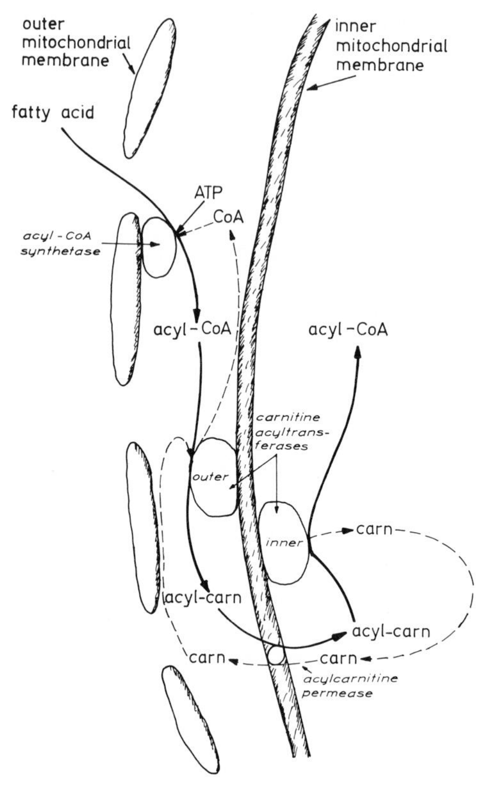

FIG. 17. Mitochondrial transfer of fatty acids. carn, Carnitine (see Section VII,F).

the mitochondria during the isolation procedure (Drahota *et al.*, 1968; Hittelman *et al.*, 1969). Maximal rates of oxidation are under these circumstances seen with about 1 m*M* carnitine, close to the tissue concentration *in vivo*.

G. MITOCHONDRIAL β-OXIDATION

The mitochondria of the brown fat cell are very well endowed for β-oxidation of fatty acids. One indication of this is the amount of the β-oxidation-specific ETF-oxidase (electron-transferring-flavoprotein: ubiquinone oxidoreductase) which has an EPR signal at g = 2.08 (Flatmark *et al.*, 1976). Relative to the amount of NADH-dehydrogenase, brown fat has four times more of the ETF-oxidase than heart, a tissue which is in itself considered to be rather well equipped for fatty acid oxidation (Ruzicka and Beinert, 1977).

The regulation of fatty acid β-oxidation in brown fat has not been studied in detail. Cannon (1971) has however followed the fate of oxidized fatty acids in brown fat mitochondria by adding labeled palmitate to the mitochondria and studying the recovery of the label in different fractions. Most of the added label was recovered in acetyl-carnitine, in citric acid cycle intermediates, and in glutamate. Under these incubation conditions a few percent was converted to CO_2 and no ketone bodies were detectable. Despite some technical differences, these results are in principle in good agreement with the results of Dryer and Harris (1975). It must be noted, however, that under alkaline analysis conditions, as used by Dryer and Harris, acetyl-carnitine is split, and the free acetate evaporates during drying of the thin-layer chromatography plate. This may explain why 80% of the added counts were not recovered by Dryer and Harris. Of those added, a few percent were found in ketone bodies.

We shall now discuss in detail the fate of acetyl-CoA produced by mitochondrial β-oxidation.

VIII. The Fate of Acetyl-CoA

The acetyl-CoA, produced by mitochondrial β-oxidation, is further metabolized by three different mitochondrial matrix enzymes: citrate synthetase, acetyl carnitine-transferase, and acetyl-CoA-hydrolase (Fig. 18). Bernson (1976) investigated the maximal activity of these enzymes; as seen in Fig. 18, citrate synthetase is by far the most effective acceptor for acetyl-CoA; furthermore this enzyme has the highest affinity for acetyl-CoA. Thus if cofactors are not limiting, all acetyl-CoA produced primarily enters the

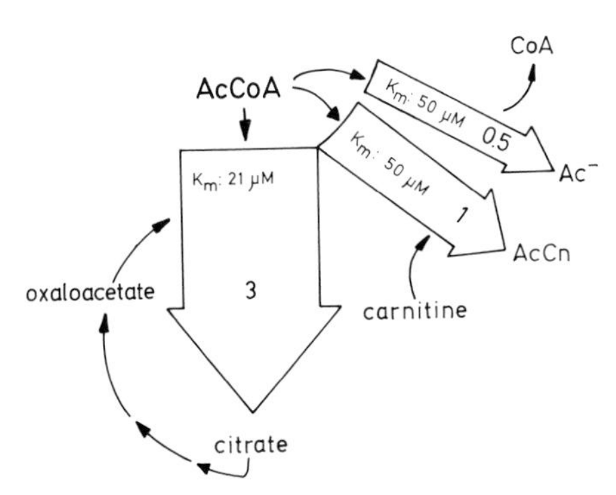

FIG. 18. Pathways for acetyl-CoA utilization in hamster brown fat mitochondria. 3, 1, and 0.5 are the rates of different systems in nanomoles per minute per milligram protein. The figure is based on the results of Bernson (1979). AcCn, Acetyl-carnitine (see Section VIII).

citric acid cycle; if there is any temporary pile-up of acetyl-CoA, acetylcarnitine is formed, and acetate is formed only if the carnitine supply is also limited.

We shall for simplicity study these pathways in an inverse order to their capacity.

A. Acetate Production

As understood from Fig. 18, acetate is only a major end product of fatty acid metabolism under conditions where carnitine and malate (oxaloacetate) are absent or at least limiting, a situation which is easily created in an experimental situation but which may seldom occur physiologically. The factors that control acetate production are thus carnitine and oxaloacetate (malate) concentrations, flux of acetyl-CoA, and the amount of, and relative activity of, the enzyme acetyl-CoA-hydrolase.

In the presence of *malate* and *carnitine*, only 3–5% of acetyl-CoA is hydrolyzed to acetate; in the absence of these aceptors, 71% of acetyl-CoA is hydrolyzed (Bernson and Nicholls, 1974). The *amount* of acetyl-CoA-hydrolase present in the mitochondrial matrix is different in different species. It is low in rat (90 nmole/mg protein/min) (Bernson, 1976) and in guinea pig (inferred from Bernson and Nicholls, 1974), and it is almost 10 times higher both in hamster (600–900 nmole, Bernson and Nicholls, 1974; Cannon *et al.*, 1977a) and in newborn lamb (500–600 nmole, Cannon *et al.*, 1977a). The reason for these marked species differencies is not known; the original hypothesis, that a high activity should be especially helpful for hibernators where the brown fat is active at low body temperatures (Bernson and Nicholls, 1974), can clearly not be the whole explanation, as lambs obviously do not fall into this group.

The *activity* of the acetyl-CoA-hydrolase is under allosteric control of NADH, ADP, and CoA (Bernson, 1976). High NADH (but not NAD^+) and high CoA inhibit the enzyme. Thus only under conditions when respiration is proceeding rapidly, and there is a lack of matrix CoA, and when carnitine and oxaloacetate are limiting, is the enzyme activated; it will then regenerate CoA for further β-oxidation.

The *temperature* dependency of the enzyme is conventional with a Q_{10} of 2 (Bernson, 1976) and thus cannot explain the finding of Bernson and Nicholls (1974) that in mitochondrial incubations, the rate of acetate production is very temperature independent. This may perhaps be secondary to a decreased malate entry at low temperatures (Cannon and Nedergaard, 1979b) rather than to "the many diffusion limited steps occurring in the matrix" (Nicholls, 1976b).

The *substrate specificity* of the enzyme has unfortunately not been investigated so far. However, Berge *et al.* (1979) have observed in guinea pig mitochondria a palmitoyl-CoA-hydrolase with an activity in excess of 20 nmole/minute/mg protein [they also measured a palmitoyl-carnitine hydrolase activity (about 1 nmole), but it cannot be ruled out that this activity in reality is due to the combined action of acyl-transferase and palmitoyl-CoA-hydrolase]. It may be speculated that this is the same enzyme as the acetyl-CoA-hydrolase, and that this enzyme is an acyl-unspecific hydrolase, which perhaps has as its main function to secure the regaining of CoA, bound up to unmetabolizable acyls. These questions await the purification of the enzyme, which both in brown fat (Bernson, 1976) and in other acetyl-CoA-hydrolase-containing tissues (see e.g., Costa and Snoswell, 1975) has proved to be a difficult task.

In conclusion, in certain species and under certain restricted conditions, acetyl-CoA may be hydrolyzed to CoA and free acetate. The reaction is irreversible and the metabolism of the free acetate is unknown; brown fat mitochondria do not seem to possess the acetate kinase necessary to recapture acetate (unpublished observations), but the brown fat cell is able to metabolize acetate (see Section XI,C,1) and so certainly are other cells in the organism, e.g., the heart.

B. Acetyl-Carnitine Production

If more acetyl-CoA is produced in the cell than can immediately enter the citric acid cycle, the acetyl groups may be sequestered as acetyl-carnitine, and the CoA liberated (Cannon, 1971). The fraction of acetyl groups sequestered as carnitine esters is [in the *absence* of malate (oxaloacetate)] very dependent upon the amount of free carnitine present. Thus if no carnitine is added, 17% of acetyls end up as acetyl-carnitine (i.e., only carnitine from the added acyl-carnitine); if 3 mM carnitine is present, 70% end up as acetyl-carnitine (Bernson and Nicholls, 1974).

In the *presence* of malate (oxaloacetate), a great amount (40%) of the acetyl groups (added as acyl-CoA) is recovered as acetyl-carnitine. If the same experiment is done with uncoupled mitochondria, less of the label is found as acetyl-carnitine, and more has entered the citric acid cycle (Cannon, 1971). One may interpret these experiments in such a way that, as long as acyl-CoA is present, the mitochondria are specifically uncoupled (Section XIII,E) and β-oxidation proceeds rapidly. When all acyl-CoA is removed, the mitochondria recouple, and citric acid cycle activity is decreased. The acetyl is then stored as acetyl-carnitine and slowly utilized, as has indeed been measured by Bernson and Nicholls (1974).

Acetyl-carnitine production can be observed *in vivo*; as much as 15–30% of all carnitine in the tissue is in the form of acetyl-carnitine (Gumpen and Norum, 1973; Alkonyi *et al.*, 1975). Acetyl-carnitine is consequently the major carnitine ester in the cell; there is no propionyl-carnitine; long-chain acyl-carnitines amount to only 3% of all carnitine, and the rest is free (Gumpen and Norum, 1973).

Thus, acetyl-carnitine is easily produced by the mitochondria in the cell, and it is easily combusted (Cannon, 1971). Whether it may leave the mitochondria—or even the cell—and take part in other processes (e.g., fatty acid synthesis) is not known.

C. Entry into the Citric Acid Cycle

Entry of acetyl-CoA into the citric acid cycle for final combustion is dependent upon the availability of oxaloacetate. Thus by the addition of 3 m*M* of the oxaloacetate precursor malate to isolated mitochondria, the fraction of acetyl-CoAs entering the cycle is much increased (Bernson and Nicholls, 1974).

However in cells, malate is not added, and the citrate synthase must use endogenous oxaloacetate. That the supply of oxaloacetate may be rate-limiting and may explain tachyphylaxis (see Section III) was clear to Prusiner *et al.* (1968b) who also stated that "oxaloacetate formation may proceed via pyruvate carboxylase or by the reversal of the malic enzyme reaction." However, neither added malate nor succinate seemed to be able to generate sufficient oxaloacetate; this is probably simply because the mitochondria are poorly permeable to these intermediates (Cannon and Nedergaard, 1979b). Rather, in the cell, the mitochondria produce oxaloacetate from pyruvate and CO_2 by the activity of pyruvate carboxylase (Cannon and Nedergaard, 1978, 1979a). The activity of this enzyme is about 40 nmole/minute/mg mitochondrial protein, and it is probably this reaction, which explains the stimulatory effect of CO_2 on brown fat cell metabolism (Pettersson, 1977). Under conditions where pyruvate carboxylase activity is stimulated, higher concentrations of malonate must be added in order to inhibit cellular respiration; this is indirect evidence that the mitochondrial level of citric acid cycle intermediates has been increased (Cannon and Nedergaard, 1979a). An inhibitor of pyruvate uptake into the mitochondria, UK5099, abolishes solely the stimulatory effect of CO_2, without affecting the respiration in the absence of CO_2, again pointing to a need for both pyruvate and CO_2 to increase oxaloacetate availability and by this respiration. Thus oxaloacetate is produced by an anaplerotic process in the cell.

It is, however, clear from the occurrence of the phenomenon of tachyphylaxis (see Fig. 5) that there is initially sufficient oxaloacetate present when cells are NE stimulated; but the successive inhibition of cellular respiration must indicate that citric acid cycle intermediates are lost after—and as an effect of—NE stimulation. In what form and why these intermediates are lost is not presently known. Dryer and Harris (1975) found that benzen-1,2,3-tricarboxylate inhibited respiration in mitochondria and concluded that the mitochondria were able to "export" citrate; this may be an indication that citrate may be lost from the mitochondria. Cannon (1971) found that a significant fraction of acyl groups could be recovered in the form of glutamate, and Cannon and Nedergaard (1979a) suggested that tachyphylaxis may be due to loss of intermediates for amino acid synthesis; they further demonstrated that the presence of amino acids in the medium during the isolation and incubation actually abolished tachyphylaxis. Why NE stimulation should lead to an increased demand for amino acids is, however, not evident.

An experimental decrease of flow through the citric acid cycle—and by this a lack of oxaloacetate—may be created by the addition of malonate (Prusiner *et al.*, 1968b) or arsenite (Williamson *et al.*, 1970) or NH_4^+ (Cannon and Nedergaard, 1979a) to the cells.

IX. The Energy Potential of Brown Fat Cells

The concentration of adenine nucleotides in brown fat cells is 10–20 nmole/million cells (Williamson, 1970; Pettersson and Vallin, 1975). The ratio ATP/ADP is 4/1 or higher in unstimulated cells; if the cells are stimulated by NE or oleate, the ratio drops down to as little as 2/1 (Prusiner *et al.*, 1968c; Williamson, 1970; Pettersson and Vallin, 1975). [The decrease is smaller when CO_2 is present (Pettersson, 1977).] After this decrease, a recovery may be observed (Fain *et al.*, 1972).

Effects of changes in energy potential have been studied by Cannon *et al.* (1973) and by Pettersson and Vallin (1976). Cannon *et al.* found little effect of change in adenylate charge on coupling of mitochondria before the charge fell below 0.3; Pettersson and Vallin found that at very low energy potentials, the respiration of endogenous substrate (fatty acids) was abolished, probably due to AMP inhibition of acyl-CoA synthetase.

The GTP/GDP ratio changes in parallel with the ATP/ADP ratio (Prusiner *et al.*, 1968c). This is probably a consequence of the existence of a high activity (2 mmole/minute/mg protein) of a nucleotide phosphotransferase in the cytosol (Skala and Knight, 1975).

The cytosolic ATP is buffered by creatine phosphate both in mouse (Berlet *et al.*, 1976) and in hamster (Cannon *et al.*, 1978). The amount of creatine is as great as or greater than the amount of adenine nucleotides. Creatine kinase specific activity is increased by 50% by cold adaptation;

the activity is then 0.2 μmole/minute/mg homogenate protein (Cannon *et al.*, 1978).

Measurement of the redox state of cytochrome b in mitochondria within intact cells may give information on the energy status of the mitochondria. Prusiner (1970) could actually measure an oxidation of cytochrome b in isolated cells after NE stimulation; as this oxidation occurred even in the presence of the ATP-synthetase inhibitor oligomycin, it is a demonstration of an NE-induced uncoupling of the mitochondria. Similarly, in intact hamsters, the redox state of NAD(P)H can be followed by fluorometric techniques; NAD(P)H is also oxidized as a consequence of NE injection (Prusiner *et al.*, 1968d).

These results, which indicate an NE-stimulated *oxidation* of the electron carriers and an uncoupling of the mitochondria, are in contrast to results obtained by Seydoux *et al.* (1977) and Seydoux and Girardier (1978). In innervated tissue, nerve stimulation (2 Hz) led to a *reduction* of NAD to NADH. However, it is only possible to get a graded response regarding oxygen consumption in this preparation by the use of frequencies below 1 Hz; 20 Hz leads to a nearly immediate total lack of oxygen within the tissue, and under such circumstances, it is naturally not possible to oxidize NADH. However, the ability to produce NAD(P)H upon stimulation is a good measure of the intactness of the neurotransmitter response and the ability of the cells in the tissue pieces to oxidize fatty acids.

X. Exports from Brown Fat Cells

The existence of a cell type within the integrated organism is justified only by the advantages to the organism as a whole of the function of the cell. We shall here discuss the products of the brown fat cell; Fig. 19 summarizes the effects of NE stimulation on the fluxes from a single fat cell.

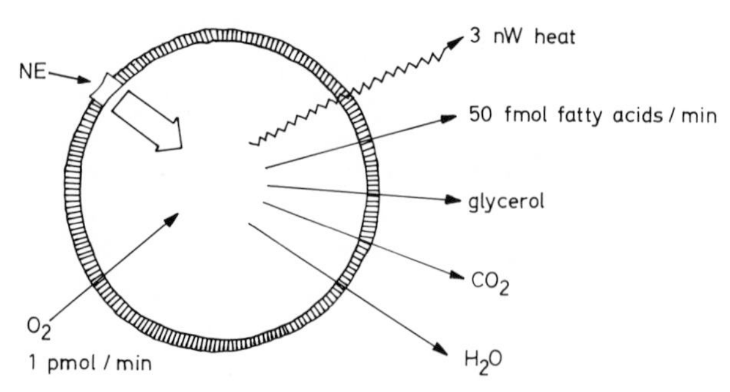

FIG. 19. Exports from brown fat cells. The values given are calculated per cell (see Section X).

A. Heat

"The energy conserved in a biological substrate can be released as heat by oxidation of the substrate to carbon dioxide and water. The caloric yield due to this complete oxidation is independent of the reaction sequence. Thus the oxygen consumption of a cell is a true reflection of its potential heat producing capability. Studies of cellular respiration can therefore be taken as an index of thermogenesis" (Prusiner *et al.*, 1968c). Nedergaard *et al.* (1977) directly examined the content of this statement by measuring both the heat output and the oxygen consumption of isolated brown fat cells. They found a heat evolution of 250 kJ/mole O; the heat production of the cell preparation was 0.8 mW/million cells. These cells, which were incubated in phosphate buffer, only had an oxygen consumption rate of 300 nmole O/minute/million cells. Rates up to three times higher have later been obtained in other buffers, and a brown fat cell may thus evolve 2.5 nW (2.5 mW/ million cells). Approximately 300 W/kg wet weight is thus produced in brown fat; this heat-producing capacity of brown fat cells can be compared with the cells in an adult human which on average only produce 1 W/kg.

Although this is a very high capacity for heat production, it may still be an insufficient capacity under certain conditions, e.g., the arousal of hibernators (Nedergaard *et al.*, 1977).

B. Fatty Acid Release

Fatty acids are released from brown fat cells after NE stimulation (Fain *et al.*, 1967), and the rate of fatty acid release from brown fat cells is—per millimole triglyceride—higher in brown fat cells than in white fat cells (Fain and Reed, 1970). Calculated per cell the releasing capacity is about the same in white and brown fat cells (Nedergaard and Lindberg, 1979), but (as pointed out by these authors) since brown fat cells contain less triglyceride and are smaller than white fat cells, the release per gram wet weight is higher in brown fat. The release of fatty acid acid does not seem to represent merely an overflow in excess of the capacity of the oxidizing systems within the cell, as it is observed even at NE concentrations which only stimulate respiration suboptimally (Cannon *et al.*, 1978).

There are some indications that fatty acids are also released from brown fat cells *in situ*. Under certain conditions, brown fat in newborn rabbits may release 20 nmole fatty acids/minute/million cells (Hardman and Hull, 1970), and the concentration of fatty acids in the blood of arousing hibernators increases, e.g., in the garden dormouse from 0.7 to 2.3 m*M* (Ambid *et al.*, 1971). The blood leaving brown fat in NE-stimulated adult rats may contain fatty acids equivalent to a release of 15 nmole/minute/million cells (Portet *et al.*, 1974) [these authors actually found that the release of fatty acids was lower in cold-adapted than in control rats, but since the glycerol release was lower, the brown fat in cold-adapted rats does not seem to have been fully stimulated; Glass (1978, personal communication) found an NE-stimulated fatty acid release also in cold-adapted rats].

In their investigation of fatty acid release from brown fat cells, Bieber *et al.* (1975) concluded that brown fat cells may have "also a secretory action" in that "in addition to generating heat, brown adipose tissue also can supply the circulatory system with free fatty acids." Even the highest current estimates of the capacity of heat production in the brown fat cells conclude that only 60% of the heat production during nonshivering thermogenesis takes place in the brown fat itself. The extra work performed by the heart and the respiratory muscles also needs substrate (and produces heat) and this substrate may come from brown fat in the form of released fatty acids. Thus, "brown fat produces some of 'its' heat by delivering substrate for combustion in other tissues" (Cannon *et al.*, 1978).

C. Glycerol

The release of glycerol from brown fat cells may be as high as 50 nmole/minute/million cells (Forn *et al.*, 1970). The glycerol released is the backbone of hydrolyzed triglycerides; it is generally assumed that all glycerol produced by lipolysis is released, but some may be phosphorylated by glycerol kinase; in rat cells the activity of this enzyme may be as high as 10 nmole/minute/million cells (calc.; Bertin, 1976) under conditions of saturating glycerol concentration; in rabbit and hamster cells, the activity is probably lower or absent.

The glycerol released by the brown fat cell *in vivo* is probably phosphorylated and metabolized in the liver (Lin, 1977).

D. CO_2 and H_2O

The end products of respiration are naturally also released from brown fat. At least CO_2 has been suggested to be more than a waste product (Malan *et al.*, 1980).

E. Other Investigated Exports

Lactate may be released from brown fat cells after insulin or catecholamine stimulation (Fain and Loken, 1969), but conditions of unrestricted oxygen supply to the cells may not have been fulfilled. Oligomycin, which inhibits oxidative ATP production, increases the conversion of glucose to lactate (Fain *et al.*, 1970), probably as a consequence of an increased demand for glycolytic ATP.

Cholesterol synthesis within, and its probable release from, brown fat cells has been suggested by Dryer *et al.* (1969). Less than 0.1% of added labeled fatty acids (or less than 0.01% of acetate) was, however, converted to cholesterol within 1 hour of incubation; the physiological significance of cholesterol synthesis in and release from brown fat cells must thus be considered to be minimal.

Steroids. Ptak (1965) found that a labeled compound was released from brown fat incubated with labeled acetate—only in brown fat from adrenalectomized and castrated mice and only if pituitary glands also were present in the incubation. This compound cochromatographed with 11-deoxy-17α-hydroxy-corticosterone. However, the compound is *not* a steroid, but a monoglyceride (monoolein or monopalmitin) (Linck *et al.*, 1975).

Monoglycerides (monoolein and monopalmitin) are synthesized from acetate in brown fat incubations and may perhaps also be released from brown fat *in vivo* since small amounts are also found in plasma (Linck *et al.*, 1975). However, the extraction procedures used favor the extraction of compounds with properties like monoglycerides, and the quantitative role of any such release must be very small.

Acetate may be produced in isolated mitochondria from hamsters and lambs (Section VIII,A), but there is so far no evidence that cells *in vitro* or brown fat *in situ* release acetate. The concentration of acetate in the blood of hamsters is around 2 m*M*, but there is no apparent correlation between the amounts of acetate and the physiological state of the hamster (Bernson, 1979). The amount of acetate in the plasma in rats (which are not able to produce acetate in brown fat mitochondria) is, however, only one-tenth of that in hamsters (Bernson, 1979).

Ketone bodies. Dryer and Harris (1975) found that 2% of carbon-label added as palmitate was recovered in the ketone bodies acetoacetate and 2-hydroxy-butyrate, and a further 0.5% in hydroxymethylglutarate, which may be a ketone body precursor. These amounts of ketone body formation are in good agreement with the statement of Bernson and Nicholls (1974) that ketone body formation must be less than 4% of added palmitate. Thus although a weak capacity for ketone body formation *may* exist in brown fat, it is clear that at least hamster mitochondria release acetate under those conditions which in liver mitochondria lead to ketone body formation.

Hormonal factors. A series of experiments by Himms-Hagen (1969, 1970, 1974) may be interpreted to show that "brown fat secretes a factor that acts upon other tissues to permit them to increase their oxygen uptake when NE-stimulated," but such a factor has not been further characterized.

Immunological factors. A series of experiments (e.g., Jankovič *et al.*, 1977) have given some evidence that may be interpreted as showing that brown fat cells are involved in immunological responses; such phenomena have not been further studied.

Vitamins. Brown adipose tissue contains rather high amounts of ascorbic acid (Pagé and Babineau, 1951). This ascorbic acid is, however, probably not associated with brown fat cells but with the catecholamine vesicles in the nerve fibers in the tissue (cf. the situation in chromaffin granules), and no further role of the ascorbic acid needs to be claimed.

F. Conclusion

As seen above there is little reason to consider anything but heat and probably fatty acids as true products of the fat cells; glycerol, water, and CO_2 can be considered waste products and no other postulated release has so far been shown to have any physiological significance.

Under certain conditions even heat can be considered to be a waste product from the brown fat cell: during diet-induced nonshivering thermogenesis, brown fat is working in order to combust excess foodstuffs and heat is then not an aim in itself, but a secondary product.

XI. Anabolic Processes

Anabolic processes in brown fat cells have received much less attention than the catabolic processes immediately leading to thermogenesis. They are naturally as important for sustained thermogenesis as is the catabolic cascade from catecholamine binding to heat production, if in no other way, then at least to supply substrate to replace combusted lipids, so that thermogenesis is not to cease due to lack of substrate.

A. Free Fatty Acid Uptake and Esterification

Fatty acid uptake as such is difficult to measure, but the rate of esterification gives a minimal rate of fatty acid uptake.

If fatty acids are added to a cell incubation, much of the palmitate is within minutes found in triglycerides (Prusiner *et al.*, 1968c; Lindberg *et al.*, 1976). The rate of esterification can be calculated to be around 5 nmoles/minute/million cells.

The esterification may be through the monoglyceride pathway or the glycerol 3-phosphate pathway. Schultz and Johnston (1971) examined these pathways and found that the cellular esterification of added fatty acid proceeded as rapidly with a mono-oleyl-ether added to provide the glycerol backbone as with glucose. These results with isolated hamster cells were in accordance with results obtained with a microsome preparation which gave the same rate of incorporation with glycerol 3-phosphate and mono-oleyl-ether (Schultz and Johnston, 1971). The relatively high capacity of the monoglyceride pathway is in contrast to white fat and to liver. Schenk *et al.* (1975a), however, found no difference in specific activity of fatty acids in different glycerides after a pulse of labeled fatty acids in rabbits, and concluded that in rabbit only the glycerol 3-phosphate pathway for triglyceride synthesis is functioning.

Schenk *et al.* (1975a) stated that "plasma free fatty acids do not serve as a substrate for thermogenesis." This statement was based on the observation that the rate of esterification of labeled fatty acids from plasma into glycerides was lower in cold-exposed than in control rabbits. However, those fatty acids which have actually been combusted cannot be detected in the glycerides; it is not even known whether—as proposed by Schenk *et al.* (1975b)—all fatty acids taken up must first be esterified (before they are burned) or whether they may pass directly to the mitochondria.

B. Uptake of Esterified Fatty Acids: Lipoprotein Lipase

The uptake of fatty acids from circulating chylomicrons and very-low-density lipoproteins is accomplished by the action of lipoprotein lipase (Fig. 20).

Although this lipase is thought to be synthesized within the fat cells themselves, it has not as yet been studied in isolated cells. The activity examined has been that of acetone-ether-extracted tissue; the activity has been tested against artificial substrates and is thus difficult to evaluate in a physiological context. Guerrier and Pellet (1979) purified the enzyme 2000 times over the

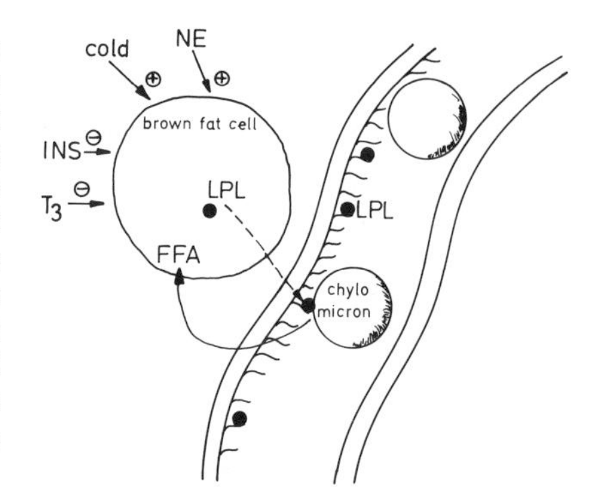

Fig. 20. The action of lipoprotein lipase (LPL) in brown adipose tissue. The lipase is probably synthesized within the brown fat cells as a response to different external stimuli; those indicated are those applied *in vivo*; no *in vitro* studies exist. The lipase is exported from the cells and binds to heparin-like glycoproteins on the epithelial cell and is thus anchored to the wall of the capillary. Here it breaks down the lipids of chylomicrons and very-low-density lipoproteins, and the free fatty acids (FFA) enter the cell (see Section XI,B).

crude acetone-ether-extracted powder by using a heparin-Sepharose affinity column. The purified lipase had a molecular weight of 67,000; it was, as would be expected of a lipoprotein lipase, activated by serum and inhibited by 0.4 *M* NaCl.

Within less than 1 day of cold exposure of rats the total activity of lipoprotein lipase is increased eightfold (Abe and Yoshimura, 1972; Radomski and Orme, 1971; Bertin *et al.*, 1978; Guerrier and Pellet, 1979); as the amount of brown fat increases during the process of cold adaptation, the specific activity decreases, but the total activity in the brown fat remains constant (Radomski and Orme, 1971).

The regulation of this impressive increase in activity is not known. Radomski and Orme (1971) found that injected NE—although leading to some increase—could not by far mimic the effect of cold stress; other hormones tested (insulin and T_3) diminished the cold-induced activity. Bertin *et al.* (1978) found that the activity of lipoprotein lipase was not well correlated with changes in tissue levels of cAMP or cGMP.

Thus the regulation of the activity of this enzyme, which under prolonged cold stress must do for the cell what the hormone-sensitive lipase does under acute cold stress, is not understood at all.

C. Fatty Acid Synthesis

1. *From Acetate*

Acetate incorporation into fatty acids by brown fat cells is "brisk" (Angel, 1969).

The enzymatic steps of the "cytosolic" (probably partly in endoplasmic reticulum) fatty acid synthesis have been followed (Fig. 21). The first step is catalyzed by acetyl-CoA-synthetase, which has a capacity close to that of liver (Hahn and Drahota, 1966). The malonyl-CoA-forming acetyl-CoA-carboxylase, the activity of which may be doubled by insulin, has an activity five times higher than liver per gram wet weight (McCormack and Denton, 1977), and fatty acid synthesis by microsomes directly from acetyl-CoA is much higher than in liver or white fat (Giacobino and Farvarger, 1971). Fatty acid synthesis from malonyl-CoA is also higher than in liver (Drahota *et al.*, 1970).

Also the mitochondrial chain elongation system, which adds acetyl-CoA directly to the growing chain, has been studied in brown fat and found to be be as active or more active than that of liver (Drahota *et al.*, 1970; Giacobino and Farvarger, 1971; Hinsch *et al.*, 1976). Hinsch *et al.* suggest that the mitochondrial chain elongation system is mainly functioning as a "pseudo-transhydrogenase"; the "real" transhydrogenase is not found in brown fat mitochondria (Lindberg *et al.*, 1967).

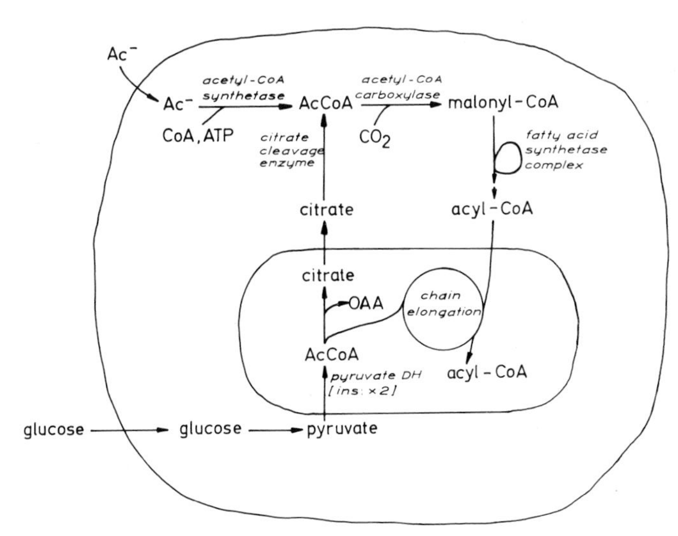

FIG. 21. Pathways of fatty acid synthesis. Acyl-CoA—the initial product of fatty acid synthesis—may be formed from 2C [acetate(Ac^-)], 3C [pyruvate (or lactate)], or 6C (glucose) precursors. pyruvate DH, Pyruvate dehydrogenase, the activity of which is doubled by insulin; OAA, oxaloacetic acid. Citrate cleavage enzyme is also known as ATP-citrate lyase.

In isolated cells the incorporation of acetate into fatty acids is inhibited by NE (Steiner and Evans, 1976). The mechanism for this antianabolic effect of NE is not known; the reported direct influence (Giacobino, 1971) of dibutyryl-cAMP on fatty acid synthesis in a cytosolic supernatant does not seem of a direct character: 0.5 m*M* is needed for half-maximal inhibition and cAMP itself, even at 5 m*M*, is without effect.

The physiological role of fatty acid synthesis from 2-carbon (acetate) or 2 × 2-carbon (ketone body) precursors in the adult animal is unknown. Regarding the fetus, Seccombe *et al.* (1977b) have suggested that in animals such as rat and man, which have a poor transfer of fatty acids over the placenta, fatty acid transfer may occur in the form of ketone bodies, and indeed these authors found a rapid uptake of 2-hydroxy-butyrate into the fetus, and the highest amount of incorporation into triglycerides was actually found in fetal brown fat. It may be suggested that acetate (Section VIII,A), produced by *maternal* brown fat, could do the same; indeed acetate can go into fetal lipids (Goldwater and Stetten, 1947).

2. *From Glucose*

Cells synthesize fatty acids from glucose (Fain *et al.*, 1967). The enzymatic steps (Fig. 21) include pyruvate dehydrogenase, the activity of which is increased by insulin (McCormack and Denton, 1977), and the citrate cleavage enzyme, which has been found in brown fat (Hahn and Drahota, 1966).

D. Glucose Uptake and Anabolism

Glucose passes the cell membrane by a "facilitated diffusion" process, the rate of which can be stimulated by insulin (Czech *et al.*, 1974a). However, the capacity of this carrier-mediated diffusion is not high, and already at incubation concentrations of 2.5 m*M*, the non-carrier-mediated entry is as rapid as the carrier-mediated.

The above observations were performed as a direct measurement of methylglucose accumulation into the cells. Commonly, the rate of glucose uptake is equated with the rate of CO_2 production from glucose, e.g., the microtubuli-inhibitor colchicine (but also the inactive analog trimethoxybenzene) halves the conversion of glucose to CO_2 by isolated cells (Rosenthal, 1979), but it is in reality not known whether this is an effect on uptake or metabolism.

Glycogen synthesis appears to occur only in the perinatal stage.

E. Gluconeogenesis

The enzymatic capacity for glucose synthesis from 3-carbon presursors is present in brown fat; the specific anabolic steps of reversed glycolysis (Fig. 22) are mitochondrial pyruvate carboxylase (Cannon and Nedergaard, 1979a), and cytosolic malate dehydrogenase, phosphoenolpyruvate carboxy-

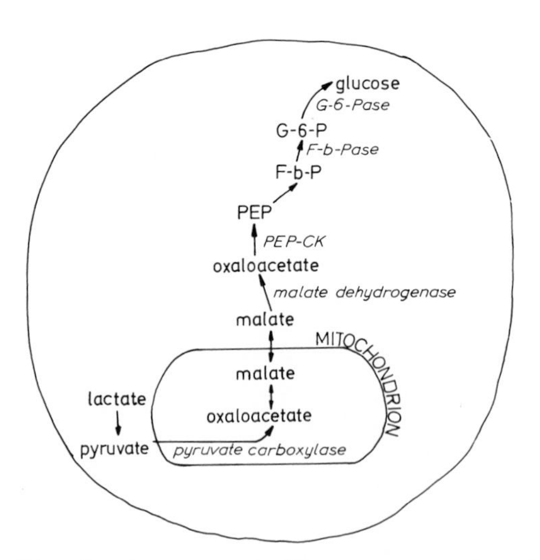

Fig. 22. The pathway for gluconeogenesis. The enzymes named are those necessary for the reversal of glycolysis. PEP-CK, Phospho*enol*pyruvate carboxykinase; F-b-Pase, fructose-bis-phosphatase (earlier known as fructose-1,6-diphosphatase, FDPase); G-6-Pase, glucose 6-phosphatase (see Section XI,E).

kinase (Frohlich *et al.*, 1976), fructose-bis-phosphatase (Seccombe *et al.*, 1977a) and glucose 6-phosphatase; some of these enzymes have been demonstrated to be present as indicated, although with very different capacities, and whether gluconeogenesis may actually occur in or be of any significance for brown fat cells is not known.

F. Amino Acid Uptake and Protein Synthesis

Only reports of a very preliminary character exist (Romert and Cannon, 1978).

XII. The Organelles and Storage Granules of the Brown Fat Cell

A. Triglyceride Droplets

The multiple triglyceride droplets within the brown fat cell are not surrounded by a unit membrane, but only a single electron-dense line; the biochemical nature of this line is not known, but it may consist of amphiphilic partial glycerides, fatty acids, and phospholipids. During acute cold stress of rats, the surface density of these droplets increases from 0.4 to 0.8 m^2/cm^3, but this process takes hours, whereas lipolysis is maximally stimulated within minutes, so the function of the increase in surface density is not known (Ahlabo and Barnard, 1974).

B. Glycogen Granules

Glycogen granules can be seen in brown fat cells from fetal brown fat (Barnard and Skala, 1970), but glycogen cannot normally be chemically detected in brown fat from nonfetal tissue (our unpublished observations).

C. Golgi Apparatus

A small Golgi apparatus may be seen in brown fat cells (Lindberg *et al.*, 1967) but nothing is known of its function.

D. Endoplasmic Recticulum

Although at first glance the brown fat cell may seem to have "an almost complete lack of endoplasmic reticulum" (Lindberg *et al.*, 1967), membrane-bound vesicles and cisternae may be seen close to the triglyceride droplets, and these formations appear to be smooth endoplasmic reticulum (Ahlabo and Barnard, 1974).

The close spatial relationship between the endoplasmic reticulum and triglyceride droplets may be related to the possibility that fatty acid synthesis takes place in the reticulum; biochemical studies of fatty acid synthesis have often been performed with a microsomal fraction (Schultz and Johnston, 1971; Giacobino and Farvarger, 1971).

E. Lysosomes

The lysosomal enzyme "acid phosphatase" has been found to increase during adaptation to cold (Nedergaard *et al.*, 1980), and this may be related to increased protein turnover (Bukowiecki and Himms-Hagen, 1971).

F. Peroxisomes

Ahlabo and Barnard (1971) found in electron microscopic studies small electron-dense particles within brown fat cells. The particles were 0.1–1 μm and often clustered; they did not have any crystalloid core, but they did stain with diaminobenzidine, and the staining was inhibited by aminotriazole; these histochemical reactions indicate that the organelles are peroxisomes, and, due to the lack of core, the organelles belong to what Novikoff and Novikoff (1972) have called microperoxisomes. Within cells from normal rats they are only about 1/50 as numerous as are mitochondria, but in cells from cold-adapted rats, their relative number increases five times.

Pavelka *et al.* (1976) examined the tissue for peroxisomal enzyme activities, but found only little 2-hydroxy-acid-oxidase and no D-alanine- or glycolate-oxidase at all. Catalase was easily detected, and in differential centrifugation, the specific activity of catalase was somewhat higher in a peroxisomal (20,000 *g*, 20 minutes) than in a mitochondrial fraction. In isopycnic centrifugations in sucrose gradients, catalase activity was highest at a density of 1.20; here also some carnitine-acetyltransferase activity was found, and as the amount was higher than would be expected from mitochondrial contamination, the authors concluded that also brown fat peroxisomes contain carnitine-acetyltransferase, as is the case for liver peroxisomes (Markwell *et al.*, 1973).

A more physiologically interesting enzyme system found in brown fat peroxisomes is the β-oxidation system, which probably primarily oxidizes long-chain unsaturated fatty acids (see Section VII,E).

Adaptation to cold leads to an increased activity of peroxisomal enzymes relative to mitochondrial (Nedergaard *et al.*, 1980); interestingly the increase in relative activity is about five times, i.e., the same change Ahlabo and

Barnard (1971) observed in the relative number of peroxisomes, and both Pavelka *et al.* (1976) and Nedergaard *et al.* (1980) conclude that for studies on the proliferation of peroxisomes, brown fat should be an interesting tissue.

G. PLASMALEMMA

Giacobino and Perrelet (1971) made a plasmalemma fraction from brown fat. They placed the "microsomal" (105,000 *g*, 90 minutes) fraction, obtained from heavily homogenized brown fat, onto a discontinuous Ficoll gradient and collected a band at a density of 1.066; the yield was naturally very small, only 1 mg protein from 3 gm brown fat. This fraction contained the plasmalemma marker enzyme 5-nucleotidase.

The plasmalemma fraction also contained NADH-diaphorase activity, the specific activity of which was half that found in the microsomal fraction on the same gradient; this is probably the result of contamination of the plasmalemma fraction with some microsomes, but it naturally cannot be ruled out that it is, as the authors discuss, a true intrinsic enzyme of the plasma membrane.

The activity of the Na^+,K^+-ATPase in the fraction was high, about 1 μmole/minute/mg protein. Notably, in the light of one hypothesis of the mechanism for thermogenesis in brown fat (Section XIII,D), the activity in a similar fraction from white fat was the same (Giacobino and Perrelet, 1971). The activity of the Na^+,K^+-ATPase is partially inhibited by ouabain (Herd *et al.*, 1970).

H. MITOCHONDRIA

The mitochondria are by far the most prominent organelles in the brown fat cell (Fig. 1). They contain at least one-third of the total amount of protein in the cell (Roberts and Smith, 1967), and occupy at least 30% of the cell volume (50% of the triglyceride-free volume) (Lindgren and Barnard, 1972); their mean volume is 0.4 μm^3, but within the mitochondrion there is so much inner membrane that each mitochondrion has 12 μm^2, about five times more than the calculated outer surface. There are 700 mitochondria per 1000 μm^3 brown fat cell, i.e., some 10,000 within one cell. The resulting high concentration of respiratory cytochromes is partly responsible for the color of the tissue.

Mitochondria are by far the most studied organelles in the brown fat cells, and the number of studies on brown fat mitochondria probably outnumbers all other studies on brown adipose tissue. These have also recently been amply reviewed (Flatmark and Pedersen, 1975; Cannon and Lindberg, 1979;

Nicholls, 1979; Lindberg *et al.*, 1981), and we shall here only review some of the most pertinent data. This mainly includes a discussion of the coupling state of the mitochondria in relation to the presence of thermogenin, and the low ATP synthetase activity of the mitochondria.

1. *The Coupling State of Mitochondria: The Presence of Thermogenin*

For the first preparations of isolated brown fat mitochondria, the same isolation procedure and the same incubation conditions as those developed for liver mitochondria were used. When traditional respiratory measurements were performed under these circumstances, high capacities for the oxidation of several substrates were demonstrated, but the mitochondria did not phosphorylate ADP by oxidative phosphorylation (Lepkovsky *et al.*, 1959; Smith *et al.*, 1966; Lindberg *et al.*, 1967). However, a series of investigators soon showed that the addition of albumin to the mitochondrial incubation made the mitochondria capable of oxidative phosphorylation (Guillory and Racker, 1968; Hohorst and Rafael, 1968) or could make them sensitive to artificial uncouplers like DNP (Joel *et al.*, 1967).

Also the combustion of "endogenous" fatty acids or fatty acid esters in the isolated mitochondria produced mitochondria which were coupled (Hittelman *et al.*, 1969). Thus the initial suggestions that brown fat mitochondria were uncoupled *in situ* seemed to have been inspired by an artifactual uncoupling during the preparation procedure.

However, Hohorst and Rafael (1968) had pointed out that even better respiratory control ratios were obtained if a nucleotide was present in the medium; GTP (at 2 m*M*) was more effective than ATP. Based on these observations, Pedersen (1970) tested the ability of different nucleotides at 1 m*M* concentrations to introduce coupling; he found that ATP, GTP, GDP, and ITP were about equally effective, whereas CTP and UDP did not have any effect. Therefore, when Cannon *et al.* (1973) observed that under all conditions were coupled mitochondria were obtained, purine nucleotides were present in the medium, it became clear that purine nucleotides affected brown fat mitochondria in a specific and unusual way. Cannon *et al.* (1973) demonstrated that the coupling effect of purine nucleotides was not influenced by oligomycin and atractylate, and it was not due to a chelating action of the nucleotides, thus it must result from the interaction of the nucleotides with the outside of the mitochondria and is not secondary to an energization of the mitochondria. A purine nucleotide binding site was also later measured on the mitochondria by Nicholls (1976c); it bound 0.7 nmole GDP/mg mitochondrial protein. Attempts to identify the binding polypeptide by affinity-labeled, tritium-labeled GTP were unsuccessful, probably due to a too low specific activity of the tritiated compound (Cannon, 1976). However, the amount of binding sites on the mitochondria was found by Rafael and Heldt (1976) to vary in parallel with the thermogenic state of the animal, and at the same time Ricquier and Kader (1976) found that a mitochondrial membrane polypeptide with a molecular weight of 32,000 also changed in parallel with the thermogenic activity of the animal. Later, Heaton *et al.* (1978) by using affinity-labeled, phosphate-labeled ATP with a high specific activity were able to conclude that the nucleotide-binding protein and the 32,000 polypeptide were one and the same.

Thus, purine nucleotides may bind to a probably brown fat unique polypeptide; this binding transfers mitochondria into a coupled state. We have introduced the name *Thermogenin* for this polypeptide (Cannon *et al.*, 1981) (see Fig. 23).

However, as pointed out by Cannon *et al.* (1973), the cytosolic purine nucleotides are always around the mitochondrion *in situ*, and as the coupling action of nucleotides is about the same for dinucleotides and trinucleotides, changes in energy potential do not influence their action until very low values are obtained, i.e., when most adenine is in the form of AMP. Thus changes in purine nucleotides are probably not a regulatory mechanism for the degree of coupling of the mitochondria *in situ*, "whereas variations in the supply of fatty acids which are known to occur during sympathetic stimulation of the tissue provide a more likely regulation for the degree of respiratory control exhibited by the tissue" (Cannon *et al.*, 1973). Later, Cannon *et al.* (1977b) demonstrated that not fatty acids themselves but *activated* fatty acids, e.g., palmitoyl-CoA, had more of the qualities required for an activator of thermogenin and, by this, for activation of thermogenesis (see Section XIII,E).

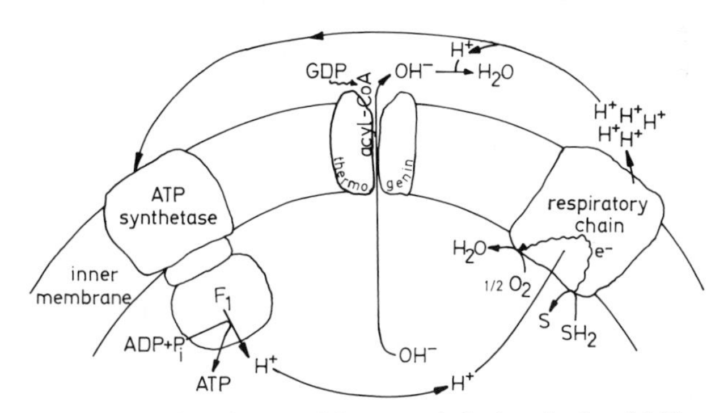

FIG. 23. Thermogenin. The existence of thermogenin in the mitochondrial inner membrane short-circuits the protonical current driven by the respiratory chain by allowing the exit of hydroxide OH^-. GDP and acyl-CoA compete for a binding site on thermogenin: in the presence of purine nucleotides such as GDP, the OH^- channel is closed; in their absence (or in the presence of acyl-CoA) the channel is open (see Section XII,H,I). F_1 indicates the ATPase unit. Adapted from Lindberg *et al.* (1981).

By using a modification of the ion-distribution method of Rottenberg (see Rottenberg, 1979, for review) for the determination of mitochondrial membrane potentials, Nicholls has been able to study the kinetics of the effects of the GDP binding on the energy status of the mitochondria in great detail; these results have been extensively reviewed (Nicholls, 1976a,b, 1979).

The existence of thermogenin also reveals itself as a high halide ion permeability of the mitochondria, i.e., the mitochondria swell in 100 m*M* solutions of Cl^- or Br^- if a permanent cation (NH_4^+ or K^+ + valinomycin) is present (Nicholls and Lindberg, 1973). As OH^- movement in the chemiosmotic theory for mitochondrial coupling (Mitchell, 1961) is indistinguishable from H^+ movement (Mitchell, 1966), Nicholls (1976a) has suggested that high halide ion permeability and high hydroxyl ion permeability (i.e., uncoupling) are two manifestations of the same property of brown fat mitochondria.

Due to the uncoupled state of brown fat mitochondria during isolation a series of "low contents" of compounds normally found in the matrix of isolated mitochondria has been reported. The compounds include water (Nicholls *et al.*, 1972), malate (less than 0.6 nmole/mg; Bernson and Nicholls, 1975), Ca^{2+} (Greenway and Himms-Hagen, 1978), and phosphate (Pedersen and Grav, 1972). All these compounds are probably lost during the isolation procedure, due to the uncoupled state, and there is no reason to think that the mitochondria *in situ* contain noteworthy low amounts of these compounds.

2. *ATP-Synthetizing Capacity*

Brown fat mitochondria from most species have a low ATP synthetase activity, relative to their amount of respiratory chain, both when measured as phosphate incorporation into ADP (or ADP stimulation of respiration) and when measured as ATPase activity (Smith *et al.*, 1966; Lindberg *et al.*, 1967; Bulychev *et al.*, 1972). This low activity is—as demonstrated by Cannon and Vogel (1977)—not due to any distinct qualities of the enzyme as such, but only to the enzyme being present in low amounts, as also shown by Houštěk and Drahota (1977) and Houštěk *et al.* (1978). As pointed out by Cannon and Vogel (1977), this low capacity for ATP production means that "significant heat production [can] only occur by uncoupling between ATP synthesis and respiration" (see Section XIII,E).

XIII. The Mechanism of Heat Production in the Brown Fat Cell

To convert foodstuff into heat in a regulated way is the main task of the brown fat cell. Heat production ability has had to be built within the same framework which in all *other* mammalian cells is constructed to work as *efficiently* as possible, i.e., per definition, by producing as *little* heat as possible. There have been several hypotheses as to how this has been accomplished; we shall discuss some of them here.

A. Substrate Supply

Based on the early observations that brown fat mitochondria were uncoupled when isolated, and implying that this was also the case *in situ*, Smith and Horwitz (1969) suggested that the combustion in brown fat cells was controlled by substrate supply. As later investigations have shown that the mitochondria may be, but are not always, uncoupled, this hypothesis cannot be maintained, although naturally, as pointed out by Horwitz (1978a) if the mitochondria were to become fully uncoupled, "oxygen consumption will be controlled by substrate availability."

B. Internal Alkalinization

Chinet *et al.* (1978) suggested that an internal alkalinization was the stimulator of thermogenesis. NE should stimulate the Na^+,K^+-ATPase and this stimulation should in some way be coupled to an effiux of protons from the cell. An increased pH within the cell then would regulate heat production, e.g., by decreasing the affinity of thermogenin protein for GDP (Nicholls, 1976c).

The experimental evidence for this hypothesis is so far limited (Chinet *et al.*, 1978).

C. K^+-Transport

Reed and Fain (1968a) concluded that the mitochondria in brown fat cells are coupled when the cells are not stimulated by catecholamines, and that the addition of fatty acids or cAMP uncouples them. Reed and Fain (1968b) then state that "rapid operation of a loosely coupled transport scheme would dissipate energy as heat and is a mechanism which could account for thermogenesis in brown fat cells." As a candidate for the transported compound they suggested a comblex involving fatty acid and K^+ (Reed and Fain, 1968b); but the molecular nature of such a complex has not been further elucidated.

D. Stimulated Na^+, K^+-ATPase

Edelman and Ismail-Beigi (1974) suggested that the increase in basal metabolism observed in hyperthyroid animals may be due to an increased

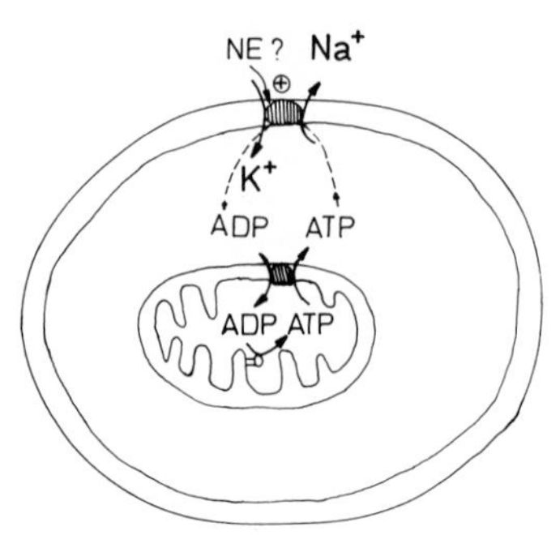

FIG. 24. The Na^+,K^+-ATPase hypothesis for heat production in brown fat (see Section XIII,D).

activity and perhaps an increased number of Na^+,K^+-ATPases in the plasmalemma of peripheral cells.

Horwitz has in several contexts discussed this possibility in brown fat (Fig. 24). NE should stimulate the Na^+,K^+-ATPase and ADP should then be returned to the mitochondria for rephosphorylation. In this form, the capacity both of the ATP-producing and of the ATP-utilizing mechanism must be high enough to account for the rate of oxygen consumption within brown fat.

As seen in Table IV, the rate of oxygen consumption in stimulated brown fat is about 100 μmole/minute/gm wet weight; if this should be used for oxidative phosphorylation, 300 μmole ATP should be synthesized and utilized per minute. Horwitz and Eaton (1975) have measured the activity of Na^+,K^+-stimulated ATPase to be (under optimal conditions) 10 μmole/hour/mg N in their supernatant fraction; if the activity in a brown fat homogenate is estimated to be as great as that in the supernatant, this activity amounts to around 5 μmole O/minute/gm wet weight, i.e., about 2% of that required. Similarly, the capacity of mitochondrial ATP synthetase in hamster can be calculated to be 0.1 μmole/minute/mg protein at 37°C (Lindberg *et al.*, 1976); if 135 mg mitochondrial protein is found in 1 gm brown fat, this rate only amounts to 14 μmole/minute, again clearly lower than expected from the rate of oxygen consumption.

Thus, both the ATP-generating and the ATP-utilizing capacities are too small to permit this futile cycle to be the heat-generating process in brown fat. This was also the conclusion of Chinet *et al.* (1977), who observed that by blocking the activity of the Na^+,K^+-ATPase by ouabain, only a 5% decrease in the rate of heat production was observed; these authors, however, did not investigate the tissue when stimulated by NE.

If, as recently discussed by Horwitz (1978a), a stimulation of Na^+,K^+ is necessary to take care of ATP produced by substrate level phosphorylation, or if it generates some kind of intracellular signal (Girardier *et al.*, 1968), the above calculations are without interest; such effects have, however, not been clearly demonstrated so far, and in any case the coexistence of another mechanism for heat production is needed.

E. Partial Uncoupling

In this hypothesis the main postulate is that the heat production of the cell is independent of ATP synthesis and breakdown, and the energy released by combustion of fat is dissipated as heat in a more direct way, or, as first stated by Lindberg *et al.* (1966): "Particular attention is being directed to the possibility that heat production may constitute an additional energy-linked process that derives energy directly from intermediates of the respiratory chain-linked energy transfer system." This hypothesis, as sketched above, has with time gained support from a series of experiments and can now be presented as shown in Fig. 23.

According to this hypothesis, the mitochondria within the cell are in a "normal" coupled state when the cell is not stimulated; the H^+ (OH^-) conducting thermogenin hole is inhibited by cytosolic purine nucleotides.

The stimulation of heat production then occurs by addition of NE; a stimulation can also be obtained by addition of fatty acids; "the mimicking of the norepinephrine effect by oleate demonstrates that a rise in the intracellular free fatty acid concentration is probably the only stimulus needed to increase the rate of respiration" (Prusiner *et al.*, 1968b). This statement is the "minimal theory" that no effects of NE other than a stimulated lipase are necessary for thermogenesis to occur (Bieber *et al.*, 1975).

The released fatty acids probably do not function as such; rather they are activated to acyl-CoA, which in isolated mitochondria can interact with GDP-binding site in a specific way. This is demonstrated by the following observations.

Palmitoyl-CoA in concentrations of 1–10 nmole/mg protein inhibits GDP binding in a competitive way with an estimated K_i of 2 μM (Cannon *et al.*, 1977b); palmitoyl-CoA can reintroduce swelling (i.e., activate thermogenin) in mitochondria in a manner competitive with GDP (Cannon *et al.*, 1977b); palmitoyl-CoA increases permeability only to halide ions (phosphate ion and cation permeability is not introduced in a specific manner) (Cannon *et al.*, 1977b; 1980), i.e., palmitoyl-CoA specifically activates thermogenin.

Thus (Fig. 25) fatty acids, produced by stimulation of the hormone-sensitive lipase, are activated in the mitochondrial outer membrane; these acyl-CoAs have a double function, in being both the substrate for β-oxidation

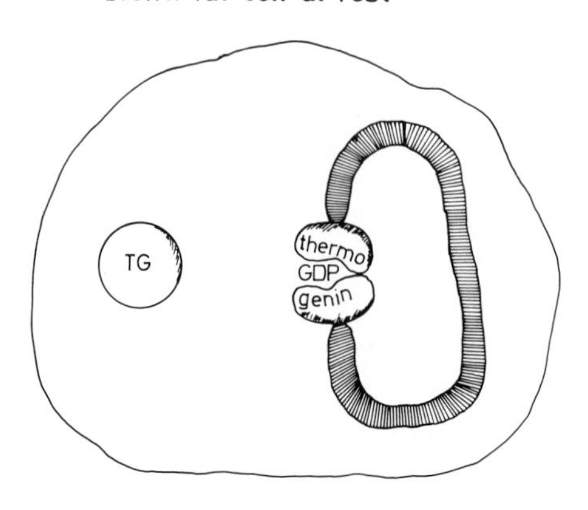

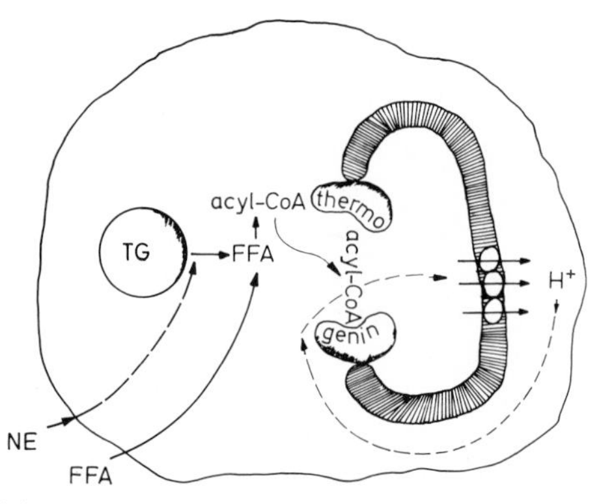

FIG. 25. The partial uncoupling theory for heat production in the brown fat cell. In the resting cell the thermogenin channel is closed by cytosolic purine nucleotides, here exemplified by GDP. In the stimulated cell, the intracellular level of free fatty acids (FFA) is increased—either by addition to the medium or by stimulation of triglyceride (TG) lipolysis with NE. Some acyl-CoA may open the thermogenin channel, and the mitochondria become uncoupled. This is here shown as a H^+ circuit; cf. however Fig. 23 (see Section XIII,E).

and, perhaps, the physiological uncoupler of brown fat mitochondria. Perhaps, because it must be stressed that although acyl-CoA fullfils the criteria which can be set for a physiological uncoupler, it still has not been demonstrated that it functions in this way within the brown fat cell itself.

In a scheme for the activation of heat production as outlined here, "an ideal metabolic state would appear to exist" where the mitochondria "maintained a high rate of respiration and a relatively low rate of phosphorylation" (Smith *et al.*, 1966) and energy dissipation as heat would easily proceed.

XIV. Conclusions and Perspectives

As seen in the preceding sections, the amount of information collected on the brown fat cells is now—only 13 years after they were first isolated—very impressive (a summary is given in Section I,C). However, as is so often the case, the broadened knowledge merely emphasizes our ignorance. Indeed, for each of the sections, major questions are still unsolved; examples should include the following:

Regarding the isolation of the cells (Section II) the yield is still uncomfortably low and the suspension not sufficiently clean. A more important problem is the inability to find in isolated cells the differences in hormonal response that would be predicted on the basis of their thermogenin content. Regarding hormonal effects (Section IV) the differentiation between α- and β-adrenergic responses is not clarified, and the physiological importance of spare receptors for adrenergic stimulation and of "spare adenylate cyclases" is far from understood. We know nothing about how the insulin message is communicated within the cell and nothing about the physiological significance of glucocorticoid effects. We know that cells from some species may respond thermogenically to glucagon but not what the *in vivo* significance of this is. Regarding second messengers (Section V) it is unknown whether the changes of membrane potential are in themselves of significance or whether they are just secondary phenomena. So far no effects of changes in cGMP levels have been demonstrated and there is no conclusive evidence that intracellular Ca^{2+} acts as a second messenger. Regarding the catabolic cascade (Section VI) the number of different protein kinases remains to be settled, and concerning question of the fate of fatty acids (Section VII) there is still only circumstantial evidence for a significant export of fatty acids from the brown fat cell to the rest of the body. The role of peroxisomal β-oxidation in relation to mitochondrial β-oxidation is not established. Although brown fat mitochondria *in vitro* can produce acetate from acetyl-CoA (Section VIII), such production has not been demonstrated in cells or *in vivo*. The regulation of fatty acid uptake (Section XI) and lipoprotein lipase activity is unknown, and there is so far no demonstration of a gluconeogenic capacity of the cells, although they seem to have the necessary enzymes. Finally, although the partial uncoupling theory for the mechanism of heat production is now established in isolated mitochondria and in part in cells, there is presently no direct evidence that acyl-CoA is the intracellular activator of thermogenin and by this of thermogenesis.

However, perhaps the most fascinating problem is not the function of the cells—but their formation. The proliferation of brown fat which is seen before birth, before hibernation, in the cold or as a response to different diets is probably caused by a common initiator. The molecular nature of this initiator is not known; there are indications that it may simply be NE (Heick *et al.*, 1973; Behrens and Depocas, 1977; see Barnard *et al.*, 1980, for review). However, all studies so far have been performed *in vivo*, and many secondary effects may occur. The time has now come when investigations into brown fat cellular proliferation must be conducted; the information collected from work with isolated mature and thermogenic cells has taught us what to look for.

There is no doubt that an understanding of the factors which regulate brown fat cellular proliferation is of more than academic interest. If we were able to manipulate the activity of brown adipose tissue, this new knowledge could probably be utilized in at least three different areas. It may be possible to make humans with serious problems of obesity less efficient in food utilization by stimulating their brown fat cells. It may similarly be possible to make animals more efficient in food utilization and thus in a world with diminishing resources to increase the production of meat without increasing the consumption of primary products. And finally, it may be possible to increase our comfort in colder surroundings by stimulation of our own heat production in our own brown fat cells and through this to depress our utilization of nonrenewable energy resources.

Thus, further work with the brown fat cell may one day make it function under our own control for the benefit of mankind.

Acknowledgments

The authors would like to thank Barbara Cannon and the rest of the brown fat group at the Wenner-Gren Institute for many valuable discussions and much practical help. The illustrations were made by Elisabeth Palmér, and the manuscript patiently typed by Renate Ahlfont and Ulla-Britt Åkerblom. Our work is supported by the Swedish Natural Science Research Council.

References

Abe, L., and Yoshimura, K. (1972). *J. Physiol. Soc. Jpn.* **34,** 81–82.

Afzelius, B. A. (1970). *In* "Brown Adipose Tissue" (O. Lindberg, ed.), pp. 1–31. American Elsevier, New York.

Ahlabo, I., and Barnard, T. (1971). *J. Histochem. Cytochem.* **19,** 670–675.

Ahlabo, I., and Barnard, T. (1974). *J. Ultrastruct. Res.* **48,** 361–376.

Ahlquist, R. P. (1948). *Am. J. Physiol.* **153,** 586–599.

Aldridge, W. N., and Street, B. W. (1968). *Biochem. J.* **107,** 315–317.

Alexson, S., Nedergaard, J., Osmundsen, H., and Cannon, B. (1980). *Adv. Physiol. Sci.* **32,** 483–485.

Alkonyi, I., Kerner, J., and Sandor, A. (1975). *Acta Biochim. Biophys. Acad. Sci. Hung.* **10,** 217–220.

Al-Shaikhaly, M. H. M., Nedergaard, J., and Cannon, B. (1979). *Proc. Natl. Acad. Sci. U.S.A.* **76,** 2350–2353.

Ambid, L., Berlan, M., and Agid, R. (1971). *J. Physiol. (Paris)* **63,** 505–522.
Angel, A. (1969). *Science* **163,** 288–290.
Barde, Y. A., Chinet, A., and Girardier, L. (1975). *J. Physiol. (London)* **252,** 523–536.
Barnard, T., and Skála, J. (1970). *In* "Brown Adipose Tissue" (O. Lindberg, ed.), pp. 33–72. American Elsevier, New York.
Bernard, T., Mory, G., and Néchad, M. (1980). *In* "Biogenic Amines in Development" (S. Parvez and H. Parvez, eds.), pp. 391–439. Elsevier, Amsterdam.
Baumber, J., and Denyes, A. (1964). *Can. J. Biochem.* **42,** 1397–1401.
Behrens, W. A., and Depocas, F. (1977). *Can. J. Physiol. Pharmacol.* **55,** 695–699.
Berge, R. K., Slinde, E., and Farstad, M. (1979). *Biochem. J.* **182,** 347–351.
Berlet, H. H., Bonsmann, I., and Birringer, H. (1976). *Biochim. Biophys. Acta* **437,** 166–174.
Bernson, V. S. M. (1976). *Eur. J. Biochem.* **67,** 403–410.
Bernson, V. S. M. (1979). "Metabolic and Bioenergetic Events in Brown Adipose Tissue in Relation to Hibernation: Studies on Cells and Mitochondria." University of Stockholm. ISBN 91-7146-027-6.
Bernson, V. S. M., and Nicholls, D. G. (1974). *Eur. J. Biochem.* **47,** 517–525.
Bernson, V. S. M., and Nicholls, D. G. (1975). *In* "Depressed Metabolism and Cold Thermogenesis" (L. Janský, ed.), pp. 165–169. Charles University, Prague.
Bernson, V. S. M., Lundberg, P., and Pettersson, B. (1979). *Biochim. Biophys. Acta* **587,** 353–361.
Bertin, R. (1976). *Biochimie* **58,** 431–434.
Bertin, R., and Portet, R. (1976). *Eur. J. Biochem.* **69,** 177–183.
Bertin, R., Gaubern, M., and Portet, R. (1978). *Experientia Suppl.* **32,** 185–189.
Beviz, A., Lundholm, L., and Mohme-Lundholm, E. (1968). *Br. J. Pharmacol.* **34,** 198–199.
Bieber, L. L., Pettersson, B., and Lindberg, O. (1975). *Eur. J. Biochem.* **58,** 375–381.
Brück, K. (1978). *In* "Perinatal Physiology" (U. Stave, ed.), pp. 493–557. Plenum, New York.
Bukowiecki, L., and Himms-Hagen, J. (1971). *Can. J. Physiol. Pharmacol.* **49,** 1015–1018.
Bukowiecki, L., and Lindberg, O. (1974). *Biochim. Biophys. Acta* **348,** 115–125.
Bukowiecki, L., Follea, N., Vallieres, J., and LeBlanc, J. (1978a). *Eur. J. Biochem.* **92,** 189–196.
Bukowiecki, L., Caron, M. G., Vallieres, J., and LeBlanc, J. (1978b). *Experientia Suppl.* **32,** 55–60.
Bulychev, A., Kramer, R., Drahota, Z., and Lindberg, O. (1972). *Exp. Cell Res.* **72,** 169–187.
Burlington, R. F., Therriault, D. G., and Hubbard, R. W. (1969). *Comp. Biochem. Physiol.* **29,** 431–437.
Busuttil, R. W., Paddock, R. J., Fisher, J. W., and George, W. J. (1976). *Circ. Res.* **38,** 162–167.
Butcher, R. W., Baird, C. E., and Sutherland, E. W. (1968). *J. Biol. Chem.* **243,** 1705–1712.
Cameron, I. L., and Smith, R. E. (1964). *J. Cell Biol.* **23,** 89–100.
Cannon, B. (1971). *Eur. J. Biochem.* **23,** 125–135.
Cannon, B. (1976). *Int. Congr. Biochem. 10th* Abstr. No. 06-7-112.
Cannon, B., and Johansson, B. W. (1980). *Molec. Aspects Med.* **3,** 119–223.
Cannon, B., and Lindberg, O. (1979). *Methods Enzymol.* **55,** 65–78.
Cannon, B., and Nedergaard, J. (1978). *Experientia Suppl.* **32,** 107–111.
Cannon, B., and Nedergaard, J. (1979a). *Eur. J. Biochem.* **94,** 419–426.
Cannon, B., and Nedergaard, J. (1979b). *Biochem. Contact Meet., 15th, Norway.*
Cannon, B., and Nedergaard, J. (1982). *In* "Biochemical Development of the Fetus and the Newborn" (C. T. Jones, ed.), Elsevier, Amsterdam (in press).
Cannon, B., and Polnaszek, C. F. (1976). *In* "Regulation of Depressed Metabolism and Thermogenesis" (L. Janský and X. J. Musacchia, eds.). Thomas, Springfield, Illinois.
Cannon, B., and Vogel, G. (1977). *FEBS Lett.* **76,** 284–289.
Cannon, B., Nicholls, D. G., and Lindberg, O. (1973). *In* "Mechanisms in Bioenergetics" (G. F. Azzone, L. Ernster, S. Papa, E. Quagliariello, and N. Siliprandi, eds.), pp. 357–363. Academic Press, New York.

Cannon, B., Polnaszek, C. F., Butler, K. W., Eriksson, L. E. G., and Smith, I. C. P. (1975). *Arch. Biochem. Biophys.* **167,** 505–518.
Cannon, B., Romert, L., Sundin, U., and Barnard, T. (1977a). *Comp. Biochem. Physiol.* **56B,** 87–99.
Cannon, B., Sundin, U., and Romert, L. (1977b). *FEBS Lett.* **74,** 43–66.
Cannon, B., Nedergaard, J., Romert, L., Sundin, U., and Svartengren, J. (1978). *In* "Strategies in Cold: Natural Torpidity and Thermogenesis" (L. L. Wang and J. Hudson, eds.), pp. 567–594. Academic Press, New York.
Cannon, B., Nedergaard, J., and Sundin, U. (1980). *Adv. Physiol. Sci.* **32,** 479–481.
Cannon, B., Nedergaard, J., and Sundin, U. (1981). *In* "Survival in Cold" (X. J. Musacchia and L. Janský, eds.). Elsevier, Amsterdam (in press).
Caswell, A. H. (1979). *Int. Rev. Cytol.* **56,** 145–181.
Chalvardjian, A. M. (1964). *Biochem. J.* **90,** 518–521.
Chan, S. S., and Fain, J. N. (1970). *Mol. Pharmacol.* **6,** 513–523.
Cheung, W. Y. (1980). *Science* **207,** 19–27.
Chinet, A., Clausen, T., and Girardier, L. (1977). *J. Physiol. (London)* **265,** 43–61.
Chinet, A., Friedli, C., Seydoux, J., and Girardier, L. (1978). *Experientia Suppl.* **32,** 25–32.
Christiansen, E. N. (1971). *Eur. J. Biochem.* **19,** 276–282.
Christiansen, E. N. (1977). *Comp. Biochem. Physiol.* **56B,** 19–24.
Christiansen, E. N., and Wojtczak, L. (1974). *Comp. Biochem. Physiol.* **49B,** 579–592.
Christiansen, E. N., Drahota, Z., Duszyński, J., and Wojtczak, L. (1973). *Eur. J. Biochem.* **34,** 506–512.
Christiansen, R. Z., and Bremer, J. (1976). *Biochim. Biophys. Acta* **448,** 562–577.
Ciaraldi, T., and Marinetti, G. V. (1977). *Biochem. Biophys. Res. Commun.* **74,** 984–991.
Colburn, R. W., and Maas, J. W. (1965). *Nature (London)* **208,** 37–41.
Costa, N. D., and Snoswell, A. M. (1975). *Biochem. J.* **152,** 161–166.
Czech, M. P., Lawrence J. C., Jr., and Lynn, W. S. (1974a). *J. Biol. Chem.* **249,** 5421–5427.
Czech, M. P., Lawrence J. C., Jr., and Lynn, W. S. (1974b). *J. Biol. Chem.* **249,** 7499–7505.
Dawkins, M. J. R., and Hull, D. (1964). *J. Physiol. (London)* **172,** 216–238.
de Beer, L., and de Schepper, P. J. (1967). *Biochem. Pharmacol.* **16,** 2355–2367.
Depocas, F., Behrens, W. A., and Foster, D. O. (1978). *Can. J. Physiol. Pharmacol.* **56,** 168–174.
Derry, D. M., Schönbaum, E., and Steiner, G. (1969). *Can. J. Physiol. Pharmacol.* **47,** 57–63.
Doniach, D. (1975). *Lancet* **1,** 160–161.
Drahota, Z., Honová, E., and Hahn, P. (1968). *Experientia* **24,** 431–432.
Drahota, Z., Coratelli, P., and Landriscina, C. (1970). *FEBS Lett.* **6,** 241–244.
Dryer, R. L., and Harris, R. R. (1975). *Biochim. Biophys. Acta* **380,** 370–381.
Dryer, R. L., Brown, D. J., Paulsrud, J. R., and Mavis, K. K. (1969). *Fed. Proc. Fed. Am. Soc. Exp. Biol.* **28,** 948–954.
Dryer, R. L., Lindberg, O., and Pettersson, B. (1971). *In* "Nonshivering Thermogenesis" (L. Janský, ed.), pp. 209–220. Academia Prague. Swetz & Zeitlinger, Amsterdam.
Dyer, R. F. (1968). *Am. J. Anat.* **123,** 255–282.
Edelman, I. S., and Ismail-Beigi, F. (1974). *Recent Prog. Hormone Res.* **30,** 235–257.
Ernster, L., and Luft, R. (1963). *Exp. Cell Res.* **32,** 26–35.
Fain, J. N. (1965). *Endocrinology* **76,** 549–552.
Fain, J. N. (1968). *Endocrinology* **82,** 825–830.
Fain, J. N. (1970). *Fed. Proc. Fed. Am. Soc. Exp. Biol.* **29,** 1402–1407.
Fain, J. N. (1975). *Methods Enzymol.* **35,** 555–561.
Fain, J. N., and Butcher, F. R. (1976). *J. Cyclic Nucleotide Res.* **2,** 71–78.
Fain, J. N., and Loken, S. C. (1969). *J. Biol. Chem.* **244,** 3500–3506.
Fain, J. N., and Loken, S. C. (1971). *Mol. Pharmacol.* **7,** 455–464.

Fain, J. N., and Reed, N. (1970). *Lipids* **5,** 210–219.
Fain, J. N., and Shepherd, R. E. (1975). *J. Biol. Chem.* **250,** 6586–6592.
Fain, J. N., Reed, N., and Saperstein, R. (1967). *J. Biol. Chem.* **242,** 1887–1894.
Fain, J. N., Loken, S. C., and Czech, M. P. (1970). *Biochim. Biophys. Acta* **197,** 40–48.
Fain, J. N., Rosenthal, J. W., and Ward, W. F. (1972). *Endocrinology* **90,** 52–59.
Fain, J. N., Jacops, M. D., and Clement-Cormier, Y. C. (1973). *Am. J. Physiol.* **224,** 346–351.
Feldman, D. (1977). *Am. J. Physiol.* **233,** E147–E151.
Feldman, D. (1978). *Endocrinology* **103,** 2091–2097.
Feldman, D., and Hirst, M. (1978). *Am. J. Physiol.* **235,** E197–E202.
Feldman, D., and Loose, D. (1977). *Endocrinology* **100,** 398–405.
Fink, S. A., and Williams, J. A. (1976). *Am. J. Physiol.* **231,** 700–706.
Flaim, K. E., Horwitz, B. A., and Horowitz, J. M. (1977). *Am. J. Physiol.* **232,** R101–R109.
Flatmark, T., and Pedersen, J. I. (1975). *Biochim. Biophys. Acta* **416,** 53–103.
Flatmark, T., Ruzicka, F. J., and Beinert, H. (1976). *FEBS Lett.* **63,** 51–55.
Forn, J., Gessa, G. L., Krishna, G., and Brodie, B. B. (1970). *Life Sci.* **9,** 429–435.
Foster, D. O., and Frydman, M. L. (1978). *Can. J. Physiol. Pharmacol.* **56,** 110–122.
Foster, D. O., Frydman, M. L., and Usher, J. R. (1977). *Can. J. Physiol. Pharmacol.* **55,** 52–64.
Friedli, C., Chinet, A., and Girardier, L. (1978). *Experientia Suppl.* **32,** 259–266.
Frohlich, J., Hahn, P., Kirby, L., and Webber, W. (1976). *Biol. Neonate* **30,** 40–48.
Giacobino, J.-P. (1971). *Life Sci.* **10,** 1089–1095.
Giacobino, J.-P., and Farvarger, P. (1971). *Biochimie* **53,** 253–259.
Giacobino, J.-P., and Perrelet, A. (1971). *Experientia* **27,** 259–261.
Girardier, L., and Seydoux, J. (1971). *Experientia* **27,** 720.
Girardier, L., Seydoux, J., and Clausen, T. (1968). *J. Gen. Physiol.* **52,** 925–940.
Goldwater, W. H., and Stetten, D. (1947). *J. Biol. Chem.* **169,** 723–738.
Greenway, D., and Himms-Hagen, J. (1978). *Am. J. Physiol.* **234,** C7–C13.
Grodums, E. I. (1977). *Cell Tissue Res.* **185,** 231–237.
Guerrier, D., and Pellet, H. (1979). *FEBS Lett.* **106,** 115–120.
Guillory, R. J., and Racker, E. (1968). *Biochim. Biophys. Acta* **153,** 490–493.
Gumpen, S. A., and Norum, K. R. (1973). *Biochim. Biophys. Acta* **316,** 48–55.
Hahn, P., and Drahota, Z. (1966). *Experientia* **22,** 706–707.
Hahn, P., and Novak, M. (1975). *J. Lipid Res.* **16,** 79–91.
Hahn, P., and Skála, J. (1972). *Biochem. J.* **127,** 107–111.
Hahn, P., and Skála, J. P. (1978). *Experientia Suppl.* **32,** 61–67.
Hahn, P., Drahota, Z., Skála, J., Kazda, S., and Towell, M. E. (1969). *Can. J. Physiol. Pharmacol.* **47,** 975–980.
Hamilton, J., and Horwitz, B. A. (1977). *Fed. Proc. Fed. Am. Soc. Exp. Biol.* **36,** 579, Abstr. No. 1625.
Hamilton, J., and Horwitz, B. A. (1979). *Eur. J. Pharmacol.* **56,** 1–6.
Hardman, M. J., and Hull, D. (1970). *J. Physiol. (London)* **206,** 263–273.
Hayward, J. S., and Davies, P. F. (1972). *Can. J. Physiol. Pharmacol.* **50,** 168–170.
Hayward, J. S., and Lyman, C. P. (1967). *In* "Mammalian Hibernation III" (K. C. Fisher *et al.*, eds.), pp. 346–355. American Elsevier, New York.
Heaton, G. M., and Nicholls, D. G. (1976). *Eur. J. Biochem.* **67,** 511–517.
Heaton, G. M., Wagenvoord, R. J., Kemp A., Jr., and Nicholls, D. G. (1978). *Eur. J. Biochem.* **82,** 515–521.
Heick, H. M. C., Vachou, C., Kallai, M. A., Bégin-Heick, N., and LeBlanc, J. (1973). *Can. J. Physiol. Pharmacol.* **51,** 751–758.
Heim, T., and Hull, D. (1966). *J. Physiol. (London)* **186,** 42–55.
Heldmaier, G., and Hoffmann, K. (1974). *Nature (London)* **247,** 224–225.

Herd, P. A., Horwitz, B. A., and Smith, R. E. (1970). *Experientia* **26,** 825–826.
Herd, P., Hammond, R. P., and Hamolsky, M. W. (1973). *Am. J. Physiol.* **224,** 1300–1304.
Himms-Hagen, J. (1969). *J. Physiol. (London)* **205,** 393–403.
Himms-Hagen, J. (1970). *Adv. Enzyme Reg.* **8,** 131–151.
Himms-Hagen, J. (1972). *Lipids* **7,** 310–323.
Himms-Hagen, J. (1974). *Can. J. Physiol. Pharmacol.* **52,** 225–229.
Himms-Hagen, J. (1979). *Can. Med. Assoc. J.* **121,** 1361–1364.
Hinsch, W., Klages, C., and Seubert, W. (1976). *Eur. J. Biochem.* **64,** 45–55.
Hittelman, K. J., Lindberg, O., and Cannon, B. (1969). *Eur. J. Biochem.* **11,** 183–192.
Hittelman, K. J., Bertin, R., and Butcher, R. W. (1974). *Biochim. Biophys. Acta* **338,** 398–407.
Hjelmdahl, P., and Fredholm, B. B. (1977). *Acta Physiol. Scand.* **101,** 294–301.
Hohorst, H.-J., and Rafael, J. (1968). *Hoppe Seylers Z. Physiol. Chem.* **349,** 268–270.
Horowitz, J. M., and Plant, R. E. (1978). *Am. J. Physiol.* **235,** R121–R129.
Horowitz, J. M., Horwitz, B. A., and Smith, R. E. (1971). *Experientia* **27,** 1419–1421.
Horwitz, B. A. (1973). *Am. J. Physiol.* **224,** 352–355.
Horwitz, B. A. (1975). *Fed. Proc. Fed. Am. Soc. Exp. Biol.* **34,** 477, Abstr. No. 1448.
Horwitz, B. A. (1976). *Isr. J. Med. Sci.* **12,** 1086–1089.
Horwitz, B. A. (1977). *In* "Drugs, Biogenic Amines and Body Temperature" (K. E. Cooper *et al.*, eds.), pp. 160–166. Karger, Basel.
Horwitz, B. A. (1978a). *In* "Strategies in Cold" (L. C. H. Wang and J. W. Hudson, eds.), pp. 619–654. Academic Press, New York.
Horwitz, B. A. (1978b). *Experientia Suppl.* **32,** 19–23.
Horwitz, B. A., and Eaton, M. (1975). *Eur. J. Pharmacol.* **34,** 241–245.
Horwitz, B. A., Herd, P. A., and Smith, R. E. (1968). *Can. J. Physiol. Pharmacol.* **46,** 897–902.
Horwitz, B. A., Horowitz J. M., Jr., and Smith, R. E. (1969). *Proc. Natl. Acad. Sci. U.S.A.* **64,** 113–120.
Houštěk, J., and Drahota, Z. (1977). *Biochim. Biophys. Acta* **484,** 127–139.
Houštěk, J., Cannon, B., and Lindberg, O. (1975a). *Eur. J. Biochem.* **54,** 11–18.
Houštěk, J., Bieber, L. L., and Lindberg, O. (1975b). *In* "Depressed Metabolism and Cold Thermogenesis" (L. Janský, ed.), pp. 154–159. Charles University, Prague.
Houštěk, J., Kopecký, J., and Drahota, Z. (1978). *Comp. Biochem. Physiol.* **60B,** 209–214.
Itaya, K. (1978a). *J. Pharm. Pharmacol.* **30,** 632–637.
Itaya, K. (1978b). *J. Pharm. Pharmacol.* **30,** 315–317.
Jankovič, B. D., Mitrovič, K., and Miloševič, D. (1977). *Clin. Exp. Immunol.* **29,** 173–175.
Janský, L., Bartunkova, R., and Zeisberger, E. (1967). *Physiol. Bohemoslov.* **16,** 366–371.
Joel, C. D., Neaves, W. B., and Rabb, J. M. (1967). *Biochem. Biophys. Res. Commun.* **29,** 490–495.
Kennedy, D. R., Hammond, R. P., and Hamolsky, M. W. (1977). *Am. J. Physiol.* **232,** E565–E569.
Kissane, J. M., and Robins, E. (1958). *J. Biol. Chem.* **233,** 184–188.
Knight, B. L. (1971). *Biochem. J.* **123,** 485–491.
Knight, B. L. (1974). *Biochim. Biophys. Acta* **343,** 287–296.
Knight, B. L. (1975). *Biochem. J.* **152,** 577–582.
Knight, B. L. (1976). *Biochim. Biophys. Acta* **429,** 798–808.
Knight, B. L., and Fordham, R. A. (1975). *Biochim. Biophys. Acta* **384,** 102–111.
Knight, B. L., and Skála, J. P. (1975). *Biochim. Biophys. Acta* **385,** 124–132.
Knight, B. L., and Skála, J. P. (1977). *J. Biol. Chem.* **252,** 5356–5362.
Knight, B. L., and Skála, J. P. (1979). *J. Biol. Chem.* **254,** 1319–1325.
Kopecký, J., and Drahota, Z. (1978). *Physiol. Bohemoslov.* **27,** 125–130.
Kramar, R., Hüttinger, M., Gmeiner, B., and Goldenberg, H. (1978). *Biochim. Biophys. Acta* **531,** 353–356.

Krebs, H. A., and Henseleit (1932). *Hoppe-Seyler's Z. Physiol. Chem.* **210,** 3–66.
Krishna, G., Moskowitz, J., Dempsey, P., and Brodie, B. B. (1970) *Life Sci.* **9,** 1353–1361.
Kumon, A., Hara, T., and Takahashi, A. (1976). *J. Lipid Res.* **17,** 559–564.
Kuo, J. F., and Greengaard, P. (1969). *Proc. Natl. Acad. Sci. U.S.A.* **64,** 1349–1355.
Kuroshima, A., and Dai, K. (1976). *Experientia* **32,** 473–474.
Kuroshima, A., and Yahata, T. (1979). *Jpn. J. Physiol.* **29,** 683–690.
Kuroshima, A., Dai, K., and Ohno, T. (1977). *Can. J. Physiol. Pharmacol.* **55,** 951–953.
Lands, A. M., Arnold, A., McAuliff, J. P., Luduena, F. P., and Brown, T. G. (1967). *Nature (London)* **214,** 597–598.
Lepkovsky, S., Wang, W., Koike, T., and Dimick, M. K. (1959). *Fed. Proc. Fed. Am. Soc. Exp. Biol.* **18,** 272 Abstr. No. 1075.
Lin, E. C. C. (1977). *Annu. Rev. Biochem.* **46,** 765–795.
Linck, G., Stoeckel, M.-E., Porte, A., and Petrovic, A. (1974). *C. R. Acad. Sci. Paris* **278D,** 87–89.
Linck, G., Teller, G., and Petrovic, A. (1975). *Eur. J. Biochem.* **58,** 511–516.
Lindberg, O., ed. (1970). "Brown Adipose Tissue." American Elsevier, New York.
Lindberg, O., de Pierre, J., Rylander, E., and Sydbom, R. (1966). *Commun. Meet. Fed. Eur. Biochem. Soc., 3rd, Warsaw.*
Lindberg, O., de Pierre, J., Rylander, E., and Afzelius, B. A. (1967). *J. Cell Biol.* **34,** 293–310.
Lindberg, O., Prusiner, S. B., Cannon, B., Ching, T. M., and Eisenhardt, R. H. (1970). *Lipids* **5,** 204–209.
Lindberg, O., Bieber, L. L., and Houštěk, J. (1976). *In* "Regulation of Depressed Metabolism and Thermogenesis" (L. Janský and X. J. Musacchia, eds.), pp. 117–136. Thomas, Springfield, Illinois.
Lindberg, O., Cannon, B., and Nedergaard, J. (1981). *In* "Mitochondria and Microsomes" (C. P. Lee, G. Schatz, and G. Dallner, eds.), pp. 93–119. Addison-Wesley, Reading, Massachusetts.
Lindgren, G., and Barnard, T. (1972). *Exp. Cell Res.* **70,** 81–90.
Loken, S. C., and Fain, J. N. (1971). *Am. J. Physiol.* **221,** 1126–1129.
Lovell-Smith, C. J., Manganiello, V. C., and Vaughan, M. (1977). *Biochim. Biophys. Acta* **497,** 447–458.
Lynch, G. R., and Epstein, A. L. (1976). *Comp. Biochem. Physiol.* **53C,** 67–68.
McCormack, J. G., and Denton, R. M. (1977). *Biochem. J.* **166,** 627–630.
McCormack, J. G., and Denton, R. M. (1979). *Biochem. J.* **180,** 533–544.
Malan, A., Rodeau, J. L., and Daull, F. (1980). *In* "Survival in Cold" (Abstracts), 71. Prague.
Malbou, C. C., Moreno, F. J., Cabelli, R. J., and Fain, J. N. (1978). *J. Biol. Chem.* **253,** 671–677.
Markwell, M. A. K., Mc. Groarty, E. J., Bieber, L. L., and Tolbert, N. E. (1973). *J. Biol. Chem.* **248,** 3426–3432.
Mitchell, P. (1961). *Nature (London)* **191,** 144–148.
Mitchell, P. (1966). "Chemiosmotic Coupling in Oxidative and Photosynthetic Phosphorylation." Glynn Research Ltd., England.
Mitchell, P. (1973). *J. Bioenerget.* **4,** 63–91.
Mohell, N., Nedergaard, J., and Cannon, B. (1980). *Adv. Physiol. Sci.* **32,** 495–497.
Moriya, K., and Itoh, S. (1969). *Jpn. J. Physiol.* **19,** 775–790.
Mosinger, B., and Vaughan, M. (1967). *Biochim. Biophys. Acta* **144,** 556–568.
Moskowitz, J., and Krishna, G. (1973). *Pharmacology* **10,** 129–135.
Muirhead, M., and Himms-Hagen, J. (1971). *Can. J. Biochem.* **49,** 802–810.
Murthy, V. K., and Steiner, G. (1972). *Am. J. Physiol.* **222,** 983–987.
Murthy, V. K., and Steiner, G. (1974). *Can. J. Biochem.* **52,** 259–262.
Nedergaard, J. (1980). "Control of Fatty Acid Utilization in Brown Adipose Tissue." University of Stockholm.

Nedergaard, J. (1981). *Eur. J. Biochem.* **114,** 159–167.
Nedergaard, J. (1982). *Am J. Physiol.* **242,** in press.
Nedergaard, J., and Cannon, B. (1979). *Methods Enzymol.* **55,** 3–28.
Nedergaard, J., and Cannon, B. (1980). *Acta Chem. Scand.* **B34,** 149–151.
Nedergaard, J., and Lindberg, O. (1979). *Eur. J. Biochem.* **95,** 139–145.
Nedergaard, J., Cannon, B., and Lindberg, O. (1977). *Nature (London)* **267,** 518–520.
Nedergaard, J., Alexson, S., and Cannon, B. (1979a). *Acta Physiol. Scand.* **105,** 68A–69A.
Nedergaard, J., Al-Shaikhaly, M. H. M., and Cannon, B. (1979b). *In* "Function and Molecular Aspects of Biomembrane Transport" (E. Quagliariello *et al.*, eds.), pp. 175–178. Elsevier, Amsterdam.
Nedergaard, J., Alexson, S., and Cannon, B. (1980). *Am. J. Physiol.* **239,** C208–C216.
Nicholls, D. G. (1976a). *FEBS Lett.* **61,** 103–110.
Nicholls, D. G. (1976b). *Trends Biochem. Sci.* **6.**
Nicholls, D. G. (1976c). *Eur. J. Biochem.* **62,** 223–228.
Nicholls, D. G. (1979). *Biochim. Biophys. Acta* **549,** 1–29.
Nicholls, D. G., and Bernson, V. S. M. (1977). *Eur. J. Biochem.* **75,** 601–612.
Nicholls, D. G., and Lindberg, O. (1973). *Eur. J. Biochem.* **37,** 523–530.
Nicholls, D. G., Grav, H. J., and Lindberg, O. (1972). *Eur. J. Biochem.* **31,** 526–533.
Normann, P. T., and Flatmark, T. (1978). *Biochim. Biophys. Acta* **530,** 461–473.
Normann, P. T., Ingebretsen, O. C., and Flatmark, T. (1978). *Biochim. Biophys. Acta* **501,** 286–295.
Novikoff, P. M., and Novikoff, A. B. (1972). *J. Cell Biol.* **53,** 532–560.
Olefsky, J. M. (1977). *J. Lipid Res.* **18,** 459–464.
Osmundsen, H., Neat, C. E., and Norum, K. R. (1979). *FEBS Lett.* **99,** 292–295.
Pagé, E., and Babineau, L.-M. (1950). *Rev. Can. Biol.* **9,** 202–206.
Pagé, E., and Babineau, L.-M. (1951). *Can. J. Res.* **28E,** 196–201.
Pavelka, M., Goldenberg, H., Hüttinger, M. and Kramar, R. (1976). *Histochemistry* **50,** 47–55.
Pedersen, J. I. (1970). *Eur. J. Biochem.* **16,** 12–18.
Pedersen, J. I., and Grav, H. J. (1972). *Eur. J. Biochem.* **69,** 453–462.
Pedersen, J. I., Christiansen, E. N., and Grav, H. J. (1968). *Biochem. Biophys. Res. Commun.* **32,** 492–500.
Pedersen, J. I., Slinde, E., Grynne, B., and Aas, M. (1975). *Biochim. Biophys. Acta* **398,** 191–203.
Pellet, H., and Lheritier, M. (1975). *C. R. Soc. Biol.* **5,** 1220–1223.
Pettersson, B. (1976). "Isolated Hamster Brown Adipocytes Studies on Catecholamine-Induced Changes in Metabolism." University of Stockholm.
Pettersson, B. (1977). *Eur. J. Biochem.* **72,** 235–240.
Pettersson, B., and Vallin, I. (1975). *In* "Depressed Metabolism and Cold Thermogenesis" (L. Janský, ed.), pp. 161–164. Charles University, Prague.
Pettersson, B., and Vallin, I. (1976). *Eur. J. Biochem.* **62,** 383–390.
Pettersson, B., Lundberg, P., and Bernson, V. S. M. (1978). *Experientia Suppl.* **32,** 101–105.
Plant, R. E., and Horowitz, J. M. (1978). *Comp. Prog. Biomed.* **8,** 171–179.
Portet, R., Laury, M. C., Bertin, R., Senault, C., Hluszko, M. T., and Chevillard, L. (1974). *Proc. Soc. Exp. Biol. Med.* **147,** 807–812.
Portet, R., Beauvallet, M., and Solier, M. (1976). *Arch. Int. Physiol. Biochim.* **84,** 89–98.
Prusiner, S. (1970). *J. Biol. Chem.* **245,** 382–389.
Prusiner, S., and Poe, M. (1968). *Nature (London)* **220,** 235–237.
Prusiner, S. B., Eisenhardt, R. H., Rylander, E., and Lindberg, O. (1968a). *Biochem. Biophys. Res. Commun.* **30,** 508–515.
Prusiner, S. B., Cannon, B., and Lindberg, O. (1968b). *Eur. J. Biochem.* **6,** 15–22.
Prusiner, S. B., Cannon, B., Ching, T. M., and Lindberg, O. (1968c). *Eur. J. Biochem.* **7,** 51–57.

Prusiner, S., Williamson, J. R., Chance, B., and Paddle, B. M. (1968d). *Arch. Biochem. Biophys.* **123,** 368–377.
Prusiner, S. B., Cannon, B., and Lindberg, O. (1970). *In* "Brown Adipose Tissue" (O. Lindberg, ed.), pp. 283–318. American Elsevier, New York.
Ptak, W. (1965). *Experientia* **21,** 26–27.
Rabi, T., Cassuto, Y., and Gutman, A. (1977). *J. Appl. Physiol.* **43,** 1007–1011.
Radomski, M. W., and Orme, T. (1971). *Am. J. Physiol.* **220,** 1852–1856.
Rafael, J., and Heldt, H. W. (1976). *FEBS Lett.* **63,** 304–308.
Rasmussen, H., and Goodman, D. B. P. (1977). *Physiol. Rev.* **57,** 421–509.
Rechardt, L., and Hervonen, H. (1976). *Histochemistry* **50,** 57–64.
Reed, N., and Fain, J. N. (1968a). *J. Biol. Chem.* **243,** 2843–2848.
Reed, N., and Fain, J. N. (1968b). *J. Biol. Chem.* **243,** 6077–6083.
Reed. N., and Fain, J. N. (1970). *In* "Brown Adipose Tissue" (O. Lindberg, ed.), pp. 207–224. American Elsevier, New York.
Revel, J. P., Yee, A. G., and Hudspeth, A. J. (1971). *Proc. Natl. Acad. Sci. U.S.A.* **68,** 2924–2927.
Ricquier, D., and Kader, J.-C. (1976). *Biochem. Biophys. Res. Commun.* **73,** 577–583.
Ricquier, D., Mory, G., and Hemon, P. (1975). *FEBS Lett.* **53,** 342–346.
Ricquier, D., Gervais, C., Kader, J.-C., and Hemon, P. (1979). *FEBS Lett.* **101,** 35–38.
Roberts, J. C., and Smith, R. E. (1967). *Am. J. Physiol.* **212,** 519–525.
Rodbell, M. (1964). *J. Biol. Chem.* **239,** 375–380.
Rodbell, M. (1965). *In* "Handbook of Physiology" (A. E. Renold and G. F. Cahill, Jr., eds.), pp. 471–482. Amer. Physiol. Soc. Washington D.C.
Romert, L., and Cannon, B. (1978). *J. Thermal Biol.* **3,** 102.
Rosak, C., and Hittelman, K. J. (1977). *Biochim. Biophys. Acta* **496,** 458–474.
Rosenthal, J. W. (1979). *Gen. Pharmacol.* **10,** 47–49.
Rosenthal, J. W., and Fain, J. N. (1971). *J. Biol. Chem.* **246,** 5888–5895.
Rothwell, N. J., and Stock, M. J. (1979). *Nature (London)* **281,** 31–35.
Rottenberg, H. (1979). *Methods Enzymol.* **55,** 547–569.
Ruzicka, F. J., and Beinert, H. (1977). *J. Biol. Chem.* **252,** 8440–8445.
Schenk, H., Heim, T., Mende, T., Varga, F., and Goetze, E. (1975a). *Eur. J. Biochem.* **58,** 15–22.
Schenk, H., Heim, T., Wagner, H., Varga, F., and Goetze, E. (1975b). *Acta Biol. Med. Germ.* **34,** 1311–1320.
Schierer, H. (1956). *Zool. Beitr.* **2,** 63.
Schultz, F. M., and Johnston, J. M. (1971). *J. Lipid Res.* **12,** 132–138.
Seccombe, D. W., Hahn, P., and Skála, J. P. (1977a). *Can. J. Biochem.* **55,** 924–927.
Seccombe, D. W., Harding, P. G. R., and Possmayer, F. (1977b). *Biochim. Biophys. Acta* **488,** 402–416.
Seglen, P. O. (1976). *In* "Methods in Cell Biology" (D. M. Prescott, ed.), pp. 30–83. Academic Press, New York.
Seydoux, J., and Girardier, L. (1978). *Experientia Suppl.* **32,** 153–167.
Seydoux, J., Constantinidis, J., Tsacopoulos, M., and Girardier, L. (1977). *J. Physiol. (Paris)* **73,** 985–996.
Sheldon, E. F. (1924). *Anat. Rec.* **28,** 331.
Shepherd, R. E., and Fain, J. N. (1977). *Fed. Proc. Fed. Am. Soc. Exp. Biol.* **36,** 2732–2734.
Sheridan, J. D. (1971). *J. Cell Biol.* **50,** 795–803.
Skála, J., and Hahn, P. (1971). *Can. J. Physiol. Pharmacol.* **49,** 501–507.
Skála, J. P., and Knight, B. L. (1975). *Biochim. Biophys. Acta* **385,** 114–128.
Skála, J. P., and Knight, B. L. (1977). *J. Biol. Chem.* **252,** 1064–1070.
Skála, J. P., and Knight, B. L. (1978). *Can. J. Biochem.* **56,** 869–874.

Skála, J. P., and Knight, B. L. (1979). *Biochim. Biophys. Acta* **582,** 122–131.
Skála, J. P., Hahn, P., and Braun, T. (1970). *Life Sci.* **9,** 1201–1212.
Skála, J. P., Drummond, G. I., and Hahn, P. (1974). *Biochem. J.* **138,** 195–199.
Slinde, E., Pedersen, J. I., and Flatmark, T. (1975). *Anal. Biochem.* **65,** 581–585.
Smith, R. E. (1961). *Physiologist* **4,** 113.
Smith, R. E., and Hock, R. J. (1963). *Fed. Proc. Fed. Am. Soc. Exp. Biol.* **22,** 341.
Smith, R. E., and Horwitz, B. A. (1969). *Physiol. Rev.* **49,** 330–425.
Smith, R. E., Roberts, J. C., and Hittelman, K. J. (1966). *Science* **154,** 653–654.
Steiner, G., and Evans, S. (1976). *Am. J. Physiol.* **231,** 34–39.
Sundin, U. (1981). *Am. J. Physiol.* **241,** C134–C139.
Sundin, U., and Cannon, B. (1978). *J. Thermal Biol.* **3,** 103.
Sundin, U., and Cannon, B. (1980). *Comp. Biochem. Physiol.* **65B,** 463–471.
Sundin, U., Herron, D., and Cannon, B. (1981a). *Biol. Neonate* **39,** 747–749.
Sundin, U., Moore, G., and Cannon, B. (1981b). Submitted.
Suter, E. R. (1969). *J. Ultrastruct. Res.* **26,** 216–241.
Svartengren, J., Mohell, N., and Cannon, B. (1980a). *Acta Chem. Scand.* **B34,** 231–232.
Svartengren, J., Mohell, N., and Cannon, B. (1980b). *Adv. Physiol. Sci.* **32,** 491–493.
Svoboda, P., and Mosinger, B. (1981a). *Biochem. Pharmacol.* **30,** 427–432.
Svoboda, P., and Mosinger, B. (1981b). *Biochem. Pharmacol.* **30,** 433–439.
Svoboda, P., Svartengren, J., Snochowski, M., Houštěk, J., and Cannon, B. (1979). *Eur. J. Biochem.* **102,** 203–210.
Świerczyński, J., Ścisowski, P., and Aleksandrowicz, Z. (1976). *Biochim. Biophys. Acta* **452,** 310–319.
Teodorn, C. V., and Grishman, E. (1959). *Fed. Proc. Fed. Am. Soc. Exp. Biol.* **18,** 510, Abstr. No. 2012.
Thomson, J. F., Habeck, D. A., Nance, S. L., and Beetham, K. L. (1969). *J. Cell Biol.* **41,** 312–334.
Williams, J. A., and Matthews, E. K. (1974a). *Am. J. Physiol.* **227,** 981–986.
Williams, J. A., and Matthews, E. K. (1974b). *Am. J. Physiol.* **227,** 987–992.
Williams, L. T., Lefkowitz, R. J., Watanabe, A. M., Hathaway, D. R., and Besch H. R., Jr., (1977). *J. Biol. Chem.* **252,** 2787–2789.
Williamson, J. R. (1970). *J. Biol. Chem.* **245,** 2043–2050.
Williamson, J. R., Prusiner, S., Olson, M. S., and Fuhami, M. (1970). *Lipids* **5,** 1–14.
Wünnenberg, W., Merker, G., and Brück, K. (1974). *Pflügers Arch.* **352,** 11–16.
Zuk, R. T., Webb, G., Clark, D. G., and Hahn, P. (1978). *Life Sci.* **23,** 1923–1928.

Index

A

B

C

D

E

F

G

H

I

L

M

N

Contents of Recent Volumes and Supplements

Volume 64

Volume 65

Volume 66

Volume 67

Volume 68

Volume 69

Volume 70

Volume 71

Volume 72

Volume 73

Supplement 10: Differentiated Cells in Aging Research

Supplement 11A: Perspectives in Plant Cell and Tissue Culture

Supplement 11B: Perspectives in Plant Cell and Tissue Culture

Supplement 12: Membrane Research: Classic Origins and Current Concepts